MANUAL OF FIELD GEOLOGY

Robert R. Compton
Professor of Geology
Stanford University

MANUAL OF FIELD GEOLOGY

JOHN WILEY & SONS, INC.
NEW YORK · LONDON · SYDNEY

Copyright © 1962 by John Wiley & Sons, Inc.

All rights reserved. This book or any part thereof must not be reproduced in any form without the written permission of the publisher.

ISBN 0 471 16698 7

Library of Congress Catalog Card Number: 61-17357

Printed in the United States of America

16

Preface

This manual was written primarily to assist undergraduate students in field training. It also offers some specialized procedures and data to advanced students and professional geologists. To keep the book a reasonable length, the more specialized materials are presented by brief descriptions, outlines, and diagrams. I have tried, however, to describe basic procedures in enough detail so that students can follow them with little if any supervision. It is assumed that the reader can identify common rocks and minerals and is acquainted with geologic structures and the basic principles of stratigraphy. Although a knowledge of petrography is desirable for some of the studies described in the last four chapters, it is not essential for the field procedures themselves.

The first eleven chapters deal primarily with techniques and basic concepts of field work. Although each of these chapters can be used independently, they are organized as a unit to help guide the student from one stage of his training to another. Chapter 1, for example, describes note taking and collecting for general field work, and also provides suggestions for students who want to collect data and specimens before starting a formal course in field geology. Chapter 2 describes the basic uses of the compass, clinometer, and hand level. Chapter 3 introduces traversing by pace and compass methods and describes a complete field project suitable for the beginner. Procedures for plotting geologic features on base maps and aerial photographs are given in Chapters 4 and 5, which also include a general discussion of rock units and contacts. The next four chapters comprise a related sequence, describing first the instrumental methods used in making maps of areas for which there is no suitable base, and then suggesting consecutive steps for plane table and aerial photograph projects. Chapter 10 introduces techniques used in detailed geologic studies, and includes a discussion of systematic sampling. Chapter 11 concludes the first group of chapters with suggestions on report writing.

The remaining four chapters have several purposes. First, I wished to emphasize the importance of observations and interpretations made at outcrops and have therefore described and illustrated a number of pertinent features and relations. Second, to help students attack specific problems, I have suggested ways to organize specialized studies and have cited references that present detailed examples of such studies. Third, the chapters bring together descriptions and classifications of rocks and structures that may be helpful in many types of field projects.

Most of the drawings in the last four chapters are based on actual outcrops. I have included (and occasionally emphasized) some unusual rocks and relations, hoping to encourage consideration of all possible hypotheses. I have also emphasized some topics because they are relatively new or, I feel, have been slighted elsewhere. Some of the references were selected to help in finding other systems and ideas.

Many persons have contributed to this manual either directly or indirectly. I appreciate deeply the initial field training I received from Randall E. Brown, A. C. Waters, and Howel Williams, who were patient and inspiring teachers. Many of the procedures were worked out during the 10 years I have taught Stanford's field geology courses, and I gratefully acknowledge the many contributions made by students and instructors. Ben M. Page, who formerly directed these courses, has been especially generous with his ideas and materials.

I wish to thank Laurence H. Nobles for reading and criticizing the entire first draft of the manuscript, and Francis H. Moffitt, who made many valuable suggestions pertaining to the chapters on surveying and aerial photographs. Valued assistance on individual chapters and sections was given by C. M. Gilbert, V. E. McKelvey, S. W. Muller, P. J. Shenon, A. C. Waters, L. E. Weiss, and Howel Williams. William R. Dickinson and Thane H. McCulloh also made important suggestions regarding several chapters, and several of my colleagues kindly discussed specific items with me. Robert G. Coleman, Allan B. Griggs, H. L. James, Douglas M. Kinney, and William R. Moran generously supplied useful materials and ideas. Permission to use or redraw illustrations was given kindly by a number of persons, who are acknowledged individually in the figure captions.

Finally, I am very grateful to my wife, Ariel, who not only helped prepare the manuscript but also read it several times and made many suggestions that improved it.

ROBERT R. COMPTON

Stanford, California
September, 1961

Contents

Chapter 1 Observing and Collecting Data and Samples   1

1-1. General basis of field geology. 1-2. Planning a field project. 1-3. Basic field equipment. 1-4. Taking geologic notes in the field. 1-5. Descriptions entered in notes. 1-6. Collecting rock samples. 1-7. Collecting fossils. 1-8. Numbering and marking specimens. 1-9. Locality descriptions.

Chapter 2 Using the Compass, Clinometer, and Hand Level   21

2-1. The Brunton compass. 2-2. Setting the magnetic declination. 2-3. Taking bearings with the compass. 2-4. Magnetic deflections of compass bearings. 2-5. Measuring vertical angles with the clinometer. 2-6. Using the Brunton compass as a hand level. 2-7. Measuring strike and dip. 2-8. Where to take strike and dip. 2-9. Measuring trend and plunge of linear features. 2-10. Care and adjustment of the Brunton compass.

Chapter 3 The Compass Traverse   36

3-1. General scheme of a geologic traverse. 3-2. Determining pace. 3-3. Selecting a course and planning a traverse. 3-4. First steps in a compass traverse. 3-5. Traversing by turning angles. 3-6. Plotting geologic features on the traverse. 3-7. Vertical profile of the traverse. 3-8. Making illustrations from the traverse data.

Chapter 4 Plotting Geologic Features on a Base Map   49

4-1. Selecting and preparing a base map. 4-2. Locating field data on a base map. 4-3. Locating geologic features by traversing. 4-4. Using a barometer (altimeter) to locate geologic features on a map. 4-5. What to plot on the base map. 4-6. Rock units for mapping. 4-7. Mapping contacts between rock units. 4-8. Mapping by the outcrop or exposure method. 4-9. Defining and mapping gradational contacts. 4-10. Using colored pencils in mapping. 4-11. Tracing and plotting faults. 4-12. Methods of approximate (reconnaissance) mapping. 4-13. Daily routine of field work.

Chapter 5 Mapping Geologic Features on Aerial Photographs 73

5–1. Kinds of aerial photographs and their uses. 5–2. How vertical aerial photographs are made and indexed. 5–3. Differences between vertical aerial photographs and maps. 5–4. Obtaining a stereoscopic image from aerial photographs. 5–5. Materials for mapping on aerial photographs. 5–6. Determining scales of aerial photographs. 5–7. Locating outcrops on photographs by inspection. 5–8. Drawing north arrows on aerial photographs. 5–9. Locating outcrops on photographs by compass and pacing methods. 5–10. Geologic mapping on aerial photographs. 5–11. Transferring geologic features from photographs to a map.

Chapter 6 The Alidade and Plane Table 88

6–1. General value of the alidade and plane table. 6–2. The peep-sight alidade and the principles of alidade surveying. 6–3. The telescopic alidade. 6–4. Adjusting the alidade in the field. 6–5. Major adjustments of the alidade. 6–6. Care of the alidade in the field. 6–7. The plane table and tripod. 6–8. Plane table sheets and their preparation. 6–9. Setting up and orienting the plane table. 6–10. Measuring vertical angles with the alidade. 6–11. The stadia method and the Beaman arc. 6–12. Stadia procedure with the Beaman arc. 6–13. The stadia interval factor and the stadia constant. 6–14. The gradienter screw. 6–15. Differences in elevation by the stepping method. 6–16. Using the stadia rod.

Chapter 7 Control for Geologic Maps 113

7–1. General nature of control surveys. 7–2. General plan of a triangulation survey. 7–3. Using existing control data for surveys. 7–4. Selecting stations for a triangulation network. 7–5. Signals for triangulation stations. 7–6. Measuring the base line. 7–7. Triangulation with the transit. 7–8. Surveying an elevation to the control network. 7–9. Observation on Polaris at elongation. 7–10. Computations and adjustments of triangulation data. 7–11. Rectangular and polyconic grids. 7–12. Triangulation with the alidade and plane table. 7–13. Surveying control by transit traverses.

Chapter 8 Geologic Mapping with the Alidade and Plane Table   135

8–1. Plane table projects and their appropriate scales. 8–2. Planning a plane table project. 8–3. Reconnoitering the control system and geologic features. 8–4. Choosing the layout of plane table sheets. 8–5. Plotting primary control on plane table sheets. 8–6. Instrument stations for stadia work. 8–7. Examining geologic features and flagging contacts. 8–8. Choosing the contour interval. 8–9. Intersecting instrument stations. 8–10. Locating instrument stations by resection. 8–11. Three-point locations. 8–12. Traversing with the alidade and plane table. 8–13. Mapping by stadia methods. 8–14. Using aerial photographs in plane table mapping. 8–15. Method of moving the plane table around the rod. 8–16. Work on the plane table sheet in the evening. 8–17. Vertical cross sections from plane table maps. 8–18. Completing plane table maps.

Chapter 9 Making a Geologic Map from Aerial Photographs   154

9–1. General value of aerial photograph compilations. 9–2. Preparations for an aerial photograph project. 9–3. Mapping geologic features on aerial photographs. 9–4. Ground control for photograph compilations. 9–5. Surveying cross-section lines for photograph compilations. 9–6. Radial line compilation. 9–7. Marking and transferring photograph points. 9–8. Plotting control on the overlay. 9–9. Compiling points from controlled photographs. 9–10. Compiling points from uncontrolled photographs. 9–11. Compiling photograph data. 9–12. Checking the map and drawing cross sections.

Chapter 10 Detailed Mapping and Sampling   170

10–1. General nature of detailed studies. 10–2. Detailed surface maps and sections. 10–3. Cleaning, excavating, and drilling. 10–4. Underground mapping. 10–5. Sampling.

Chapter 11 Preparing Geologic Reports   185

11–1. General nature of geologic reports. 11–2. Organizing and starting the report. 11–3. Clarity of the report. 11–4. Use of special terms. 11–5. Front matter for the report. 11–6. Form of the report. 11–7. Planning illustrations for the report. 11–8. Kinds of illustrations. 11–9. Drawing methods. 11–10. Detailed geologic maps and cross sections. 11–11. Stratigraphic illustrations.

Contents

Chapter 12   Field Work with Sedimentary Rocks   208

12–1. Interpreting sedimentary rocks. 12–2. Lithologic and time-stratigraphic units. 12–3. Naming and describing sedimentary rocks. 12–4. Beds and related structures. 12–5. Surfaces between beds. 12–6. Unconformities. 12–7. Tops and bottoms of beds. 12–8. Measuring stratigraphic sections. 12–9. Sampling for microfossils. 12–10. Logging wells and drill holes. 12–11. Surficial deposits and related landforms.

Chapter 13   Field Work with Volcanic Rocks   250

13–1. Volcanic sequences and unconformities. 13–2. Cartographic units of volcanic rocks. 13–3. Naming volcanic rocks. 13–4. Structures of basic lavas. 13–5. Structures of silicic lavas. 13–6. Pyroclastic and closely related deposits. 13–7. Volcanic feeders and related intrusions.

Chapter 14   Field Work with Igneous and Igneous-Appearing Plutonic Rocks   272

14–1. Concepts of plutonic geology that apply to field studies. 14–2. Cartographic units of plutonic igneous rocks. 14–3. Naming plutonic rocks. 14–4. Contacts of plutonic units. 14–5. Planar and linear fabrics in plutonic rocks. 14–6. Inclusions and related structures. 14–7. Layers and schlieren in plutonic rocks. 14–8. Fractures and related structures in plutons. 14–9. Alterations of plutonic rocks.

Chapter 15   Field Work with Metamorphic Rocks   296

15–1. Studies of metamorphic rocks. 15–2. Cartographic units of metamorphic rocks. 15–3. Naming metamorphic rocks. 15–4. Premetamorphic lithology and sequence. 15–5. Studying metamorphic deformation. 15–6. Foliations and lineations. 15–7. Geometric styles of folding. 15–8. Foliations and lineations related to folds. 15–9. Deformation structures in massive rocks. 15–10. Joint and vein patterns in deformed rocks. 15–11. Analyses on spherical projections. 15–12. Oriented samples for microscopic studies. 15–13. Mapping metamorphic zones. 15–14. Migmatites and related rocks.

Appendixes   327

Index   363

MANUAL OF FIELD GEOLOGY

1
Observing and Collecting Data and Samples

1-1. General Basis of Field Geology

To geologists, the *field* is where rocks or soils can be observed in place, and *field geology* consists of the methods used to examine and interpret structures and materials at the outcrop. Field studies are the primary means of obtaining geologic knowledge. Some studies may be as simple as visiting a single outcrop or quarry, making notes and sketches on the relations between certain rocks, and collecting a suite of specimens. Others may require weeks or months of geologic mapping, systematic sampling, and careful integrating of field and laboratory measurements.

Geologic mapping is so essential to many field studies that it is sometimes considered synonymous with "field geology." Maps are used to measure rock bodies, to plot structural measurements, and to relate many kinds of data. Frequently, they permit interpretations of features that are too large to be studied in single exposures. Many folds and faults, for example, can be discovered only by geologic mapping, and even if they can be seen in outcrops, they must be mapped over large areas to be understood. Geologic maps are also used to construct such important projections as cross sections. Together, maps and sections are an ideal means of presenting large amounts of information to other persons.

Although mapping may be essential to a field study, observations made at individual outcrops are fundamental. Rocks must be identified before they can be mapped. Moreover, many genetic relations can be understood only after exposures are examined in detail. No amount of mapping can supplant these crucial observations. For example, a detailed map of an igneous body might show only that it is a concordant layer between sedimentary formations. Relations at one or two outcrops, however, could demonstrate that the body is a lava flow rather than a sill. Once this is established, obscure features associated with the flow might be identified and then utilized in further interpretations.

Field studies as a scientific method. Because the geologist is continuously observing relations and making interpretations in the field, his general methods may be compared with the classical scientific method. In its conventional form, the scientific method consists of several steps. The investigator first observes and collects data. He then formulates a hypothesis to explain these data. Next, he tests the hypothesis in all possible ways, particularly by studying additional relations that can be predicted on its basis. If the hypothesis survives all tests, it is considered tentatively verified; if not, another hypothesis must be formulated and tested.

In geologic studies it is often most effective to consider all possible hypotheses together. G. K. Gilbert (1886, p. 286) explained this procedure as follows:

> There is indeed an advantage in entertaining several (hypotheses) at once, for then it is possible to discover their mutual antagonisms and inconsistencies, and to devise crucial tests—tests which will necessarily debar some of the hypotheses from further consideration. The process of testing is then a process of elimination, at least until all but one of the hypotheses have been disproved.

If only one hypothesis is utilized, or if one is adopted too quickly, there may be a tendency to overlook evidence that would disprove it. T. C. Chamberlin (1897) discussed this problem thoroughly in his description of "The method of multiple working hypotheses," a paper recommended to all geologists.

One reason why many hypotheses should be considered in the field is that outcrops cannot be revisited to test every new idea. Furthermore, different kinds of data may be so interrelated that all must be studied together to be understood. Field studies must thus go far beyond a mere mapping and collecting of individual rocks or structures. Even a simple rock specimen will lack potential meaning if not selected in light of all associated features.

Because the geologist uses the scientific method continuously as he works, he must be well armed with hypotheses when he goes into the field. He must also use imagination and ingenuity at the outcrop. If none of his initial ideas withstands testing, several possibilities should be considered. Some hypotheses seem to fail because data were collected on the basis of incorrect assumptions, or because information was classified inconsistently on a map. In other cases, tentative hypotheses are too simple to account for natural events, and more complex possibilities should be considered.

Interpreting complex relations. Geologic features as simple as those diagrammed in textbooks are rarely found in the field. Even those features that at first appear simple will commonly have small but important complexities. These complexities are not impediments; they are often the keys to understanding an association of features.

One type of complex relation is formed where two or more processes act at the same time. The weathering of a given rock, for example, is typically a composite effect of several mechanical and chemical processes. To interpret the overall effect, it is first necessary to understand how each process would work alone and then to consider how the processes would modify each other. By examining many outcrops it is generally possible to find places where single processes have been dominant. To interpret the weathering of a sandstone, for example, mechanical changes might be studied at exposed rock ledges, whereas chemical changes could be seen in valley-bottom soil profiles. The effects of combined processes could then be studied at various intervening localities.

A second type of complex relation is caused by overprinting of two or more geologic events. The determination of relative ages from these outcrops provides a means for interpreting geologic history. Relative ages are determined by noting how older features are modified relative to younger ones. An analysis should be started with the age relation that is shown most clearly and then extended, step by step, either backward or forward in geologic time. It is often necessary to know the shapes or positions of features in intermediate stages, and this can be done by analyzing and then removing the effects of later events. Where a vein has been folded, for example, it may be possible to establish its original shape and orientation by first measuring folds in surrounding beds, and then unfolding the beds back to simple planar forms. Events that are obscure in one outcrop may be obvious in another; by plotting many outcrops on a map, one event can be related to the others, both in space and time.

A serious error that can be avoided by studying age relations thoroughly is that of concluding that features have a causative relation merely because they are closely associated. This error is especially easy to make if the association fits a well-established theory. If, for example, quartz veins are associated with hornfelses throughout a given area, it might be tempting to conclude that they were both caused by one agent—an intrusive magma. A thorough study might show, however, that the veins also formed in much younger rocks and were therefore not caused by the contact metamorphism. Perhaps they formed abundantly in the hornfelses because these rocks fractured readily.

Outcrops are sometimes so complex or so sparse that no certain conclusions can be reached from a field study. The study may still have great value, however, if the observations are described fully in a report. In many cases, the geologist can state the relative probability of several alternate hypotheses. The uncertainties may be resolved when more data become available or when laboratory or mathematical tests are made.

1-2. Planning a Field Project

Geologic field projects generally proceed in three stages: the stage of planning; the stage of mapping, observing, and collecting; and the stage of preparing a report. The effectiveness of a project is determined largely during the planning stage. There are few cases where the geologist can study all kinds of data and relations thoroughly, and he must therefore select the scope of his study and plan his work so that the most pertinent data are collected. It is essential to know clearly the purpose of a given project. The purpose may be simple and clear from the outset, as in some studies of economic deposits. In many purely scientific studies, the purpose should be reviewed after considerable reading, discussion, and thought. In addition to the principal objective, there may be several subsidiary ones, which are sometimes difficult to evaluate until field work is well underway.

It is wise to plan a project in such a way that its scope can be increased or changed to a reasonable degree during the field season. For this reason, moderate amounts of extra equipment and supplies should be taken to the field. Although specific steps vary, the following recommendations should be considered in planning.

1. Determine if other geologists are working in or near the area of interest by inquiring at state and federal agencies and appropriate companies and colleges. Correspond with them to determine whether a new study would duplicate their work needlessly.

2. Accumulate and study reports and maps of the region in order to gain an understanding of the broader features of the geology and geography. Determine what is known, specifically, about the problems and relations that fall within the scope of the study being planned.

3. Visit the area in order to reconnoiter its topography and geology and to obtain permission for camping, mapping, and collecting. If a visit is not possible, do these things as completely as correspondence and discussions will permit.

4. Determine the scales and quality of maps and aerial photographs of the area. If these will not provide an adequate base for geologic

Observing and Collecting Data and Samples

mapping, consider what means will be used to construct a map. Will a topographic (contour) map be needed? What is the smallest map scale that will be useful? What are the most efficient surveying methods which will give adequate precision to the work?

5. Evaluate the probable schedule and costs of the project. To do this effectively, consider not only the mapping procedures but also how well the rocks are exposed, how accessible the area is from a base camp or office, and to what degree the weather is likely to interfere with field work.

6. Order maps, aerial photographs, and various other field and office equipment, allowing plenty of time for delivery. The check lists of Appendix 1 may help in doing this.

7. Reread critically all reports that pertain to the area, as well as books or papers that present basic ideas and methods pertinent to the project. Accumulate as complete a field library as possible, and photograph, copy, or abstract items that cannot be taken to the field.

1-3. Basic Field Equipment

The basic equipment needed for examining, describing, and collecting rocks is modest in amount and need not be expensive. It consists of a hammer with either a pick or chisel point at one end, a hand lens, a pocket knife, a notebook or looseleaf clip folder, a 2H or 3H pencil or a good ballpoint pen, a 6-inch scale, heavy paper or cloth bags for collecting samples, and a knapsack for carrying lunch and field gear.

The field notebook must be considered and selected with care because the notes recorded in it will become part of a permanent record. The paper should be of the highest quality, for thin and inexpensive papers tend to disintegrate in dry climates, and heavily filled ("slick") papers wrinkle permanently when dampened. Top-quality engineer's level-books (standard $4\frac{1}{2} \times 7\frac{1}{4}$ in.) with water-resistant paper and waterproof covers are an excellent choice. They are small enough to fit in a trouser pocket and are bound so that they can be opened flat or folded back cover-to-cover. The U.S. Geological Survey uses a somewhat larger ($5\frac{1}{4} \times 8\frac{1}{2}$ in.) book that is bound at the upper edge and has perforated pages that can be removed easily and reorganized during the report-writing stage.

The possibility of using standard ($8\frac{1}{2} \times 11$ in.) looseleaf sheets for notes should also be considered. These sheets give extra space for sketches, accessory maps, and sections, and they may be carried in

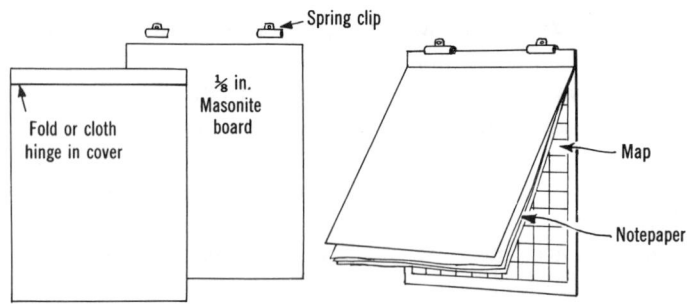

Fig. 1-1. Clip folder for 9 × 9 in. aerial photographs and 8½ × 11 in. note sheets.

a clip folder that is large enough to carry field maps or aerial photographs as well. This has the advantage of reducing the number of items of field gear. The individual sheets, however, must be labeled and stored with care so that they will not be misplaced. If the climate in the field area is either very dry or quite wet, the sheets should be made up from the heaviest ledger stock available.

The clip folder must measure at least 9¼ × 11 in. in order to hold 9 × 9 in. aerial photographs. An excellent one can be made by using ⅛-inch Masonite for a baseboard and a sheet of heavy plastic or cardboard for the cover (Fig. 1-1). Some geologists prefer an aluminum folder with a steel spring hinge, and this type of carrier is indeed good for wet or dirty working conditions because of the tight fit of the hinge and cover. For average mapping, however, its extra weight and rather sharp edges are undesirable. It also tends to become very hot when left in the summer sun for a few minutes and exceedingly cold when carried in freezing weather. Moreover, the common 8½ × 11 in. size is not large enough for 9 × 9 in. aerial photographs.

For note taking, various drawing pencils and ballpoint pens should be tested. A good ballpoint pen gives dark, legible copy that will not smear on good paper, will erase reasonably well, and is more or less waterproof. It may clog, however, under dusty working conditions, and even the best pen may run out of ink in the middle of a day's work. The ideal pencil produces legibly dark copy that will not smear easily. The optimum hardness of its lead will depend on the paper, the individual, and the climate (damp paper is gouged by even moderately hard pencils). Generally, something from H to 3H is best. A cap eraser and a pocket clip should be put on each pencil, and at least one extra set should be available in camp in case one is lost in the field. A fountain pen should be used only if the ink is known to be waterproof and permanent.

Hand lenses used in most field studies should magnify 10 to 15 times, although additional higher or lower-power lenses may be needed for some determinations. Relatively high-powered lenses have a small field, a short depth of focus, and must be held very close to the object being examined. They are therefore more difficult to use, especially where lighting is poor. Regardless of its magnifying power, the quality of a lens should be tested by focusing on both smooth and rough surfaces. The magnified image should be sharp throughout, even near its edges. Three-element lenses generally give undistorted images.

An ordinary geologist's hammer is adequate for most rocks, but a hammer weighing about 2 lb may be needed to collect hard rocks (tough sandstones, lavas, hornfelses, and so forth). Rectangular corners and edges on the striking end of a new hammer should be hammered or filed to a bevel because they may send off steel chips at dangerous speeds when struck against a tough rock.

Preparations for wet weather. Wet weather should be anticipated; it can upset a field schedule seriously. Maps and aerial photographs can be waterproofed completely by painting or spraying them on both sides with transparent plastic. Notepaper can be treated in the same way, or it can be dipped in a solution of paraffin and a volatile solvent. Paraffin should not be used on maps or photographs because they cannot then be marked with ink. The most serious damage to maps and notes is caused by slipping on steep muddy slopes, and this can be remedied by wearing nailed boots. In order that field work in rainy areas will be thorough and consistently precise, the geologist should clothe himself so that he will be comfortable (though not necessarily dry).

1-4. Taking Geologic Notes in the Field

Brief descriptions are recorded in the notebook as rocks and other geologic features are discovered and studied in the field. This must always be done directly at the outcrop, for it is difficult to remember accurately the host of facts and ideas developed during a day's work. Field notes serve as a basis for writing a report after the field season; furthermore, they may be an important record for other geologists who become interested in the area. They also serve to make the geologist think more critically and observe more carefully in the field.

Most of the items recorded in the notebook will be factual data on the geology. Many of these facts are best recorded by word descriptions, but drawings or diagrams should be used wherever they save time and space or add clarity. In many cases, small accessory maps and cross sections serve to record large amounts of data briefly and clearly. With exceptions that will be noted elsewhere, the items recorded should not duplicate data that can be recorded on field maps; for example, strike and dip of beds are generally plotted by a symbol on the map, not recorded in the notes. Where a photograph is taken, a simple diagram noting the direction of the view and labeling its important features is likely to prove valuable. The notes shown in Fig. 1-2 include examples of various kinds of entries.

The discontinuous nature of rock exposures makes it necessary to base some geologic relations on inferences rather than on observable facts. It is important to enter these inferences in the notes, though each note must show clearly that it is an inference. Hypotheses should also be stated and reviewed in the notes because interpretation must go hand in hand with observation. Even gross speculations and vague ideas have a valued place in field notes, as long as they can be identified for what they are.

Each person will develop a somewhat different style of note taking, but all notes must be legible, accurate, and as brief as clarity will allow. Use of engineering-style (slant) lettering rather than longhand writing will contribute greatly to each of these requirements. Field notes are not rapid scratches or memoranda made merely to aid the memory, but rather a record that is taken *to be used* at some future date, perhaps not by the person who made it. Common abbreviations should be used (Appendix 2), but vowels should not be extracted at random to save a few inches of note space. Highly abbreviated copy may be more or less intelligible to its originator, but it is likely to require a lengthy and unsure translation by anyone else. Explanatory lists of all but the commonest abbreviations should be added to the notes before they are filed away after the field season.

If notes are taken on looseleaf sheets, the top of each sheet must bear the date, the geologist's name, a brief geographic title or description of the area covered by the page, and the name or number of the base map or aerial photograph used. If a permanently bound book is used, these data must be recorded on the first page of each day's entries. A margin of about 1½ in. should be reserved on the left-hand side of each sheet for notations that call attention to specimens, photographs, or special problems (Fig. 1-2). Consecutive numbers are placed along the right side of this margin as each outcrop or relation is described.

Just where and how often notes should be taken will vary greatly with the study, and only experience will allow the geologist to set an optimum pace. It may be helpful for the beginner to review and summarize his notes each evening; by doing this he will more quickly get a feel for the relative value of his various entries. In general, it is better to record too much data than to have to revisit an area.

1-5. Descriptions Entered in Notes

Exactly what data should be recorded in field notes will vary with the project. In most geologic surveys, the notes concentrate on descriptions of rocks and structures, especially those features that indicate the origin of the rocks or their relative ages. As a field project gets under way, these descriptions typically cover one outcrop at a time, in the order in which they are found and examined. As the geologist becomes more familiar with the rock formations and structures, his notes should record critical descriptions of features that have been traced through a series of outcrops. This organization is very important because it keeps notes from becoming repetitive and difficult to use.

Before the geologist leaves the field for the season, he should make sure that his field notes include full descriptions of rock units and structures in all parts of the area he has surveyed; otherwise he will not be able to write an accurate report on them.

In Fig. 1-2, note 1 is an example of a rock description made on the basis of a single outcrop, while note 8 is an example of a more complete lithologic description relating to a number of outcrops. Lithologic descriptions are more usable if recorded in a fairly systematic way, as by the outline that follows.

1. Name of unit and/or brief rock name.
2. Specific locality or area to which description applies.
3. Thickness and overall structure or shape of unit in this area.
4. Main rock types and their disposition within unit.
5. Gross characteristics of area underlain by unit (topographic expression, color and type of soil, vegetation, nature of outcrops).
6. Characteristic structures of unit.
    a. Range of thicknesses and average thickness of beds or other layered structures.

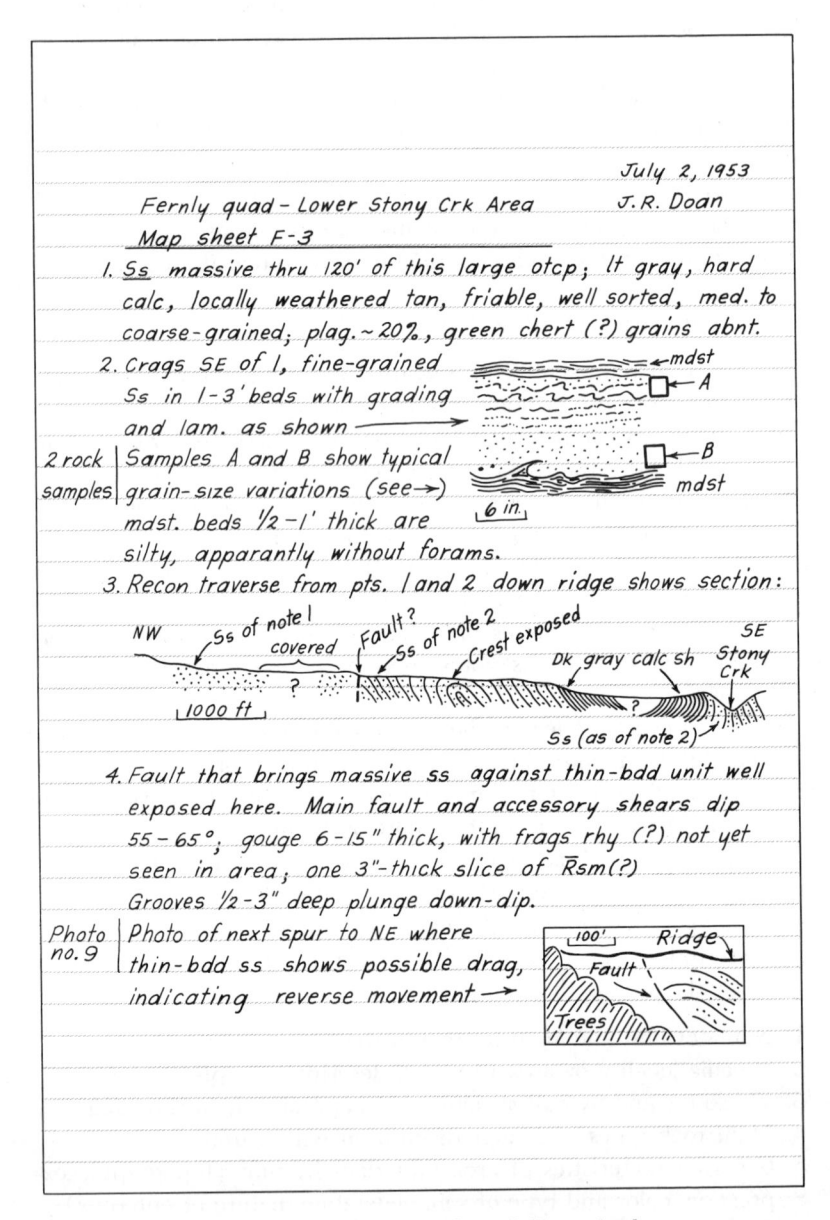

Fig. 1–2. Page from a geologic field notebook.

5. Major syncline with dark calc sh at core apparently missing on this ridge; check Silor and Mill Crks for
Check major fault.
Fossil 6. <u>Trigonia</u> collected 10 ft. above base massive ss unit. Exc. preservation indicates not re-worked.
   Loc: Stony Crk area, on 3<u>rd</u> SE-flowing trib. above Jones Mill, 1170' (paced along stream) above confluence with Stony Crk.; prominent fluted gray otcp on NE bank crk. Approx: center NE 1/4 Sec 3, T3S, R4E.
7. Unconf between massive ss and deeply weathered gr, reddish arg gr looks like old soil profile under ss; very crude x-bdng in ss suggest trough-sets (and fluviatile origin?). Poor exp.
8. P.M. spent examining gr over main ridges SW of 7. It is prob. what Smith called Pine Mt. gr in region to SW. Area is craggy, covered by low brush and small pine groves. Reddish mica-bearing soil characteristic. No lg met otcps but scattered incls of qzt and bio-feld granulite; incls locally in crudely planar swarms (see map) but no other planar or linear structures. Stream otcps show fresh pink rock, with comp: qz, 35; K feld, 40; plag, 20; bio (much altered to chl), 5 to 8; musc, <1. Grains 1/3 - 1/2 in., locally with lg euhedra K feld. (1.5% of rock); texture hypid. to allotriomorphic. Rock appears very similiar over this area.
Gravels Ridge crests have patches of gravel with well-rounded pebbles of qz and aplite (prob from gr area); these support Smith's idea of uplifted fluviatile surface on NE flank Pine Mt.

Fig. 1–2. (Continued)

b. Shapes of beds or other structures (tabular, lenticular, lineate, etc.).
   c. Primary features within beds or other structures (grading, laminations, cross-bedding, channeling, distorted flow banding, inclusions, etc.).
   d. Characteristic secondary structures, especially cleavage and prominent weathering effects.
7. Fossils (especially if a lithologic characteristic of unit).
   a. Distribution of fossils.
   b. Special characteristics of fossiliferous rocks.
   c. Position and condition of fossils (growth position, fragmental, rounded, pitted or fluted by solution, external or internal molds, etc.).
8. Description of rocks, with most abundant variety described first.
   a. Color, fresh and weathered (of wet or dry rock?).
   b. Induration (of weathered or completely fresh rock?).
   c. Grain sizes (range of sizes and principal or median size).
   d. Degree of sorting or equigranularity.
   e. Shapes of grains.
   f. Orientations or fabric of shaped grains, especially in relation to rock structures.
   g. Nature and amount of cement, matrix, or groundmass, if any.
   h. Nature and amount of pores (porosity), and any indications of permeability (is this of truly fresh rock?).
   i. Constitution of grains (mineral, lithic, fossil, glass) and their approximate percent by volume.
9. Nature of contacts.
   a. Sharp or gradational, with descriptions and dimensions of gradations.
   b. All evidence regarding possible unconformable relations.
   c. Criterion or criteria used in tracing contact in field.

Care must be used in determining the colors, induration, and mineralogy of units that are weathered almost everywhere, otherwise their "typical" recorded lithology can be totally unlike descriptions of the same unit in drill cores or mine samples. This does not mean that weathered materials should not be examined, for weathering may make it possible to see structures and minerals that cannot be seen readily in fresh rock.

Faults, unconformities, and intrusive contacts are examples of structures that are likely to require systematic and thorough descriptions

Observing and Collecting Data and Samples 13

in the field notes. Note 4 of Fig. 1–2 illustrates a note on a major fault. In all cases, even the most minor things that may relate to the origin of the rocks or structures must be described with care. Photographs or drawings showing details of structures are likely to prove very useful, and the locations of outcrops where relations are especially clear should be described fully.

1–6. Collecting Rock Samples

Even though fairly thorough lithologic descriptions are made in the field, rock samples must be collected for a number of reasons. Many rocks can be identified more exactly in the camp or office where help may be had from a microscope or a more experienced geologist. Accurate determinations of porosity and permeability, which must be made in the laboratory, require especially thorough and careful sampling of fresh materials. Important mineral ratios, such as that between dolomite and calcite in carbonate rocks, are best determined in the laboratory. Where several people are studying an area jointly, rock samples must be used to standardize names and descriptions. Even when a geologist is working alone, representative samples are useful in making comparisons between widely separated parts of an area, particularly when the final report on the area is written.

The most important specification for a sample is that it be truly representative of the unit studied; this means that an outcrop, or preferably several outcrops, should be examined carefully before a sample is selected. Where rocks are variable, suites of small samples of the principal types may be more useful than a single "average" specimen.

Specimens should be broken directly from the outcrop, and if their exact source can be marked on the outcrop with a bit of cloth, colored tape, paint, or colored crayon, so much the better. Unweathered samples are generally preferred to weathered ones; the ideal specimen has one weathered side but is otherwise unweathered. Size specifications for samples vary with the grain size and homogeneity of the rock. For homogeneous rocks with grains smaller than $\frac{1}{16}$ in., samples measuring about $3 \times 4 \times 1$ in. are generally adequate, but rocks with grains as large as $\frac{1}{8}$ in. are likely to require samples about twice that size. Still larger samples should be collected if rocks are coarser grained or show such small-scale structures as thin beds, primary igneous layers, or coarse metamorphic layers or veins. The comments on sampling given in Section 10–5 may also be helpful, especially if samples are being collected for the purpose of determining the bulk composition of a rock or rock unit.

If bedding or other foliate structures are not obvious in the sample, it should be fitted back to the outcrop and the direction of the planar structure marked on it with a felt-tip pen, a crayon, or a piece of tape. If the direction of the top of the specimen is also marked, and the structural attitude at the locality is recorded on the map, the specimen may later be studied relative to the structures of the map. These oriented specimens are particularly valuable for permeability tests and all types of petrographic studies (Section 15–12).

The only reason to size or trim samples with any care is to make them fit collection drawers of limited size; otherwise the time spent in trimming should be spent at the next outcrop. In many cases, a fairly flat and well-shaped sample can be obtained in a few blows by striking a spall off an angular edge or corner of an outcrop. This must be done with care, however, because these spalls can seriously injure one's shins, face, or eyes. There are two other warnings regarding use of a hammer: (1) heavy blows should never be struck on a hard rock when other persons are watching nearby, and (2) if one hammer is used to strike another heavily, where the latter is being driven in as a wedge, steel chips can be thrown off at dangerous speeds.

1–7. Collecting Fossils

Fossils are collected for three basic reasons: to determine the geologic age and sequence of rocks, to correlate rock units with other fossiliferous rocks, and to help in determining the environment of deposition of sediments. Each of these reasons is so important that fossils should be sought in every kind of sedimentary and pyroclastic rock and when found should be collected with care. Nor should it be assumed that metamorphic rocks are necessarily unfossiliferous. A few poorly preserved fossils from slates, phyllites, quartzites, or marbles may do more toward solving the major structures of a metamorphosed terrain than hundreds of structural readings.

Before beginning a field study, it should be determined what kinds of fossils will be particularly useful. A geologist preparing for a survey of late Paleozoic rocks, for example, should acquaint himself with the appearance of fusulinids, ostracods, brachiopods, and various key molluscs. He can do this by examining collections and discussing the possibilities with a paleontologist, or he can examine photographs and detailed drawings in books and papers. If possible, he should take to the field a small library or collection of drawings and photographs illustrating the main forms.

Fossils are so scarce in some areas that finding them may be a considerable problem. The initial search should be concentrated on float and weathered outcrop surfaces because fossils that are nearly invisible on a freshly broken surface generally weather so as to contrast with their matrix. Typically, weathered fossils are light gray, pale tan, or white, though some may be colored dark gray or black by carbonaceous materials. The geologist should make a practice of glancing at as much fragmental material as possible, no matter how many times he has walked across an area. Where fossiliferous float is found, it should be traced uphill or upstream to its source. Even if the source cannot be found at once, the fossils should be numbered, their location marked on a map, and a note entered to describe their location and the manner of their occurrence. They should be saved, for they may eventually prove more valuable than anything broken from an outcrop. If the search is unfruitful, it is worthwhile to get advice from a paleontologist since he can generally point out the sorts of rocks that are likely to be fossiliferous.

In sequences of clastic noncalcareous rocks, beds containing such fossils as molluscs, echinoderms, and large foraminifera typically stand out as relatively resistant calcareous ribs. The fossils may also occur in calcareous concretions. Weathered outcrop surfaces of these rocks are irregularly fluted and here and there a particularly well-preserved fossil may be etched into relief. In sequences of limestones, sandstones, and calcareous shales, such fossils as algae, molluscs, corals, echinoderms, brachiopods, and foraminifera are likely to form massive reef-like accumulations (bioherms) or thin, richly fossiliferous beds. These structures should be sampled critically because large and spectacular specimens consisting mainly of algae and corals may not be diagnostic. Well-preserved fossils of floating or swimming invertebrates like ammonites and foraminifera, and thin-shelled animals like trilobites, are best sought in bedded shales and fissile or thin-bedded limestones, while the delicate and important graptolites are generally found in laminated shales or limestones that can be split easily along bedding planes.

Fossil bones and teeth occur in a great variety of sedimentary and volcanic rocks, but remains of the more valuable terrestrial vertebrates occur most frequently in nonmarine lacustrine, fluviatile, or deltaic deposits. These rocks are commonly varicolored in shades of red, green, maroon, or gray. Small bones and teeth that weather from friable sandstones and siltstones may accumulate in loose material under steep ledges. These fragments can be found most quickly by sifting the loose materials; because of their weight and size, they may also be concentrated in nearby rivulet channels. Fossil bones from unconsolidated Tertiary and Quaternary sediments can be distinguished from modern bones by their greater weight and by the fact that they do not give off a strong odor when held to a flame. If bones or teeth are already loose, they may be packed in a great deal of soft paper and transported, but if they are found in an outcrop, they should not be removed from their matrix until a vertebrate paleontologist has seen them. He will use special techniques to remove and pack the fossils; furthermore, if he can see the position of the fossils in the outcrop, he may be able to locate more parts of the same animal. If only a few vertebrate fossils can be collected from a remote locality, the teeth, skull, and foot bones are likely to be most valuable.

Fossil leaves, flowers, and seed pods that show structural details are valuable fossils, though paleobotanists may need a large collection from one locality in order to determine their age or to suggest a correlation with other rock units. Leaves and flowers generally occur as brown impressions or gray to black carbonaceous films on the bedding planes of shales, slates, or thinly bedded water-laid tuffs. The carbonaceous films are fragile, thus the rock slabs must be handled carefully and wrapped individually in soft paper. Well-preserved fossil wood is occasionally useful for approximate age determinations or for correlations; it may provide useful indications of climatic conditions of the past.

Where fossils occur in a firm matrix, it is better to collect them with some matrix rather than to try to work them free in the field. This saves the fossils from being spoiled by crude trimming techniques and also protects them during transport. Parts that tend to disintegrate can be painted with shellac, daubed with cellulose cement, or sprayed with an adhesive. All but the toughest specimens should be padded with enough paper, rags, grass, or leaves to prevent abrasion by other samples in the knapsack. Each specimen should then be placed in a sample bag and marked as described in the next section. In camp, specimens may be cleaned and trimmed if a small hammer, sand bag, cold chisels, shellac, and brushes are available; however, specimens that are scarce, unusually well preserved, or likely to be critical in age determinations should be left for the more expert hands of a paleontologist.

Microfossils. Small fossils customarily called *microfossils* are frequently of great value because: (1) they can commonly be separated from rocks that are otherwise unfossiliferous, (2) their large numbers and widespread distribution make them ideal for biostratigraphic studies, and (3) they may be separated from drill cores and cuttings. The larger microfossils can be seen readily with the unaided eye and can be identified approximately with a hand lens. Examples of these are the large foraminifera (fusulinids, nummulites, and orbitoids), the ostracods, and the larger conodonts. These fossils are most likely to occur in limestones, shales, and cherts (both bedded and nodular), but they also occur in various sandstones. In friable clastic rocks, the large foraminifera may be concentrated in more firmly cemented calcareous layers or concretions.

Smaller microfossils range from sizes that can be seen readily with a hand lens down to truly microscopic forms. The commonest of these are spores, pollen, small foraminifera, radiolaria, conodonts, and diatoms. Even where these fossils cannot be seen with a hand lens, it is usually worthwhile collecting samples of shales, mudstones, siltstones, chalks, and friable siliceous or tuffaceous rocks. These samples are washed and sieved in the laboratory, and a binocular microscope is used to pick the fossils out of the other clastic materials. Bulk samples of nonfriable limestones and calcareous shales may be collected for conodonts, diatoms, and radiolaria because these noncalcareous fossils can be separated by decomposing the rocks in acid. Microscopic spores and pollen can also be separated in this way; they are most likely to occur in fresh, dark gray (carbonaceous) shales and limestones.

Foraminifera are leached quite rapidly from porous rocks; therefore a pick, mattock, or some other entrenching tool must be used to cut down to fresh (typically gray) rock. Shales with gypsiferous crusts on their fractures are likely to be barren of usable calcareous microfossils. Suggestions regarding systematic sampling for microfossils are given in Section 12–9.

1–8. Numbering and Marking Specimens

Each rock or fossil specimen must be marked with a number matching that used in the notes and on a map or aerial photograph. Most samples can be marked directly with a felt-tip pen. The number may also be written on a piece of adhesive tape fixed firmly to the sample in the field. If the sample is wet, the number can be written on a piece of paper that is secured to the sample with string or a rubber band. The sample should then be put in a paper or cloth bag on which its number is marked clearly so that it can be identified without being unpacked.

The last figure of the specimen number is generally the same as the note number and the locality number on the map. If more than one sample is collected at one locality, lower-case letters may be used after the locality number. The locality number should be preceded by letters or numbers that will identify it with the particular base map or aerial photograph on which its location is marked. Where a survey or series of surveys is being made by a group of geologists, it is also well to write the initials of the collector before all the other numbers so that supplementary information on the specimen can be obtained easily. As an illustration of such a number, the specimen of note 6, Fig. 1–2, is JRD-F3-6.

After each day's field work, rocks and fossils other than fragile and well-packed specimens may be laid out in order and checked against the numbers of the field notes. If the label applied in the field does not appear permanent, a daub of paint can be put on the rock and the number lettered on it with waterproof ink. The quick drying autoenamels that come in small bottles with a brush applicator are ideal for this purpose. Another way to apply a permanent number is to letter it with waterproof ink on a small strip of paper, which is glued to the specimen with cellulose or plastic cement. A little cement applied over the patch protects it from moisture and abrasion.

Packing samples for shipping. Small numbers of relatively light samples may be shipped in strong cardboard boxes (preferably not in bags), but wooden boxes or nail kegs should be used where large numbers of samples or very heavy samples are being shipped long distances. Samples should be dried thoroughly before being packed. Average rock specimens will not be damaged if wrapped in a double sheet of newspaper and packed firmly into place against other samples. Soft rocks, fragile minerals, and most fossils must be wrapped first in soft paper and then in tougher paper so as to be completely padded. Straw, grass, rags, or crumpled paper should be stuffed into the box since almost all damage during shipping is caused by movements of loose samples. The boxes should be nailed or wired strongly enough to withstand rough handling.

1–9. Locality Descriptions

Data and specimens that are worth collecting are worth locating with care; this can be done either by marking the point directly on a map or aerial photograph or by recording a description of the locality in the field notes. When a map has a scale of 1:62,500 or less, or is known to be generalized or obsolete, important rock and fossil localities must be described fully in the notes. The purposes of locality descriptions are: (1) to assist others in finding the locality, (2) to assist the geologist in revisiting the locality, and (3) to provide the means of plotting the localities on a suitable base map when it becomes available.

Locality descriptions are best based on points or features that are permanent, can be pinpointed on the ground, and are generally shown accurately on maps. Perhaps the best of these are triangulation stations and bench marks of the various government surveys or the township-section system of the Bureau of Land Management. These points generally carry a marker with the name or numbers of the station stamped in a brass plate, though some section corners are marked only by a pipe or a roughly marked stone. From these points, a locality may be measured and described by means of a compass bearing and a paced distance, as described in Chapters 2 and 3. In areas covered by the township-section system of the Bureau of Land Management, roads or fences commonly lie along the north-south and east-west borders of sections and quarter sections, and this gives a handy refer-

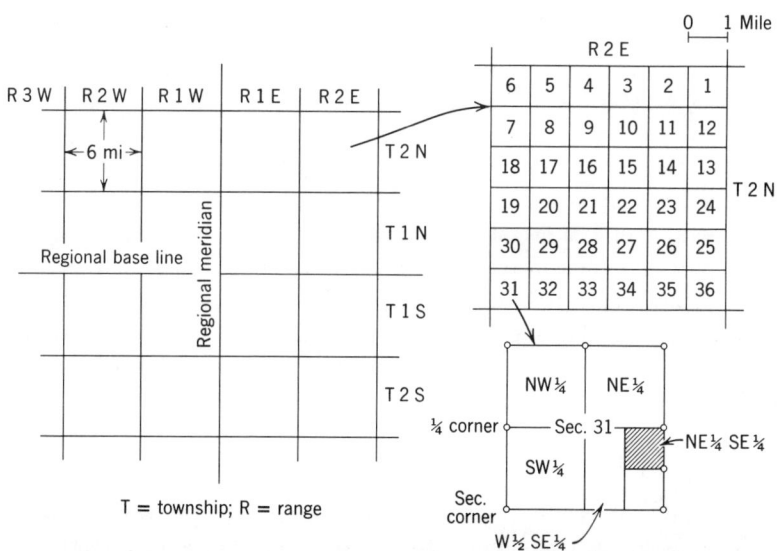

Fig. 1–3. Bureau of Land Management cadastral system of numbering townships, sections, and parts of sections. After Bureau of Land Management, 1947.

ence grid from which to pace out locality positions (Fig. 1–3). Other suitable landmarks for locality descriptions are sharply defined hilltops, stream intersections, road or railroad crossings, solid buildings, and similarly permanent features. Where localities are so far from these features that a compass bearing and paced distance cannot be used to describe them in the field notes, they may be located by taking bearings on several prominent points (Section 4–2).

Locality descriptions should begin with the name of the quadrangle or some large and well-established geographic feature, proceed to smaller and more local features, and, finally, describe the immediate landmarks and appearance of the outcrop itself. The field notes shown in Fig. 1–2 include an example of a locality description (note 6).

References Cited

Bureau of Land Management, 1947, *Manual of instructions for the survey of the public lands of the United States, 1947:* Washington, D.C., U.S. Government Printing Office, 613 pp.

Chamberlin, T. C., 1897, The method of multiple working hypotheses: *Journal of Geology*, v. 5, pp. 837–848.

Gilbert, G. K., 1886, The inculcation of scientific method by example, with an illustration drawn from the Quaternary geology of Utah: *American Journal of Science*, v. 31, pp. 284–299.

2

Using the Compass, Clinometer, and Hand Level

2–1. The Brunton Compass

A compass, clinometer, and hand level can be used to make a great variety of surveys and to measure the attitudes of various geologic structures. These three basic instruments are combined in the Brunton Pocket Transit, which is commonly called the Brunton compass. This compass is held by hand for most routine procedures, as those described in this chapter; however, it can be mounted on a tripod for more precise measurements, or can be used with a special ruler on a plane table (Section 6–2). Although the detailed instructions given in this chapter pertain especially to the Brunton compass, the same general procedures can be adapted readily to other kinds of compasses, clinometers, and hand levels.

The various parts of the Brunton compass are shown in Fig. 2–1. The compass is made of brass and aluminum—materials that will not affect the magnetized compass needle. When the compass is open, the compass needle rests on the pivot needle. The compass needle can be braked to a stop by pushing the lift pin, which is located near the rim of the box. When the compass box is closed, the lift pin protects the pivot needle from wear by lifting up the compass needle.

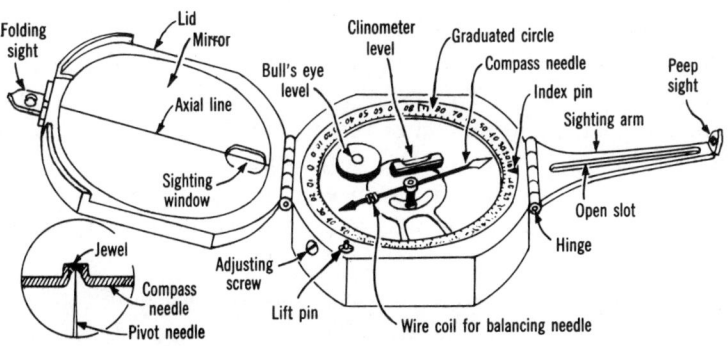

Fig. 2–1. The Brunton compass. Insert at lower left shows enlarged section through needle bearing.

The round bull's eye bubble is used to level the compass when a bearing is read, and the tube bubble is used to take readings with the clinometer. The clinometer is moved by a small lever on the underside of the compass box (not shown in the figure).

A compass should be checked to ascertain that: (1) both levels have bubbles, (2) the hinges are tight enough so that the lid, sighting arm, and peep sights do not fold down under their own weight, and (3) the point of the sighting arm meets the black axial line of the mirror when the mirror and sighting arm are turned together until they touch. Other adjustments that may be required are described in Section 2–10.

2–2. Setting the Magnetic Declination

The graduated circle of the compass can be rotated by turning the adjusting screw on the side of the case. The 0 point of the graduated circle is brought to the point of the index pin to measure bearings from magnetic north. To measure bearings from true north (the usual case), the graduated circle must be rotated to correct for the local magnetic declination. The local declination and its change per year are given in the margin of quadrangle maps; however, the correction for annual change will be only approximate if the map is more than about 20 years old. The declination can also be determined from an isogonic chart (Appendix 6). Finally, the declination at any given point can be determined by setting the compass on a firm, level surface and sighting on Polaris, the North Star. This reading should be corrected approximately for elongation (Section 7–9).

Because the east and west sides of the compass circle are reversed, it may be momentarily confusing as to which way to turn the circle. Each setting should be reasoned out and checked. A declination of 20° east, for example, means that magnetic north is 20° east of true north, and therefore the circle is turned so that the index points to 20 on the E side of 0. To check this, the compass is held level and oriented so that the white end of the needle comes to rest at 0. The entire compass is then rotated 20° in the direction known (geographically) to be east of north. If the needle then points in the direction of the sighting arm, which is magnetic north, the declination has been set correctly.

2–3. Taking Bearings with the Compass

A *bearing* is the compass direction from one point to another. A bearing always has a unidirectional sense; for example, if the bearing

Using the Compass, Clinometer, and Hand Level

from A to B is N 30 W, the bearing from B to A can only be S 30 E. Using the Brunton compass, the correct bearing sense is from the compass to the point sighted when the sighting arm is aimed at the point. The white end of the needle gives the bearing directly because the E and W markings are transposed. To read accurate bearings, three things must be done simultaneously: (1) the compass must be leveled, (2) the point sighted must be centered exactly in the sights, and (3) the needle must be brought nearly to rest. When the point sighted is visible from the level of the waist or chest, the following procedure should be used.

1. Open the lid about 135°; turn the sighting arm out and turn up its hinged point (Fig. 2-2A).
2. Standing with the feet somewhat apart, hold the compass at waist height with the box cupped in the left hand.
3. Center the bull's eye bubble, and, keeping it approximately centered, adjust the mirror with the right hand until the point sighted and the end of the sighting arm appear in it.
4. Holding the compass exactly level, rotate the whole compass (on an imaginary vertical axis) until the mirror images of the point sighted and the tip of the sighting arm are superimposed on the black axial line of the mirror.
5. Read the bearing indicated by the white end of the needle, which should be nearly at rest.
6. After reading the bearing, check to make sure the line of sight is correct and the compass is level.
7. Record or plot the bearing at once.

When the point sighted is visible only at eye level or by a steep downhill sight, the following instructions apply.

1. Fold out the sighting arm as above, but open the lid only about 45° (Fig. 2-2B).

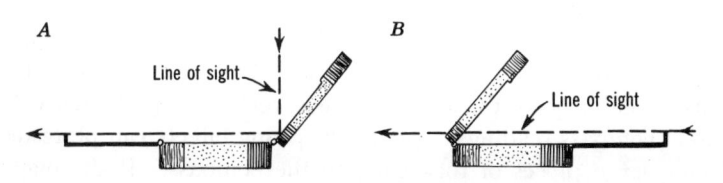

Fig. 2-2. Compass set for taking a bearing at waist height (A) and at height of eye (B).

2. Hold the compass in the left hand at eye level, with the sighting arm pointing toward, and about 1 ft from, the right eye.

3. Level the compass approximately by observing the mirror image of the bull's eye bubble, and, holding the compass approximately level, rotate it until the point sighted appears in the small sighting window of the lid.

4. Holding the compass exactly level, rotate it until the point sighted and the point of the sighting arm coincide with the axial line of the window.

5. Read the bearing in the mirror, double checking for alignment and level.

6. Transpose the direction of the bearing before recording or plotting it (the compass was pointed in reverse of its bearing direction).

With practice, bearings can be read to the nearest $\frac{1}{2}°$ provided the needle is steady. When holding the compass at waist level, the largest errors result from sighting the wrong point in the mirror. In the second method, a good deal of patience is required to level and read the compass from a mirror image. In either method, the compass must be leveled accurately to give good results on inclined sights. If the swing of the compass needle cannot be damped by the lift pin, the bearing must be read as the center of several degrees of swing. Unless much patience is used, these readings are likely to have errors of 1 or 2°.

2-4. Magnetic Deflections of Compass Bearings

The compass will give incorrect bearings if there is any local deflection of the earth's magnetic field. Objects containing iron, such as knives, hammers, and belt buckles, should be kept at a safe distance while a reading is made. This distance can be determined by placing the compass on a level surface and bringing the object toward it until the needle is deflected. A strong pocket magnet must never be carried near a compass. Steel fences, railroad rails, and steel pipelines should be avoided if possible.

Rocks and soils rich in iron, especially those containing the mineral magnetite, can cause deflections that are difficult to detect. Bodies of basalt, gabbro, skarn, and ultrabasic rocks are particularly likely to affect compass readings. Relatively strong effects can be tested by bringing large pieces of rock close to the compass. If the magnetic mass is small compared to the distance between two stations, foresights and backsights between the stations will give inconsistent re-

Using the Compass, Clinometer, and Hand Level

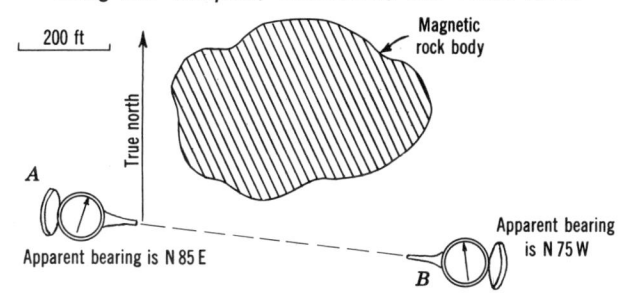

Fig. 2-3. Deflection of compass needles at two ends of a line that passes near a small magnetic rock body.

sults (Fig. 2-3). Where larger masses are involved, the deflection can be measured by pinpointing two stations on an accurate map, measuring the bearing between them with a protractor, and comparing this bearing with a compass bearing taken between the same stations in the field. This measurement will correct the declination for that part of the area. The same result may be achieved by taking readings on Polaris at a number of points within the area to be mapped. If the magnetic disturbances are moderate and vary systematically over a given area, a local isogonic map can be constructed from which corrections of compass readings may be made. If magnetic deflections are large and distributed irregularly, mapping must generally be done with other instruments, as the peep-sight alidade (Section 6-2) or the *sun compass*, a nonmagnetic instrument operated on the basis of the time of day and the direction of the sun's rays. It is also possible to make a compass traverse in such a way that the deflections are accounted for (Section 3-5).

2-5. Measuring Vertical Angles with the Clinometer

Vertical angles can be read to the nearest quarter of a degree with the clinometer of the Brunton compass. Instructions for this procedure are:

1. Open the lid about 45° and fold out the sighting arm, with its point turned up at right angles.
2. Hold the compass in a vertical plane, with the sighting arm pointing toward the right eye (Fig. 2-4). The compass must be about 1 ft from the eye so that the point sighted and the axial line in the sighting window can be focused clearly.

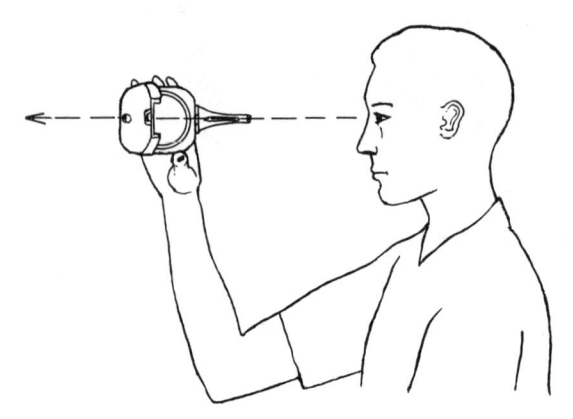

Fig. 2–4. Using the Brunton compass as a clinometer.

3. Look through the window of the lid and find the point to be sighted, then tilt the compass until the point of the sighting arm, the axial line of the window, and the point sighted coincide.

4. Move the clinometer by the lever on the back of the compass box until the tube bubble is centered, as observed in the mirror.

5. Check to make sure the sights are still aligned, then bring the compass down and read and record the angle.

Computing difference in elevation. The approximate difference in elevation between the point occupied and the point sighted can be computed in the field if the slope distance is paced and if a small table

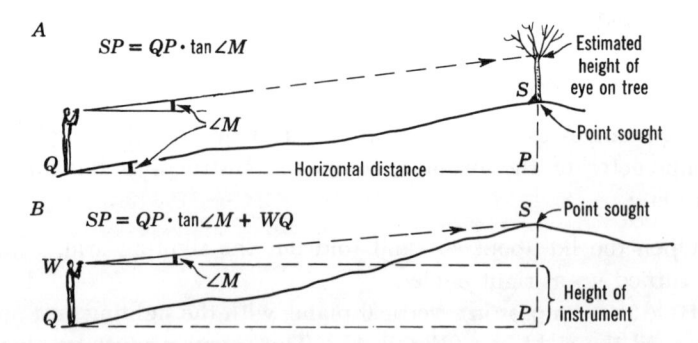

Fig. 2–5. Finding difference in elevation (SP) from a vertical angle (M) and horizontal distance (QP). (A) Relation used when sighting on a point at height of eye. (B) Relation used when sighting uphill to point itself.

Using the Compass, Clinometer, and Hand Level

of sines of angles is available (difference in elevation = slope distance × sine of vertical angle). The difference in elevation can also be determined if the horizontal distance between the points can be scaled from a map or aerial photograph, the difference in elevation then being computed from a table of tangents. The height of the instrument above the point occupied is taken into account either by (1) sighting to a point that is at an equal distance above the ground (Fig. 2–5A) or (2) sighting directly to the point on the ground and correcting the difference in elevation by adding the height of the instrument for uphill sights and subtracting it for downhill sights (Fig. 2–5B).

2–6. Using the Brunton Compass as a Hand Level

The Brunton compass is converted to a hand level by setting the clinometer exactly at 0, opening the lid 45°, and extending the sighting arm with the sighting point turned up. The compass is held in the same way as when measuring vertical angles. It is tilted slowly until the mirror image of the tube bubble is centered. Any point lined up with the tip of the sighting arm and the axial line of the sighting window is now at the same elevation as the eye of the observer. By carefully rotating the entire instrument with a horizontal motion, a series of points that are at the same elevation can be noted.

Difference in elevation by leveling. The difference in elevation between two points can be measured by using the Brunton compass as a hand level. The measurement is started by standing at the lower of the two points and finding a point on the ground that is level with the eye and on a course that can be walked between the

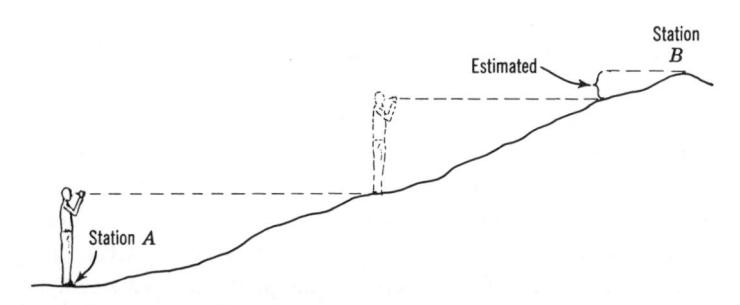

Fig. 2–6. Measuring the difference in elevation between two stations by using a hand level and counting eye-level increments.

two end points. As the first level sight is made, an object such as a stick, leaf, or stone is marked mentally and kept in sight while walking to it. Standing on this marker, another point at eye level is chosen farther uphill, and the procedure is repeated until the end point is reached (Fig. 2–6). The number of moves is tallied and multiplied by the height of the surveyor's eye, with the last fractional reading estimated to the nearest even foot or half a foot. If the country is reasonably open, the traverse can be made both quickly and accurately. The serious error of miscounting the tally can be prevented by keeping a pencil tally on the cover of the compass or by using a tally counter.

2–7. Measuring Strike and Dip

The strike and dip of planar geologic structures, such as bedding, faults, joints, and foliations, can be determined by several methods with the Brunton compass. Strike is generally defined as the line of intersection between a horizontal plane and the planar surface being measured. It is found by measuring the compass direction of a horizontal line on the surface. Dip is the slope of the surface at right angles to this line. The best method for measuring a given strike and dip depends on the nature of the outcrop and the degree of accuracy desired. The amount of the dip, too, may affect the choice because steeply dipping planar structures can be measured far more accurately and easily than gently dipping ones. Special methods are needed to measure dips of less than 5° accurately.

In the section on taking bearings (Section 2–3), it was noted that a bearing has a unidirectional sense and that the white end of the compass needle must be read in all cases. Traversing and locating points by intersection require this strict usage. In the case of a strike line, however, there is no reason for such a distinction. It is recommended that for measuring strike only the *north half* of the compass be used, regardless of which end of the needle points there. Strikes would thereby be read as northeast or northwest, never southeast or southwest. This helps eliminate the occasional serious error of transposing a strike to the opposite quadrant when reading, plotting, or recording it. These errors can occur easily where two men are working together and calling out structural data from one to the other.

The instructions that follow refer to the strike and dip of a bed or beds, but the same methods can be used for measuring other planar structures.

Using the Compass, Clinometer, and Hand Level

1. Most accurate method for one outcrop. This method requires an outcrop that shows at least one bedding surface in three dimensions. If the bedding surface is smooth and planar, no more than a square foot or so of surface need be apparent, but if it is irregular, several square feet must be visible. Where road cuts or stream banks truncate beds smoothly, a hammer can be used to expose and to clean off a bedding surface. The measurement is made by stepping back 10 ft or so from the outcrop to a point from which the bedding surface can be seen clearly. The observer then moves slowly to the right or left until he is in that one position where the bedding surface just disappears and the bedding appears as a straight line (Fig. 2–7). In this position, his eye is in the plane that includes the bedding surface. Using the Brunton compass as a hand level, the point on the edge of the bedding surface that is level with the eye is found. This horizontal line of sight is the strike of the bed, and its bearing is determined and plotted.

To measure the dip, the observer opens the lid and sighting arm of the compass and holds it in the line of sight used to measure the strike. The compass is then tilted until the upper edge of the box and lid appear to lie along the bedding plane (Fig. 2–7B). The clinometer lever is rotated until the tube bubble is centered, and the dip is then read and recorded to the nearest degree. The intersection of the strike line and the dip line on the map is customarily taken to be the point at which the reading was made.

If the bedding line contains no distinctive feature that marks the point on a level with the observer's eye, it is necessary to mark the point with a pebble, stick, or some other object; otherwise, the reading will be approximate only.

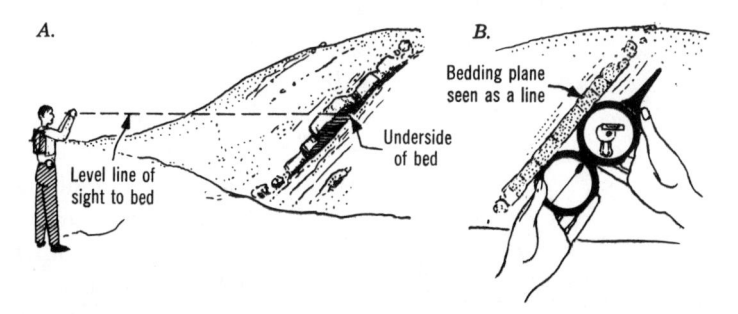

Fig. 2–7. Measuring strike and dip. (*A*) Sighting a level line in the plane of a bedding surface. (*B*) Measuring the dip of a bedding surface.

II. Method for steeply dipping beds. Where beds dip more than 60°, a level line of sight to a bedding plane can be found by a somewhat less precise method than that just described. After the observer is in a position to see a bedding surface as a line, he takes a bearing by the eye-level method (second method of Section 2–3), being careful to center the bull's eye bubble exactly and to sight on the trace of the bedding. If he prefers to take the bearing by the chest-level method, he may move to such a position that *the compass will be in the plane of the bedding surface.* By either method, the compass can be held level enough to define the strike line within a few degrees. The dip is measured in the same way as described above.

III. Method of leveling between two outcrops. This is the best method for measuring the attitude of gently dipping beds. It can be used where a well-marked bed crops out on the opposite sides of a small valley, gully, or excavation. The observer stands in front of one outcrop so that his eye is at the level of the top or base of the bed. He then uses the Brunton compass as a hand level to find a level line to the same bedding surface in the opposite outcrop. The bearing of this line is the strike of the bedding, and the dip is measured by sighting across at the opposite outcrop, just as described above for method I.

IV. Method of holding compass against bedding surface. Methods I and II cannot be used where brush, rocks, or trees make it impossible to get in a position to see a bedding surface as a line. It may then be necessary to take a reading by holding the compass against a bedding surface. The surface chosen must be smooth, clean, *and representative of the outcrop.* The compass is opened and one of the lower edges of the compass box is held firmly against the bedding surface; the compass is then rotated until the bull's eye bubble is centered (Fig. 2–8A). The bearing in this position is the approximate strike.

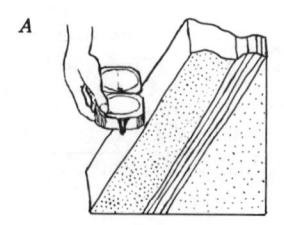

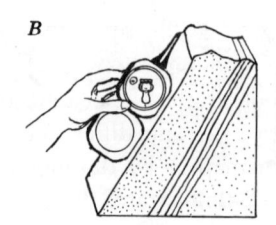

Fig. 2–8. Measuring approximate strike and dip by holding compass against a bedding surface.

The dip is read by placing the side of the compass box and lid directly on the bedding surface and at right angles to the direction of the strike (Fig. 2–8B). The clinometer is then turned until the tube bubble is centered.

The simplicity of this method makes it appealing; however, the base of the compass is only 2 in. long, and therefore large errors may result unless judgment is used in selecting the surface. On a fairly large bedding surface, a clip board or map case (with no steel parts) may be held firmly against the surface to help average out its irregularities. The compass is then held against the board to make the reading.

V. Methods for nearly horizontal beds. Where dips are less than 10°, small irregularities in the bedding will cause major local variations in strike. Method III will give superior results, but it requires unusually good exposures. Method I will give reasonably good results down to dips of about 4° if a large single section of a bedding surface is exposed. To obtain the strike of beds dipping still more gently, it is often easier to determine the direction of maximum slope (dip) first and then to take the bearing of a line at right angles to it. A quick way to do this is to pour a little water on the surface, stand over the streak it forms, and measure the strike as the bearing at right angles to the streak. Another way to find the maximum slope is to place the compass on a bedding surface as though measuring dip, then center the clinometer bubble while rotating the compass slowly back and forth against the surface. Because of the irregularities on most bedding surfaces, however, this method is generally unreliable.

VI. Three-point method. The strike and dip of gently dipping beds can be determined accurately from three points that lie at different elevations on one bedding surface. The bedding surface must be identified with certainty at each outcrop, and therefore the top or base of a distinctive rock unit or bed should be used. The construction requires that the distances and directions between the three points be known, as well as the differences in elevation between them. The distances and directions can generally be measured from a map or aerial photograph; they can also be determined by a compass traverse (Chapter 3). The differences in elevation can be determined from a contour map or by the methods described in Sections 2–5 and 2–6.

Figure 2–9 illustrates three outcrops (A, B, and C) and the construction used to determine strike and dip from them. The strike is found by locating a point, as D, which has the same elevation as the inter-

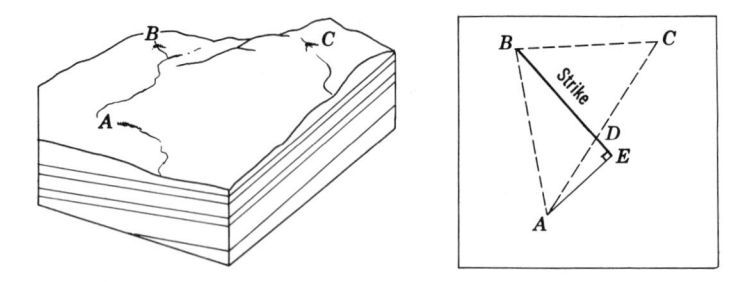

Fig. 2-9. Determining average strike and dip by making a construction based on a map of three points on a bedding surface (see text for explanation).

mediate point, *B*, and is on the line joining the highest and lowest points, *C* and *A*. Point *D* can be located by solving the relation

$$AD = AC \frac{\text{difference in elevation between } A \text{ and } B}{\text{difference in elevation between } A \text{ and } C}$$

The level line *BD* is the direction of strike. If the dip is small, it is best to determine it by measuring the distance *AE* (perpendicular to the strike) and solving the relation

$$\text{tangent of the angle of dip} = \frac{\text{difference in elevation between } A \text{ and } B}{AE}$$

The dip is then read from a table of tangents. The dip may also be determined graphically by constructing a view at right angles to the strike, as explained in books on engineering drawing and structural geology.

VII. Method for beds truncated by nearly level surfaces. The strike of beds exposed on horizontal or nearly horizontal surfaces of unpaved roads, trails, stream beds, terraces, or bare ridges may be measured quickly and accurately. This is done by standing directly over the outcrop, cupping the compass in both hands at waist height, and aligning it with the trace of bedding that passes underfoot. The long slot in the sighting arm is useful for aiming the compass in this way. In order to determine the dip, it is generally necessary to hammer out and clean off a bedding surface, upon which the compass is placed directly.

2-8. Where to Take Strike and Dip

Before measuring strike and dip, it must be determined whether the attitude will reliably represent bedding. Some "outcrops" are not in place at all, being large boulders, blocks of float, or segments of landslides. A general survey of the slopes around outcrops will generally resolve such problems. If there is still some question as to the reliability of a measurement, a question mark may be entered next to the plotted symbol or the strike line may be broken (Appendix 4).

Downhill creep of soil and mantle commonly bends planar structures that occur on slopes. Where this is likely, readings are best restricted to the bottoms of valleys, to the tops of ridges, and to cuts that are deep enough to get beneath the zone of deformation. Thick beds of massive rocks, such as sandstone, quartzite, and lava, are not likely to be bent by downhill creep.

Outcrops should also be examined to make certain that what is taken for bedding or foliation is not jointing, bands of limonite staining, or some other kind of discoloration. Changes in texture (especially grain-size) or changes in mineral composition are the best indicators of bedding. In massive sandstones, bedding may be shown only by the approximate planar orientation of mica flakes, platy carbonaceous or fossil fragments, shale chips, or platy and elongate pebbles. The possibility that bedding features in sandstones are only local cross-bedding must be considered. The identification of bedding in metamorphic rocks may be still more difficult, and there are a number of planar structures of igneous and metamorphic rocks that should be identified carefully wherever they are measured and plotted (see appropriate Sections in Chapters 12, 13, 14, and 15).

2–9. Measuring Trend and Plunge of Linear Features

Trend and *plunge* are used to define the attitudes of linear features. The trend of a linear feature is the compass direction of the vertical plane that includes the feature. If the feature is horizontal, only the compass direction is needed to define its attitude. If it is not horizontal, the trend is taken as the direction in which the feature points (*plunges*) downward. The plunge is the vertical angle between the feature and a horizontal line.

To measure the trend of a linear feature, the observer stands, if possible, directly over a surface that is parallel to the linear feature (Fig. 2–10). This surface is sometimes described as "containing" the feature or as the surface on which its maximum length is seen. The observer faces in the direction in which the linear feature points downward. He determines the bearing of this direction (the trend) by holding the compass at waist height and looking down vertically on the feature through the slot of the sighting arm. When the slot

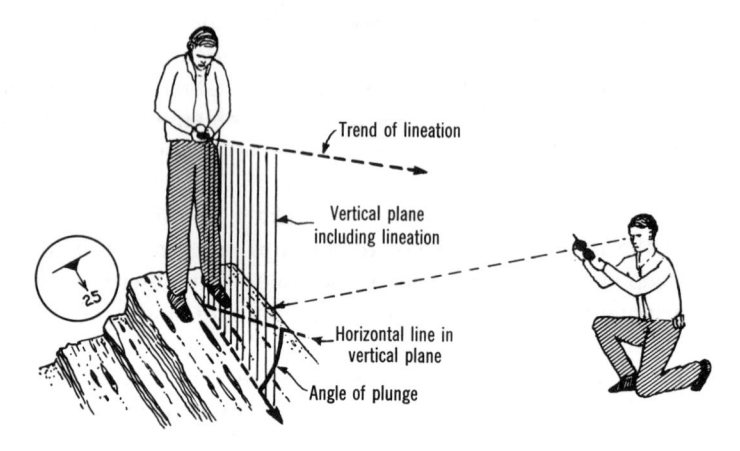

Fig. 2-10. Measuring the trend of linear structures (upright figure) and the amount of their plunge (kneeling figure). The map symbol (insert) shows how the lineation arrow can be combined with the strike and dip symbol of the foliation surfaces.

is parallel to the trend of the feature, the bearing at the white end of the needle is read. The trend is then plotted on the map as a line originating at the point occupied by the observer.

To measure the plunge of the feature, the observer moves so that he is looking at right angles to its trend (Fig. 2-10, right). The reading is taken on the trace of the linear feature seen from this position, exactly as in taking the dip of the trace of a bed. An arrow point is then drawn on the map at the downward plunging end of the trend line, and the amount of plunge is lettered at the end of this arrow. For horizontal linear features, an arrow point is drawn at both ends of the line.

2-10. Care and Adjustment of the Brunton Compass

The compass should never be carried open in the hand while walking over rough or rocky ground. If an extra mirror and glass cover are included in the field gear, these can be replaced in the field, but if the hinges are bent or the level vials broken, the instrument must be sent to the manufacturer for repair.

If the compass is used in the rain, or if it is accidentally dropped in water, it should be opened and dried because the needle will not function properly when its bearing is wet. The glass cover can be removed by forcing the point of a knife blade under the spring washer that holds it in place. This should be done at a point opposite the dove-tail join of the washer, using care not to crack the glass cover along its edge. With the washer off, the glass cover can be lifted from the box, and the needle taken off its bearing. The cone-shaped pit of the jewel bearing should be cleaned and dried with a sharpened toothpick and a bit of soft cloth or soft paper. The needle lift is then removed and the inside of the compass dried and cleaned. Care should be taken not to push so hard against the clinometer level vial that it turns on its axis. After the needle-lift arm, needle, and glass cover are reassembled, the spring washer should be placed on the glass cover by joining its dove-tailed ends first, placing them firmly against the compass box and then forcing the rest of the washer down by moving the fingers in both directions away from the dove-tail join. If this cannot be done with the fingers, two small pieces of wood can be used.

The mirror can be removed by tapping in the small retaining pin and removing a spring washer like that on the box. The new mirror must be inserted so that its black sighting line is at right angles to the hinge axis of the lid. This can be done approximately by rotating the mirror until the sighting line bisects the sighting window in the lid. The setting should then be checked by closing the lid against the upturned point of the sighting arm and determining if the point of the arm meets the sighting line of the mirror.

Before a new or a borrowed compass is used in the field, it should be checked to make certain the clinometer level is correctly set. To do this, the clinometer is set at 0, and the compass is placed on a smooth board that has been leveled exactly with an alidade or a good carpenter's level (a bull's eye level is not sufficiently accurate for this). If the tube bubble does not come to center, the compass is opened as described above and the clinometer level vial moved appropriately. Ordinarily this can be done without loosening the clinometer set screw. The new setting is checked by placing the compass on the board again, and the procedure repeated until the bubble is centered exactly.

In starting work in a new field area, one may find that the dip of the earth's magnetic field is so great as to cause the compass needle to rub against the glass lid when the compass is held level. To correct this, the glass cover is removed and the copper wire coil on the needle moved one way or the other until the needle lies level.

3
The Compass Traverse

3-1. General Scheme of a Geologic Traverse

In a *traverse* a series of points are surveyed by measuring the direction and distance from one point to a second, and from the second to a third, and so on to the last point. The directional course of this series of measurements is generally irregular; if it is eventually brought around to the starting point, the traverse is said to be *closed*. Each of the points of the traverse is called a *station*, while the measured distance between two stations may be called a *leg* of the traverse. The traverse is used as a skeletal map on which geologic data are plotted along or near the traverse legs. These data may be compiled with those from other traverses to make a complete geologic map. They may also be used to construct a vertical cross section and columnar section showing the rock units and structures traversed. The traverse is commonly used to measure thicknesses of rock units, to compile detailed descriptions of sedimentary or volcanic sequences, and to study deformation in complexly faulted or folded rocks. If a topographic map of suitable scale and accuracy is available, it should be used as a base on which to plot the traverse. Published topographic maps, however, have scales of 1:24,000 to 1:62,500, and many geologic studies require scales larger than 1:6,000 (1 in. = 500 ft). Most detailed studies therefore require constructing a map from the traverse data.

The scale of the traverse is chosen so that the smallest units that must be shown to scale can be plotted easily. If the purpose of the traverse requires, for example, that beds 10 ft thick must be shown to scale, the best scale for the work is 1 in. = 100 ft. In general, anything that cannot be plotted as a feature $\frac{1}{10}$ in. wide cannot be mapped easily to scale, although features that plot $\frac{1}{20}$ in. wide can be shown accurately if great care is used.

The method by which the traverse should be surveyed is determined by the accuracy required and the time and equipment available. For many projects, a compass is adequate for determining the bearings of the legs, and the legs can be measured with a tape or by pacing. An advantage of pacing is that it can be done by one man; its accuracy over reasonably even ground is adequate for most projects whose working scale is 1 in. = 100 ft or more. For more exacting and detailed traverses, a tape should be used to measure distances. The alidade and plane table are commonly used where magnetic variations are large, where terrain is rough, or where distances of several miles must be traversed accurately (Section 8–12). Traverses requiring still greater precision may be made with a transit and steel tape (Section 7–13).

3–2. Determining Pace

In geologic field studies, distances are commonly measured by pacing. It should be possible to pace over smooth terrain with errors of only a foot or two in each 100 ft. The pace should be calibrated and tested for slopes and rough ground as well as for smooth terrain. This can be done by walking a taped course of 200 ft or more and dividing the distance by the number of paces. One or two other courses should then be paced to test the calibrations. To keep the mental tally of paces at a reasonable rate, a pace is generally counted at every fall of the right foot. The calibrated courses should be walked in the same manner as that used normally in the field; a pace should never be forced to an even 5 or 6 ft.

Smooth slopes of low or moderate declivity may be measured by taking normal steps and then correcting the slope distance trigonometrically (map distance = slope distance × cosine of the vertical angle of slope). Slopes that are too steep to walk with normal strides can be paced by establishing one's pace for measured uphill and downhill courses. Pacing over rocky, brushy, or irregular ground requires patience and practice. It can often be done by correcting the count to normal as pacing proceeds. Right-angle offsets can be made around obstacles in some places; in other places, the number of normal paces through an obstacle can be estimated by spotting the places where steps would fall if the obstacle were not there.

The greatest source of error in pacing is miscounting, especially dropping 10 or 100 paces in a long measurement. For this reason traverse legs should be paced in both directions. For much pace and compass work it is desirable to use a *tally counter,* a small odometer-like counter operated by a lever. A *pedometer,* which tallies automatically each time it is jolted by a footfall, is useful for open ground but does not permit corrections for offsets and broken paces.

3-3. Selecting a Course and Planning a Traverse

As in all geologic projects, reconnaissance and comparison of the possible courses for a survey save time in the long run. Where possible, the traverse course should cross the strike of beds or other structures at about 90°. The course must have adequate rock exposures and be accessible enough to permit an efficient survey. These requirements are met ideally by roads that give a series of cuts across the strike of the rocks. Railroads have the drawback of affecting bearings taken with a compass. Many stream courses cross the strike directly and have abundant rock exposures, but pacing along streams may be difficult. Open ridges and beaches commonly afford excellent traverse courses.

In addition to locating the traverse, the reconnaissance should provide answers to the following questions:

1. Is the sequence of rocks monotonously the same, or can the rocks be grouped into two or more units, each characterized by a certain lithology?

2. What scale must be used to show the thinnest rock units that seem significant to the study?

3. If the data from the study will be presented in a geologic cross section or columnar section, how will the sizes of the illustrations limit the data collected and the working scale of the traverse?

Generally, more data must be collected than will be used in the final illustrations, but information must not become so detailed that the principal objectives of a survey are obscured. Field notes will be difficult to use, for example, if they present voluminous, random descriptions of every bed observed.

3-4. First Steps in a Compass Traverse

The procedures of a typical compass traverse include: (1) surveying the stations and legs of the traverse, (2) measuring a profile along the traverse course, (3) plotting the stations and geological features on a field sheet, and (4) describing the geologic features in field notes. The equipment, most of which is described in Section 1-3, should include a geologist's hammer, Brunton compass, clip board or notebook, pocket knife, medium-hard pencil with clip and eraser for plotting, ballpoint pen or medium pencil for taking notes, hand lens, knapsack, specimen bags, protractor, and 6-inch scale. The scale

> Arroyo Seco Area, Monterey Co. Calif.
> Pace-compass traverse on Gila Rd. SW of Lytle Crk.
> R.L. Jeems 4-21-60

Sta. 1   is SW corner S abutment Lytle Crk. bridge
         Elev. = 467 ft. Brng to sta. 2: S2W
         Vert. $L = +1°20'$. Dist. = 92 ft.

Sta. 1 to 29 ft: Gray mdst with distinctive spheroidal frac;
         no sign bdg; weathers pale tan; sand grains of qz,
         mica, feld total 20% (?); both silt and clay abnt
         in matrix. A few forams seen but appear
Forams   leached (sample).

@ sta. 1 + 29 ft: Ctc ss/mdst; sharp, much glauconite
         suggests disconf. but beds parallel. Current grooves
         in mdst trend down-dip.

Sta. 3   is 1 x 2" stake 18 ft S of rd. Brng. sta 2 → 3 =
         S 48 W, Dist. = 147 ft.

Sta. 1 + 29 to sta. 3: Ss; gray (weathering tan), in beds
         1-3 ft thick; interbeds carb silty mdst are 1-6 in.
         thick; ss mainly med-grained but base of
         thicker beds coarse, locally pebbly. Minls ss: ang
         qz (60%), white feld (35), bleached bio (~5).

Fig. 3–1. First note page for geologic compass traverse.

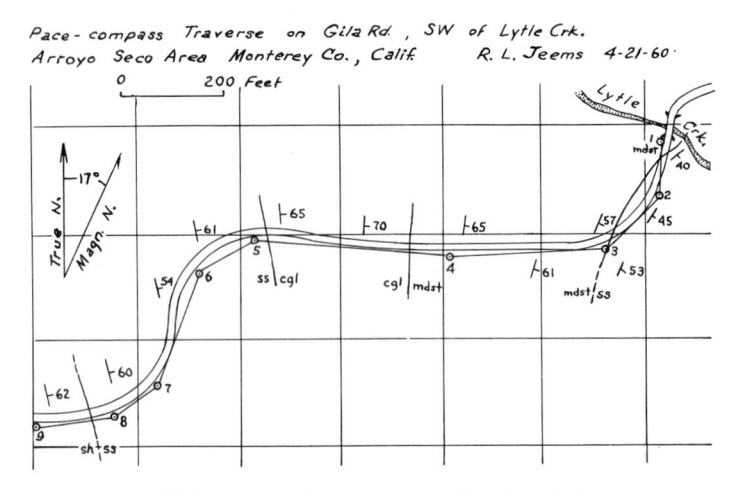

Fig. 3–2. Skeletal map of the traverse, plotted as it is surveyed.

should be divided in tenths of an inch or in other divisions that allow direct conversion to the scale of the field plot. Several sheets of cross-ruled paper are needed for the field plot.

The traverse should be begun and/or ended at permanent markers that can be found readily by other persons. Bench marks, triangulation stations, highway or railroad turning point markers, culvert posts, solid fences, or property corners are all suitable. If these are lacking, solid 2 × 2 in. stakes can be used. In any case, these points must be described in the notes.

The traverse is begun by standing at one of the end points and sighting along the general line of the traverse to the farthest point visible along a course that can be paced. The distance to this tentative station is paced, and if the point proves to be a good choice, it is marked with a stake or other device, and a bearing is read back to the starting point. If possible, the distance should then be checked by pacing back to the starting point. The bearing to the station is then taken, and if it agrees within a degree (preferably, a half a degree) with that of the backsight, the field map is started by plotting this first leg. The bearing and distance are entered in the notes, as shown in Fig. 3–1.

If the field plot of the traverse is made on cross-ruled paper, the ruled lines can be used as a north-south and east-west grid for plotting bearings. The sheet should be given a title, as that shown in Fig. 3–2.

The Compass Traverse

Before the first leg is plotted on it, the general layout of the traverse must be planned so the plot will not go off the sheet on the second or third leg.

Generally, the scale of this map need be only half as large as that of the final illustration. Its purpose is to develop a continuous picture of both the geographic features and the geologic structures as they are traversed. This permits checking the continuity of faults and contacts that cross the traverse course more than once. Bearings and structural attitudes that have been misread or transposed can be detected by orienting the map in the field and comparing the plotted symbols with the outcrops.

3-5. Traversing by Turning Angles

If magnetic disturbances are appreciable (Section 2-4), traverses should be surveyed by turning angles between adjacent legs. Before starting the traverse, it is necessary to determine the bearing of a line that can be sighted from the first traverse station. This can generally be done by: (1) pinpointing the first traverse station on a map, (2) drawing a line on the map from this point to any other point that can be seen from the first traverse station, and (3) measuring the bearing of this line with a protractor. If no usable point can be seen, or if there is no base map, the bearing of the initial leg can be determined by taking a sight on Polaris (Section 2-2); otherwise the compass bearing of the first leg must be assumed to be correct.

The traverse may then proceed as follows:

1. Read and record the compass bearing to the point used to establish a bearing from the first station, as line 1-M in Fig. 3-3.

2. Read and record the compass bearing to the first forward station of the traverse, as line 1-2 in the figure.

3. Determine the angle between these lines ($\angle a$) from the bearings. Compare the computation with the field map to make certain the correct quadrants have been used.

4. Plot the first traverse leg by using this angle.

Fig. 3-3. Surveying a traverse course by turning angles.

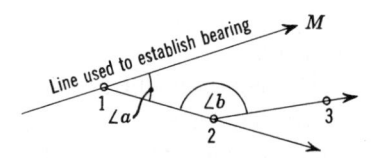

5. Read and record the bearing from Station 2 to Station 1.

6. Read the bearing to Station 3, and compute the angle ($\angle b$) *from the bearings read at Station 2.*

7. Plot the leg 2-3 on the basis of this angle, and continue in a similar way through the other stations.

Although this method will correct for magnetic deflections at each traverse station, local variations between the stations will not be corrected. Structure symbols plotted at or near stations will therefore be correct, but others may be incorrect.

3–6. Plotting Geologic Features on the Traverse

The following geologic structures should be plotted on the field map: (1) contacts between rock units, (2) strike and dip of bedding and other planar structures required for a given project, (3) faults, with dip arrow and, if possible, note of upthrown and downthrown sides, (4) axial trace of folds, with trend and plunge of axis, (5) other linear structures required for a given project, and (6) names or brief notes labeling the more important rocks and features. The plotting should be done with a sharp medium-hard pencil and in the following way: (1) the distance from the nearest station is scaled off along the traverse leg; (2) offset, if any, is plotted to give the location of the structure; (3) the strike or trend of the structure is drawn with a protractor; (4) a dip line or arrowhead is added and the amount of dip or plunge lettered in; (5) *the traverse plot is oriented relative to the terrain, and the attitude shown by the symbol is compared with the outcrop;* (6) if there is doubt about its accuracy, the structure is remeasured. No amount of intuitive thinking in the office can improve on symbols plotted in this way.

Notes should include descriptions of lithology and all small-scale structures that may be helpful in interpreting the history of the rocks. Some notes will be detailed descriptions of specific localities; others will be overall descriptions of rock units. Unit descriptions should include the lithologic characters that make the unit distinctive, the nature of the contacts, and the range of lithologic variations within the unit (Section 1–5). If a detailed cross section is to be prepared from the traverse, drawings of entire road cuts or other outcrops may be helpful.

The notes should be recorded in the order in which features are encountered on the traverse. The stations are generally given consecutive numbers while the notes between them may be listed conveniently by the number of feet from the last station. If parts of the traverse are revisited and supplementary notes are taken, these numbers offer a simple means of fitting new data into place.

3-7. Vertical Profile of the Traverse

The vertical profile of the traverse must be surveyed in order to measure rock units accurately and to project geologic features to a vertical cross section. If the traverse course runs irregularly up and down, the profile should be surveyed as the traverse proceeds. This can be done by measuring vertical angles between stations and to points where the grade of the profile breaks, and computing differences in elevation (Section 2-5). It can also be done by using the Brunton compass as a hand level and finding differences in elevation directly (Section 2-6). If the traverse course is on an evenly sloping road, stream, or ridge, the profile need not be measured at every station; instead, an overall survey of the gradient can be made after the traverse is completed.

Structures must be plotted on the map at the point where they intersect the profile of the traverse or where they intersect an arbitrary datum surface above the profile. Along most road cuts it is convenient to plot structures at waist height (3½ ft) above the road. Structures that cannot be projected reliably to this level must be surveyed individually, and their vertical distances above the datum must be recorded. Figure 3-4 illustrates why this must be done.

The line of section of a geologic cross section will almost always be offset somewhat from the traverse course. The profile of the traverse will therefore not give the true ground profile for the section.

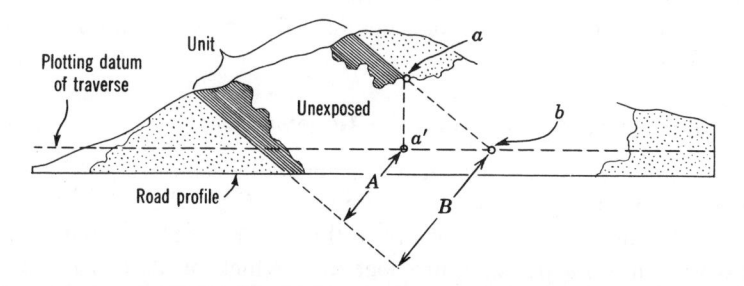

Fig. 3-4. Large road cut, exposing a contact at a. This contact must either be plotted at b, or the elevation at a must be recorded. If it were plotted at a', the thickness of the unit would be measured by A instead of B.

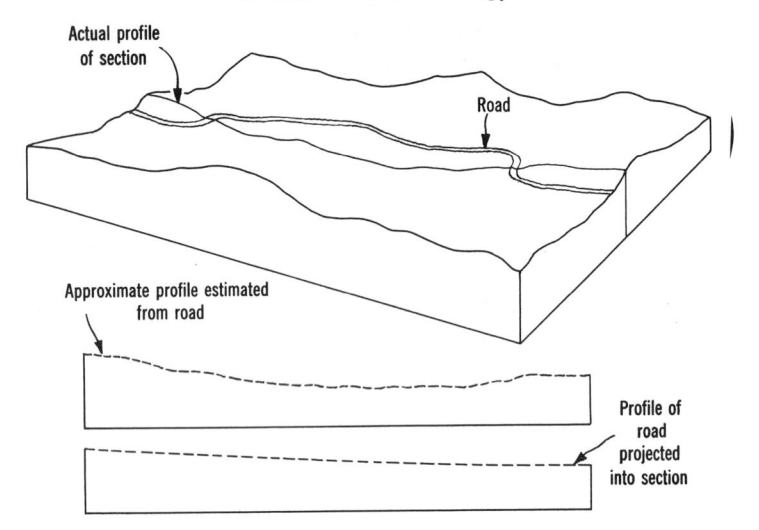

Fig. 3-5. Three possible profiles for a cross section. The traverse course is along the road.

As shown in Fig. 3-5, this problem may be met in three ways. First, the actual ground profile may be surveyed along the line of the cross section. This requires that the section line be chosen first, as described in Section 3-8. The survey of this profile may take considerable time and effort if the terrain is irregular or covered by trees and brush. The second possibility is to estimate an approximate profile along the line of section by reading vertical angles from the traverse course and sketching the terrain between. If this is done, the profile on the final illustration must be dashed and labeled as approximate. The third possibility is to ignore the profile along the line of section and to use the profile of the traverse on the final illustration, dashing it and labeling it suitably.

3-8. Making Illustrations from the Traverse Data

Data from a detailed traverse are generally compiled into a map, a vertical cross section, and a columnar section. These figures may become the only permanent record of the traverse. They are generally drawn in ink on a transparent sheet from which prints can be made. The materials required are a drawing board, T-square, 10-inch or 12-inch triangle, medium-hard pencil, eraser, ruling pen, a piece of tracing linen or heavy drawing paper of appropriate size, crow-quill pen and holder, black waterproof ink, 12-inch scale, and an accurate protractor with a base of 6 or more inches. Additional items that may be useful are a contour pen, a drop-circle compass, lettering guides, and a pen cleaner.

The traverse should be plotted accurately in pencil before any ink is used. This reduces ink erasures and provides a means of arranging the figures in a single illustration. The instructions that follow suggest consecutive steps for this procedure.

1. Using the exact bearings and distances recorded in the traverse notes, plot the traverse course accurately in pencil at the same scale that will be used for the inked illustration.

2. Add all geologic and geographic features, making strike lines about ⅜ in. long; lettering of numbers need be legible only.

3. Draw the line of the cross section, choosing it so that it will pass as close as possible to the outcrops covered by the traverse and at the same time cross the strike of the rocks as nearly to 90° as possible.

4. Project contacts, faults, and strike lines to the line of the cross section. This must be done by special methods if the rocks are folded. In Fig. 3–6A, for example, the units are pinched in an unnatural way if the contacts are projected straight to the line of section. The area is on the limb of a plunging fold, and the contacts should curve into the line of section as shown in B. To make these projections accurately, it is necessary to know the general shape or kind of fold involved and to use methods suitable for projecting folded beds from scattered structure symbols. These methods have been described by Badgley (1959, especially Chapter 3) and in part by Busk (1957); they are also presented briefly in most textbooks on structural geology.

5. Draw a tentative base line for the vertical cross section; this line should have exactly the same length as the section line of the traverse map and *be exactly parallel to it.* The cross section on the final plate should be oriented so that its right-hand end is either its more easterly end or is oriented due north.

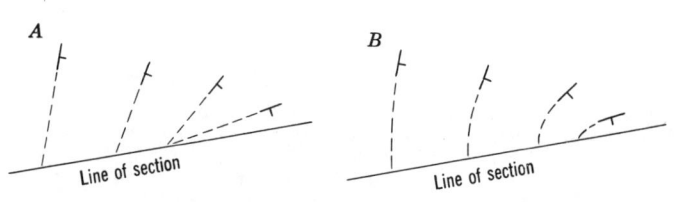

Fig. 3–6. Projecting folded beds into the line of section.

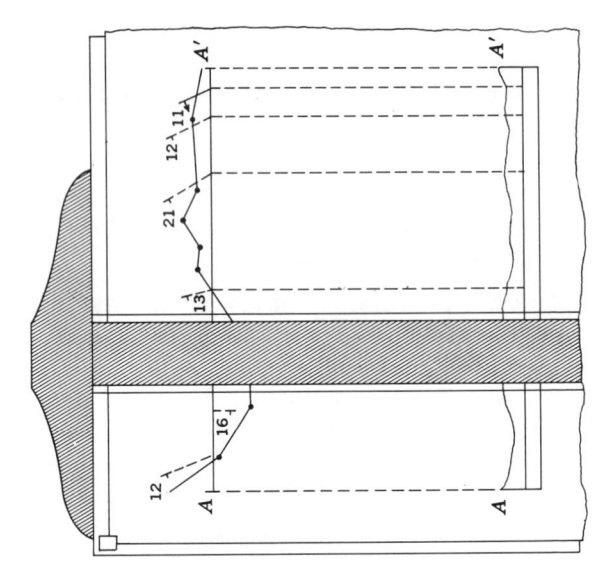

Fig. 3–7. Using a T-square to project features from the line of section AA' to the cross section.

6. About ¾ in. above the base line of the section, plot a profile line that represents the datum at which all or most of the structure readings were taken in the field.

7. Tape the sheet to a drawing board so that the section line is parallel to an edge of the board. Using a T-square on this edge, project all structures from the section line of the traverse to the profile of the cross section or to whatever elevation was used for each structure reading (Fig. 3–7).

8. Draw bedding lines where structure readings have been projected to the profile, making each about ¼ in. long. Where the line of section crosses the strike obliquely, the bedding shown on the cross section must dip less steeply than the dip measured in the field. This dip, called the *apparent dip,* can be determined from the true dip by using the diagram of Appendix 9.

9. Complete structures under the profile line, projecting them down to whatever depth the geologic data will permit. In many cases this will require shifting the base of the section that was drawn provisionally in step 5. Folded beds must be constructed accurately, as noted in step 4. Examine the structures mapped on either side of the line

Fig. 3–8. Constructing an accessory vertical section to determine where a folded unit will plunge into a cross section. Projection lines are dashed. The distances XY and XZ are transferred to the main cross section. These distances can also be found by trigonometric calculations; for example, $XY = XM \cdot \tan 30°$.

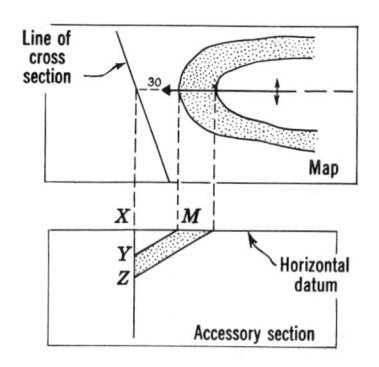

of section to determine if some may cut the section below the ground profile. Folded units, for example, may plunge into the section, and faults that strike parallel to the section may dip into it below the surface (Fig. 3–8). An example of a cross section is shown in Fig. 11–5.

10. Determine the true thicknesses of the rock units. They may be scaled from the cross section if the section cuts the strike of the units at about 90°. If this is not the case, calculate the thicknesses trigonometrically, as described in Section 12–8.

Age	Formation	Thickness	Lithology	Description
Eocene–?	Linder Formation	800'+	Eroded / Poorly exposed	Gray silty mudstone intercalated with 2-6 ft beds of gray calcareous sandstone
Cretaceous	Rojas Sandstone	1050'		Poorly sorted red sandstone with lenses of aplite and quartzite pebbles; 3 ft beds of conglomerate near base; most beds 30-100 ft thick
				Unconformity
Jurassic	Miler Shale	1200'+	Faulted	Brown calcareous shale intercalated with 12 in. beds of fine-grained sandstone — Inoceramus bed — Incomplete

Fig. 3–9. Columnar section, showing a sequence that has been divided into three rock units.

11. Using a pencil and a separate piece of cross-section paper, build up a columnar section of the rocks traversed, starting with the youngest at the top. This may be done at a larger scale than the other figures. Signify breaks in the sequence caused by unconformities, faults, or lack of exposures (Fig. 3-9).

12. Prepare very brief but informative lithologic descriptions for each unit. Letter them in pencil beside the trial column in order to space them correctly.

13. Draw in the lithology of each unit on the column by making a diagrammatic but accurate record of the kinds, proportions, and positions of the rocks. The symbols of Appendix 5 can be used for most units. Check the symbols against the lithologic descriptions.

14. If the map and the sections are to be drawn in one illustration, arrange the separate penciled sheets so that the bases of the cross section and columnar section are parallel. Space the separate drawings so that they are legible but take up as little space as possible. Trace them on a single transparent sheet, or transfer them to a sheet of opaque paper. Add a title that gives a geographic name to the project, the geologist's name, and the date of the survey. Add a bar scale and a north arrow (showing the magnetic declination).

References Cited

Badgley, P. C., 1959, *Structural methods for the exploration geologist:* New York, Harper and Brothers, 280 pp.

Busk, H. G., 1957, *Earth flexures, their geometry and their representation and analysis in geological section, with special references to the problems of oil finding:* New York, William Trussel, 106 pp. (an unabridged republication of the 1929 edition).

4

Plotting Geologic Features on a Base Map

4–1. Selecting and Preparing a Base Map

Maps used for plotting geologic features and note numbers in the field are called *base maps*. *Planimetric* base maps show drainage, culture (man-made features), and perhaps scattered elevations; *topographic* base maps show contours as well. Accurate topographic maps are ideal base maps, for cross sections can be made from them in any direction, and their contours provide several means of plotting outcrops accurately.

To be useful, a given map must have a suitable scale, must have been made recently enough to show existing culture and drainage, and must have contours that delineate the topography accurately. The 7½-minute (1:24,000) topographic maps of the U.S. Geological Survey are excellent base maps. Their scale is such that features 200 ft wide (0.1 in. on the map) can be plotted easily, and features 100 ft wide can be shown to scale if drawn with care (an average pencil line covers 5 to 10 ft of a feature at this scale). Because most of these maps have been made from aerial photographs, however, trails, minor roads, and buildings that lie under trees may be shown inaccurately.

Many quadrangle maps with a scale of 1:62,500 are useful base maps, but some older maps having this scale may be generalized or obsolete. Small features must be plotted carefully on them because features 500 ft across on the ground are only 0.1 in. wide on the map. Moreover, it make take considerable time in the field to locate individual points on these maps, for many topographic and man-made features are not shown on them. It is often preferable to map on aerial photographs with a larger scale and then transfer the geologic features to a topographic map (Section 5–11).

It may be desirable to have a base map enlarged two or three times to give more space for plotting symbols and note numbers. Enlarging does not increase the accuracy of the base map, but it may improve the geological mapping because of the greater ease of plotting. Enlargements must be made so as not to distort the scale of the map and must be printed on paper that will take ink and pencil marks. Ordinary photostat prints are not suitable for most mapping.

Although topographic maps are made by several federal and state agencies, the U.S. Geological Survey prepares and distributes editions that are available for sale to the public. It also publishes index maps that show the location, scale, and edition date of published maps. In addition, regional offices of the Topographic Branch of the Geological Survey publish index maps of topographic work in progress. These indexes should be consulted, since preliminary maps are often available long before final editions. Addresses of regional offices as well as information on available maps and aerial photographs can be obtained from the Map Information Office, U.S. Geological Survey, Washington 25, D.C. When making inquiries, the exact area of interest should be described and also outlined on a small-scale map of the area.

Planimetric maps are valuable bases in areas where roads, buildings, and waterways are spaced closely enough to permit accurate locations of geologic features. In many suburban areas, detailed, modern planimetric maps may be far preferable to generalized or outdated topographic maps, particularly where there is relatively little relief. Detailed planimetric maps are generally held by city or county surveyors and land assessors, by irrigation districts, or by local harbor or river authorities. Modern planimetric maps with intermediate scales are made by certain government agencies, for example, the U.S. Forest Service and the U.S. Bureau of Land Management. Generally, these various maps are not for sale but can be copied or borrowed on a reasonable request.

At least two copies of a base map are needed, one for plotting features in the field and the other for making a geologic compilation as field work progresses. Extra copies can be used for plotting locations and numbers of rock and fossil specimens. The copy used for field work should be cut into sheets that will fit the map case. Each field sheet may be mounted with waterproof cement on a larger piece of cross-ruled paper so that the cross-rule lines form a north-south and east-west grid around the map. Alternately, a grid may be inked directly on the map. A title, lettered in waterproof ink on the margin of each sheet, should show the name of the quadrangle (or other) map, a number or index key to designate the sheet's position on the original map, and the geologist's name.

Base maps for scribing techniques. Base maps can be printed photographically on colored plastic sheets that are marked with a stylus rather than a pencil. This method has been used for a number of years by the Topographic Branch of the U.S. Geological Survey, and it promises to be adopted widely for geologic mapping. The base material (for example, Stabilene Scribe Coat Film of the Keuffel and Esser Co.) is a tough plastic coated by a colored surface film. The film can be cut off with a sharp stylus to make lines, points, or symbols. Errors are corrected by painting over scribed areas with a quick-drying fluid (like the Touch-up Kit of Keuffel and Esser Co.) and then rescribing as necessary. The base map can be transferred to sensitized scribing sheets at most blueprint laboratories, though a photographic negative must first be made from the map. The advantages of the method are that the materials are waterproof and very durable, lines of precise and even width can be made, and errors can be corrected quickly and completely. Contact prints can be made inexpensively from the scribed sheets.

4–2. Locating Field Data on a Base Map

A geologic map is made by locating many points, lines, and other data on a base map. Its value will depend a good deal on the accuracy of these locations. Points on the ground can be located on a map by a number of methods, the most suitable of which must be chosen for a given situation. If there is any question about the accuracy of one method, another should be used to check it. The methods given here can generally be used where terrain and vegetation afford average visibility; Sections 4–3 and 4–4 describe methods that can be used under more limiting conditions.

Location by inspection. Data can be plotted directly by inspection when the configuration of features makes it possible to identify them positively on the map. Examples of such points are distinctive turns or intersections in streams, roads, or ridges.

Location by inspection and a bearing line. Data along linear features, such as ridges, roads, or streams, can often be located by taking a bearing to a point that can be identified exactly on the map, then plotting the reverse bearing from that point to intersect the linear feature on which the observer stands (Fig. 4–1A). The location will be most reliable if the bearing line intersects the linear feature at about 90°. The procedure should be repeated with another sight to check the location.

Location by inspection and pacing. Where visibility along a linear feature such as a road or stream does not permit using the method just described, a tally of the paces taken along the feature can be

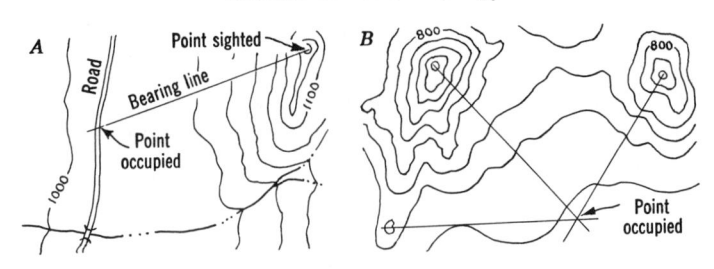

Fig. 4-1. (*A*) Locating a point along a road by taking a bearing to a nearby hilltop. (*B*) Locating a point by drawing three lines from nearby features.

used. The data are plotted by scaling distances from a positively identified feature along the linear course. A tally counter is an asset for this method because the paced distances are often long.

Location by bearing and pacing. Where geologic data do not lie along an identifiable map feature, they may be located by reading a bearing to a nearby point that can be identified on the map. The distance to that point is then paced, and after a backsight is taken to check the bearing, the distance should be checked by pacing back to the outcrop. The average bearing and distance are then used to plot the outcrop on the map. If it is not possible to pace in both directions, a tally counter should be used to eliminate errors in counting.

Location by intersection of bearing lines. Often, points that can be identified on the map are too distant for pacing, and intersection methods must be used. Three points are found that can be identified exactly on the map, and the bearings to these points are measured with a compass. When the reverse bearing lines are plotted on the map with a protractor, they should intersect at the point occupied (Fig. 4-1*B*). The intersections of the three lines are more likely to coincide if the angles of intersection are large; they should never be less than 30°. If the lines form a triangle rather than a single intersection, the bearings and the identifications of the points should be checked. Bearings to other points may be used as further checks. A residual error could be caused by local variation in the magnetic declination, and this will be difficult to correct unless a long reverse bearing is read (Section 2-4).

Location by intersection of bearing and contour lines. It may be possible to find only one distant point to which a bearing can be read. If the elevation of the point occupied can be determined, however, the intersection of the bearing line and the appropriate contour line on the map will locate the point. The elevation of the point occupied can be found with a Brunton compass as follows:

1. Set the compass for use as a hand level (Section 2–6).
2. Sight along level lines to nearby ridges until a crest, saddle, or break of slope is found that is at the same level as the point occupied.
3. Identify this feature on the map and read its elevation from the contours; this is the elevation of the point occupied.

The elevation can also be determined with an accurate altimeter or barometer (Section 4–4).

The precision of this method depends a good deal on how closely the contours are spaced on the map—locations on steep slopes are likely to be more exact than those on gentle slopes. Modern maps are made precisely enough so that almost all contours are no more than one-half contour interval off their correct position.

Using control signals for locations. When mapping featureless plains, marshland, or tundra, it may be necessary to set up control signals before geologic features can be plotted accurately. Rock cairns, flags on poles, or distinctive trees or rocks are commonly used as signals; they should be visible over as large an area as possible. The signals must first be located accurately on the base map by triangulation methods (Section 6–2). They can then be used as a basis for locating outcrops, as by the intersection or pacing methods described earlier.

Locating points by estimates. The geologist should become as proficient as possible at estimating distances in the field. Estimates should be used to check locations made by other methods; they can also be used to make fairly accurate locations. The steps that follow suggest one way to do this.

1. Read the bearing to a distinctive feature and plot the bearing line on the map.
2. Estimate the distance to the feature and plot a tentative point based on this distance.
3. Read a vertical angle to the feature and calculate the difference in elevation between the feature and the point occupied (difference in elevation = horizontal distance × tangent of the vertical angle).
4. Read the elevation of the feature from the map and use the calculated difference in elevation to determine the elevation at the point occupied. A second tentative point can then be plotted by using the contour lines.

5. If the two tentative points do not coincide, re-estimate the distance to the feature and calculate the difference in elevation again. A few approximations like this should bring the two points to coincidence, provided the contours are spaced fairly closely.

4–3. Locating Geologic Features by Traversing

To locate geologic features in wooded areas, traverses must be made from whatever features can be identified accurately on the map. Compass-pace methods are generally suitable for these traverses, though a peep-sight alidade and traverse board may be preferred where magnetic variations are appreciable (Section 6–2). The mechanics of the traverse are similar to those described in Chapter 3 except that a profile is not required, and the map scale is typically smaller.

Preliminary reconnaissance of a traverse course in heavily wooded country may take nearly as much time as the traverse itself, but in fairly open country a reconnaissance may profitably indicate where outcrops occur, what they look like from a distance, and the spacing of traverse lines needed to locate enough of them. In wooded or brushy areas, the traverse should follow the course of least resistance by making use of open ridges, stream courses, paths, and clearings that permit relatively long and clear courses for bearings and pacing. Additional outcrops are located by offsets from this skeletal framework, either by single sights or by short accessory traverses. Where vegetation is too thick to permit long sights, the traverse becomes intricately winding, and bearings and distances must be determined carefully to keep the plot from deviating widely from its correct position.

A "blind" pace and compass traverse of several miles is likely to introduce errors that are greater than errors of plotting alone; therefore, traverses must be tied as frequently as possible to points shown accurately on the map. A traverse must always be started at such a point and must be ended at the same point or at one equally accurate. In this way the traverse's overall *error of closure* can be seen. This error may be treated as follows:

1. Check the plot of the traverse against the recorded bearings and distances, and correct it as necessary.

2. If there is a residual error, consider whether it is large enough to affect appreciably the alignment and position of geologic features; if not, leave the traverse plot unchanged.

3. If the error is so large that important features may be displaced, re-do the traverse to the point where the error is detected, or is mini-

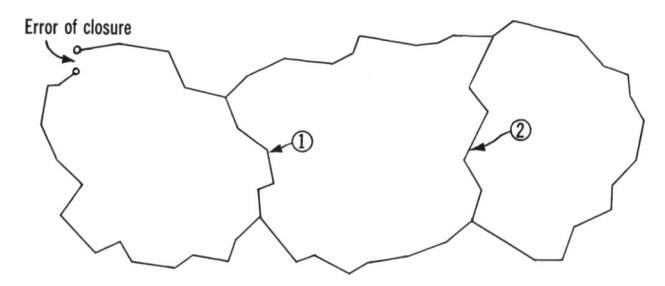

Fig. 4–2. Locating errors in a traverse by surveying accessory traverses at 1 and 2.

mized to meet the requirements of the project. On traverses that close, or that curve in broad arcs, try checks across the course to locate major errors more quickly (Fig. 4–2).

4. If it cannot be determined whether or not it is worthwhile to re-do a traverse to correct an error, leave the traverse as plotted and reconsider it when the map is more complete. Subsequent traverses may cross the traverse in question and locate the error.

A large error of closure should not be altered on the basis of intuition. It may be distributed proportionately among the legs of the traverse by methods described in texts on surveying; however, such a buried (and easily forgotten) error may introduce errors into subsequent traverses or may affect conclusions regarding the geology.

4–4. Using a Barometer (Altimeter) to Locate Geologic Features on a Map

In wooded or brush-grown areas where pacing is difficult and where there are only occasional open views of the surrounding country, a barometer or altimeter can be used effectively to locate points on a topographic map. The three prerequisites are: the contours of the map must be accurate, the contour interval must be such that the contours are spaced fairly closely, and the instrument must permit readings to within about 5 ft.

The instrument should be tested before the field season to see if it will record small differences in elevation quickly and reliably. This can be done by taking successive readings on two or more floors of a building. The barometer should also be tested to see whether normal handling and light tapping will appreciably affect readings taken at one elevation, for some instruments respond unfavorably to normal handling. The instrument must be handled carefully in the field and must be used well below the maximum elevation for which it was constructed. Details of the construction, maintenance, and use of these instruments should be obtained from the manufacturer and studied carefully (see, for example, Hodgson, 1957).

A map point is located with a barometer by plotting a line along which the point is known to lie and then comparing the elevation read from the barometer with the contours crossed by the line. The line on which the point lies can be determined in several ways. If the point lies on a ridge, along a stream, on a trail, or on any other linear feature that is well defined on the map, an elevation reading from the barometer is all that is needed to locate the point. If it is possible to see out to a feature that can be identified on the map, the bearing line to the feature can be plotted and the point located on this line by comparing the contours with the elevation read from the instrument. Where a point must be located in a completely blind situation, as on a heavily wooded hillside, a bearing line to it can be determined by a short traverse from any point that has been located nearby. The barometer reading is then used to locate the point at the appropriate contour. This method is essentially one of traversing by vertical measurements of position rather than by horizontal ones, and on very steep or heavily wooded slopes it is more accurate than pacing. When possible, both methods should be used.

The barometer is sensitive to changes in atmospheric pressure other than those caused by changes in elevation; consequently, it must be adjusted frequently so that all readings will refer to one datum. There are three principal causes for barometric variations at one elevation: temperature changes, overall weather changes, and changes in local wind velocity or pattern. The correction for temperature is approximately 0.0020 ft per °F for each foot read from the instrument. For example, if an instrument has been adjusted at a point of known elevation at a temperature of 50° F, and has then been moved to an unknown point where, at a temperature of 70° F, it reads 100 ft higher, the 100 ft should be increased by $0.0020 \times (70 - 50) \times 100$, or 4 ft. Where the temperature at the second point is the lesser, the correction is subtracted from the reading. Tables and graphs to facilitate these temperature corrections can generally be obtained from the manufacturers of the instrument.

Probably the best way (commonly, the only way) to keep an instrument adjusted to barometric changes caused by weather and wind is to check in frequently at points of known elevation. The most accurate checks can be made at bench marks or other surveyed control points shown on the map. Somewhat less precise checks can be made wherever an elevation can be read from the contours of the map. These accessory points must, of course, be located independently of the barometer. In heavily wooded country the points can be located most easily along stream courses, ridges, roads, or other distinctive linear features shown on the map. Where openings permit, hillside points can often be located by intersecting with a compass or by leveling to nearby ridges with a hand level (Section 4–2). The optimum timing for making these locations and adjusting the instrument varies greatly with the situation. In most areas they should be timed at every 30 to 60 minutes, and the geographic course of a day's mapping must be planned accordingly. During periods of high or gusty winds, it may be impossible to keep an instrument adjusted suitably.

4–5. What to Plot on the Base Map

Contacts and faults are the most important geologic features plotted on the map; they will be considered in detail in Sections 4–6 to 4–12. Folds are generally depicted on the map as a line showing the trace on the ground of the axial plane (Appendix 4). If the crest or trough is plotted, rather than the trace of the axial plane, this must be noted in the explanation of the map. Although these trace lines can be located and plotted directly in some places, symbols for most large folds can be drawn only after rock units and bedding attitudes have been plotted over large areas. Where folds are well exposed, it is important to observe and record: (1) the trend and plunge of the axis, (2) the strike and dip of the axial plane, (3) the plunge of small-scale folds associated with the axial region and limbs, and (4) the strike and dip of secondary cleavages, or the plunge of the intersections between these cleavages and the bedding. These data will be especially valuable if drawings can be entered in the field notes showing the actual shapes and interrelations of small-scale folds, cleavages, or other secondary structures. If possible, these drawings should show the fold in *profile*—that is, in a cross section drawn at right angles to the fold axis. Small-scale folds and related structures are described in Chapter 15.

A large number of planar structures can be plotted as symbols that show their strike and dip; the more common of these are bedding, compositional layering ("banding") in igneous and metamorphic rocks, various cleavages and mineral foliations (including flow structures in igneous rocks), veins, and joints. These various structures must be classified as accurately as possible and plotted with distinctive symbols that show clearly what kinds of features have been mapped. Planar and linear structures are described and illustrated in Chapters 12, 13, 14, and 15.

Symbols should be as standard as possible (Appendix 4). Where they must show unusual structures, a small letter or abbreviation can indicate their special meaning.

Numbers for specimen localities and field notes can be plotted by placing a dot at the location and lettering the number next to it. To keep plotted numbers to a minimum, data that do not refer to specific points can be located in the notes by reference to geographic features or to other plotted numbers.

4–6. Rock Units for Mapping

A rock unit or *lithologic unit* is a rock body distinctive enough to be delineated from adjacent rock bodies along surfaces called contacts. Mapping the traces of these contacts is the most important single procedure of most field projects; thus the nature of contacts and rock units should be considered carefully both before and during the field season. Rock units large enough to plot on maps (mappable or *cartographic* units) provide the practical, demonstrable basis for a geologic survey. In many regions such units have already been described and named formally in the literature. In regions where units have not been designated, new units must be mapped, studied, and ultimately defined and named. Rules for classifying and naming sedimentary units are given in Section 12–2, and suggestions regarding units of igneous and metamorphic rocks are given in Sections 13–2, 14–2, and 15–2.

Rock units are sometimes called *lithogenetic units* because each forms under nearly uniform conditions. There is a great variety of primary (genetic) characteristics that may make a given rock body distinctive. The more important of these are listed in Section 1–5. Some units consist of one rock only, as shale or granite; others are intercalations of two or three closely related rocks, as thinly bedded shale and limestone, or interbedded sandstone, siltstone, and mudstone; still others are characterized by several different rocks that typically form in one environment, as a mixture of pillow lavas, mafic tuffs, volcanic sandstones, red chert, and dark shales.

In addition to being lithologically distinctive, useful rock units must have contacts that can be located and traced out in a reasonable amount of time. Unconformities are ideal contacts because they are sharp and they clearly express the changes in conditions that ended the formation of one rock unit and began another.

Key beds. In structurally confusing or lithologically monotonous terrains, distinctive beds that are too thin to plot to scale should be mapped as single lines. Examples of such *key beds* are tuff or ash layers, carbonaceous beds, glauconite-rich layers (greensands), distinctive clay beds, thin fossiliferous beds (biostromes), and fossil soils. In many cases, a given unit of this type is nearly synchronous over a large area, and it can therefore serve as an approximate time horizon.

Matching or correlating rock units. Before mapping in a region where rocks units have been established, published definitions and descriptions of the units and their contacts should be consulted. If possible, the type areas of the units should be visited to see exactly what the rocks and contacts look like and to collect samples for later comparisons. In the area to be mapped, units can sometimes be identified by tracing them continuously from the type area or from other mapped areas. This is not practicable in most cases, however, and the rocks in the new area must then be compared with the type units by examining them where they are well exposed. Lithologic details can often be used to match the units; however, these details become less reliable as the distance from the type area increases. Moreover, nearly identical rocks can occur at more than one stratigraphic position or in more than one igneous body. Correlations should then be checked by matching similar sequences or associations of units. One of several shale units, for example, might be recognized because it is the only one underlain by a thick sandstone and overlain by a thin limestone. Sometimes entire sequences of similar units can be matched on the basis of one or two key beds or persistent unconformities.

Fossils provide another means of identifying units, although they must be used with caution. Ideally, a given unit is characterized by an assemblage of fossils that are closely related ecologically. A sandstone, for example, might be recognized widely by its suite of neritic fossils. Fossils that are not ecologically equivalent do not disprove the identity of two sandstone outcrops, for a single unit may have been deposited in more than one environment, or fossils from different environments may have been transported to it. If units are identified at widely separated localities by the geologic age of the fossils they contain, two kinds of situations can cause errors. First, the age of a given unit may vary laterally; thus fossils from different areas might seem to disprove continuity of the unit. Second, some fossils do not permit sufficiently limited age determinations to prove or disprove lithic correlations. Two outcrops of shale, for example, are not necessarily correlative simply because each bears Early Cambrian fossils. Further suggestions regarding units based on time rather than on lithologic characteristics are given in Section 12–2.

Selecting new units. It is often impossible to identify a complete suite of established units in a new area, even after considerable reconnaissance. New units must then be selected and defined, or perhaps established units should be subdivided or redefined. First, however, the established units should be evaluated fairly and completely, for redefined names tend to confuse the published stratigraphic record. New units must be of practical value; thus they should be as distinctive and genetically meaningful as possible. Some units cannot be defined exactly until adjacent units have been mapped. Mapping should, consequently, start with the more clearly delimited of the tentative units.

4–7. Mapping Contacts between Rock Units

Tracing and plotting contacts between rock units are basic procedures of geologic mapping. This is the most efficient way to map units at small and intermediate scales (1:24,000 or less), especially where rocks are well exposed. The method of plotting each exposure to scale (Section 4–8) should be considered for larger-scale mapping.

Mapping is best started along a sharp contact between two distinctive rock units. The contact should be mapped by walking along its trace and plotting points on the map where the contact can be seen or where its position can be inferred closely, as within $\frac{1}{25}$ in. on the map. The number of points that must be located accurately will vary with the degree of irregularity of the contact. If the points are spaced $\frac{1}{4}$ to $\frac{1}{2}$ in. apart on the map, the contact can generally be plotted between them with little error. A sharp, well-exposed contact is drawn as a solid thin line. This line should show the actual location of the contact, except where minor irregularities must be generalized because of the scale of the map. The strike and dip of the contact should be measured and plotted where possible, and all data that may be pertinent to its origin should be described in the notes. Lines and symbols must be plotted neatly and exactly. Inking in the office will improve the appearance of the lines but it cannot improve their accuracy; in fact, inking has the unfortunate effect of making sketchy lines appear exact.

Many contacts are exposed at only a few places; some are not exposed at all in natural outcrops. Such contacts can be mapped by walking a zigzag course between outcrops of the two rock units that lie on either side and by plotting a line that passes between the limits thereby established. These limits are often so narrow that the line will be located as exactly as a well-exposed contact (at a scale of 1:24,000, for example, within about 100 ft). More approximate contact lines should be drawn with dashes, and although the limits for these symbols are not standard (varying with the project and the personal feelings of the geologist), it is desirable to set limits for each project. Dashes about ⅛ in. long, for example, might be used for contacts located within ¹⁄₁₀ in. on the map, while dashes about ¹⁄₂₀ in. long might be used for less accurate contacts. Question marks can be added to dashed lines when the existence of the contact itself is uncertain. Such admissions of inexactitude do not commonly appear on published maps, but they are an important part of the field record.

There are several means for locating traces of poorly exposed contacts, and the success of a field project may depend on how skillfully they are used. Rock fragments in the soil or mantle often delimit the area underlain by a given unit. Even where downslope creep has displaced and mixed the float, the upslope limit of fragments from a unit can be used to locate the upslope contact of that unit (Fig. 4–3). When rocks such as shales, tuffs, and soft siltstones are covered by a soil that carries no rock fragments, burrowing animals commonly bring up fragments of rock from the subsoil. Units of soft clastic rocks can often be identified by scattered concretions or hard fossils in clayey soils. Where no large residual fragments can be found, the composition, color, and texture of the soil itself may be used to trace

Fig. 4–3. Using the uphill limit of float fragments (here, pebbles from a conglomerate) to locate the upper contact of a unit. Note that downslope movement of the pebbles makes it impossible to use the same criterion to locate the lower contact.

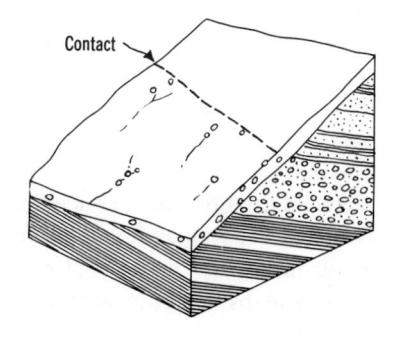

contacts between many rock units. Contacts between such units as sandstone and shale, sandstone and limestone, or granite and gabbro, can be located by the distribution of abundant quartz grains in the soil. More subtle differences between soils will become apparent as mapping progresses and the soil profiles of the various rocks are examined (Section 12–11). Vegetation commonly varies from one bedrock unit to another, particularly in areas of moderate rainfall and high summer temperatures. When these variations cannot be seen on the ground, they can often be distinguished on aerial photographs. Still another aid in locating unexposed contacts is the break in slope at the junction of units that differ in their resistance to weathering and erosion. Some units can so control the topography that they lie entirely in valleys, while others may form low ridges or lines of hills.

When there is no indication of the trace of a contact, it can be projected on the basis of strike and dip symbols measured nearby. This can be done by standing where the contact is exposed, setting the clinometer of the compass for the dip of the beds, and sighting in the direction of the strike as if measuring the dip. Several points are located accurately on the map where the imaginary projected surface intersects the ground surface. The trace of the contact is then dashed in between these points to conform naturally to the topography. This method is particularly useful in mapping sedimentary rocks, but it can also be used where metamorphic rocks have persistent compositional layering, or where igneous rocks have planar structures parallel to their contacts.

The contacts can also be constructed in the office by projection methods (see, for example, Billings, 1954, p. 427; Nevin, 1949, p. 359).

Covered units. Most of the methods described above must be used with caution where bedrock units are covered by transported surficial deposits such as glacial drift. Outcrops in these areas are likely to be scarce, but can usually be found by careful searching along stream courses, ridges, and road or railroad cuts. Data from water wells, drill holes, quarries, and other excavations should be used wherever possible. Bedrock features can generally be found and interpreted more effectively if the surficial deposits themselves are mapped and studied (Section 12–11). After the surficial units are mapped, for example, bedrock float and topographic forms can often be used to locate concealed bedrock contacts. Even well-exposed bedrock structures should be interpreted in light of surficial deposits. Folds in bedrock outcrops, for example, could be caused by glacial drag as well as by tectonic forces. Whether surficial deposits are mapped or not, the methods described in Section 4–8 should be used to plot bedrock outcrops.

4–8. Mapping by the Outcrop or Exposure Method

Tracing contacts is an efficient method of mapping at small and intermediate scales, but if the map scale is large (1:12,000 or more) the outcrop or exposure method is often preferable. In this method, each exposure is plotted to scale by drawing its contact with the surrounding surficial materials (Fig. 4–4). Although small outcrops must be generalized, the contact lines should show the shapes and sizes of the outcrops as exactly as possible. Contacts between bedrock units are drawn as solid lines within outcrop areas and as dotted lines across covered areas. The outcrops should be mapped by working systematically across an area; otherwise the continuity and significance of bedrock structures can be missed.

Letter symbols or colors can be used to designate the units within the outcrop areas. Some geologists use colored ink for the lines around exposed areas to distinguish them from other contacts. Others draw the lines in black ink, but then add a colored line on the side away from the exposure; this clarifies the positions of large, irregular exposures.

One advantage of the outcrop method is that observed facts are separated clearly from inferences. Another is that other geologists can find isolated or hidden outcrops easily, and can themselves evaluate the evidence on which concealed contacts have been drawn. A further description of this method has been given by Greenly and Williams (1930, especially pp. 189–208).

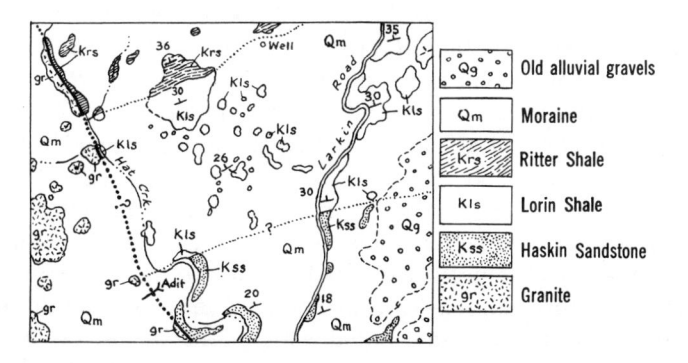

Fig. 4–4. Fragment of an outcrop map. See Appendix 4 for explanation of structure symbols.

4–9. Defining and Mapping Gradational Contacts

Where two rock units grade into one another, criteria for mapping a contact must be established with care and used consistently. The gradation should be examined at several places where it is well exposed so that its genetic significance can be considered. The criteria for drawing a line should have genetic meaning and be such as to persist laterally. If there is no clear genetic basis for drawing a contact, an arbitrary physical criterion can be used for a particular area and project. Criteria must be evaluated as to the amount of exposure, the scale of the base map, and the schedule of the project, for mapping gradational contacts may be difficult or time consuming.

In one type of gradation, rock units grade into one another through a gradual change of texture or composition. Such gradations probably form when conditions of genesis change slowly and continuously. Examples of these gradations are sandstone units that grade through silty sandstones to argillaceous siltstones, mafic granitic rocks that grade to light-colored granitic rocks within a single pluton, and slates that grade through phyllites to increasingly coarser-grained schists. Contacts between these units may be drawn at the center of the zone of gradation, or at the place where a particular mineral or textural feature (average grain size, pebbles, phenocrysts, etc.) is first noted. If the zone of gradation is narrow relative to the scale of the map, and the criteria chosen can locate it quite exactly, a solid line can be used on the map, as described in the last section. Because gradational contacts may pass laterally into sharp contacts, however, the line will be more meaningful if plotted with some special symbol. The hachured symbol suggested in Appendix 4 and Fig. 4–5 is not standard, but it expresses the sense of gradation as well as the indefiniteness of the contact. A line consisting of short ($\frac{1}{20}$ in.) dashes has been used more widely for gradational contacts; however, this symbol has also been used for inferred contacts and for poorly located sharp contacts.

In a second type of gradation, units grade into one another through a zone in which rock types are intercalated or mixed. Such relations indicate that genetic conditions oscillated between extremes as they changed or possibly that materials were mixed grossly, as by intrusion. Examples are sandstone that grades to shale through a zone of interbedded sandstone and shale, granite that grades to schist through a zone of schist that is impregnated with granite veins, or peridotite that grades to gabbro through a zone of interstratified basic and ultrabasic layers. These contacts may be mapped by locating midpoints in the gradation as described above; however, some may be mapped more exactly by drawing a solid line at the top or bottom of a bed or layer of a certain minimum thickness or a certain lithology. The contact between a unit consisting mainly of sandstone and a unit consisting mainly of shale, for example, could be drawn at the top of the uppermost sandstone bed that is more than 10 ft thick. This choice might have been made because a thick sandstone bed is likely to crop out whereas other components of the gradational zone are not. On the basis of such a criterion, the contact has the sense of a sharp contact. Nonetheless, mapping such contacts may become quite involved. In particular, units that grade through interstratified zones in a vertical (stratigraphic) sense are also likely to grade into one another in a lateral sense. Figure 4–5 shows a case of intertonguing sedimentary units, along with three contact symbols used to plot the gradation. If outcrops are scattered, a contact started at *M* in the figure might be traced to *N* and then jumped down to the next bed of suit-

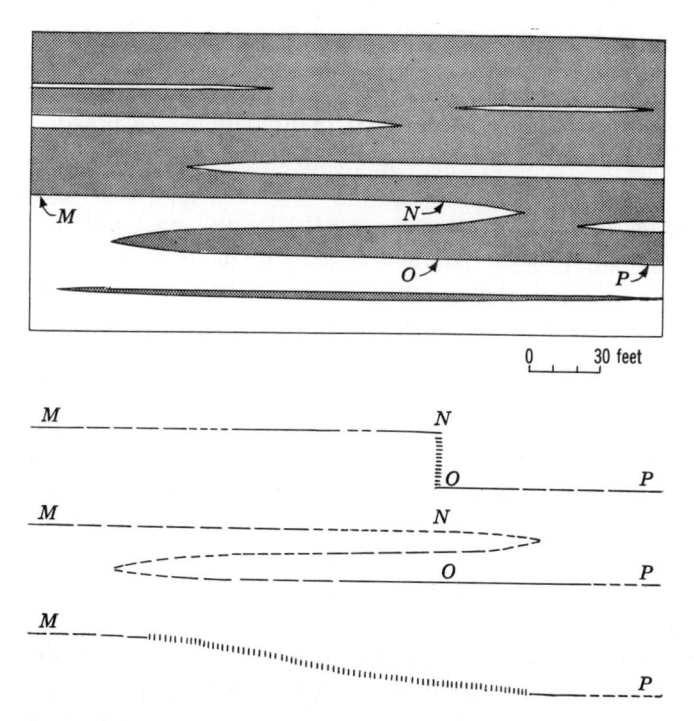

Fig. 4–5. Gradational zone between a sandstone (unpatterned) and a shale unit, showing three ways of plotting a contact (below).

able thickness at point O. A hachured symbol may be used to bridge the gap (expressing the lateral gradation), and the contact can then be continued to P and beyond. In the second case, outcrops are so abundant that the individual wedge-shaped beds can be traced out completely. Mapping of this sort may be so tedious, however, that it deteriorates into a sketching of relations that are not as reliable as the plotted lines would seem to indicate. In the third case, outcrops are so scarce that the top of the sandstone bed can be located only here and there, and a hachured symbol may be used to join the points in a generalized way. Although the three contacts in the figure show different amounts of detail, each expresses the gradational nature of the contact.

As mapping progresses, criteria should be reviewed from time to time and both the dimensions and the nature of the gradations should be described in the notes. Samples should be collected for petrographic study of the gradations. These data will be needed for definitions and interpretations of gradational relations in the final report.

4–10. Using Colored Pencils in Mapping

Waterproof colored pencils can be used effectively to map rock units where outcrops are scattered or gradations are broad. A distinctive color should be chosen for each unit and a mark made on the map at each outcrop of the unit. Most outcrops must be shown diagrammatically by small spots, but the larger ones should be drawn approximately to scale. The marks must be made so lightly that contours show through them and structure symbols can be plotted over them. As an area is mapped, the colored spots will show not only where contacts must pass but also the dimensional accuracy of their location.

Where colors are used to plot outcrops of certain rock types, gradational zones will appear as uncolored bands, and a gradational contact symbol may then be located within this band. This method of finding gradational rock boundaries is particularly useful for internal contacts of plutonic igneous bodies, zones of alteration, or contacts between metamorphic zones, as all of these boundaries may be irregular and unpredictable.

Traces of key beds and small intrusive bodies such as thin dikes and sills will be more distinct if drawn with colored pencils. Important variants within a major lithologic unit, like conglomerate lenses in a sandstone formation, can be plotted as lines and dots of a distinctive color. Colored pencils may also be used for faults or fold axes where structural data are crowded on a map.

Pencil marks must be moderately erasable and waterproof. Since pencils are lost easily in the field, a piece about 2 in. long may be cut from each and carried in a pocket. A color can be selected quickly from these stubs, and if one is lost it can be replaced from the supply in camp.

4-11. Tracing and Plotting Faults

Faults must be sought out and mapped with care. Supposed faults have been plotted on the basis of a few slickensides or a valley that lies suspiciously across the strike of beds; whereas some actual faults have gone unmapped because they could not be seen in many outcrops. This occurs partly because most faults form zones of crushed or altered rock that erode easily and therefore are rarely seen in natural exposures. Furthermore, where faults are exposed, minor ones may appear as impressive as major ones. Superficial shears formed by landslides, for example, may be prominent in shallow cuts.

The geologic map provides the most important evidences of faulting: (1) offsets of contacts between rock units, (2) repetition of rock units, or (3) cutting out of parts of rock units. The mapping of contacts and the determination of detailed stratigraphic sequences will therefore show the positions of most faults.

During the early stages of mapping, when the stratigraphic sequence is known only in part, unexposed faults (or possible faults) can be recognized by: (1) nearly straight scarps that trend somewhat uphill and downhill, (2) nearly straight valleys that cross structures obliquely or that appear to offset other valleys, (3) lines of ponds, springs, or water seeking plants, (4) linear zones of alteration, strong cementation, or discoloration, (5) abrupt changes in attitude of bedding or foliations, (6) abrupt termination of folds, dikes, or faults, and (7) blocks of float that are foreign to the underlying rock units. When such features are seen clearly from a distance or on aerial photographs, they should be sketched tentatively on the map. These possible faults are then examined and either proved by mapping of adjoining rock units or disproved by tracing contacts or beds across them. The trace of a fault should not be projected as a straight dashed line without good evidence, for some faults turn abruptly, and many end suddenly in a flexure or a maze of minor faults and joints.

When a fault has been detected by the mapping of rock units, it

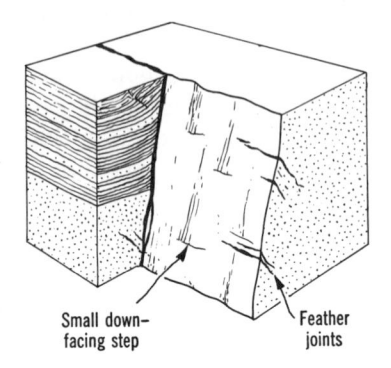

Fig. 4–6. Some useful features that may be found on and near fault surfaces. Note that the thin beds are bent by drag but the thick ones are not.

should be sought out on steep valley walls, in road cuts, quarries, mine workings, or any place where data on its dip and structural details might be obtained. Where the fault is exposed poorly or in only two dimensions, it should be excavated so that the actual fault surface or fault zone can be examined. After the average strike and dip have been read and plotted, the surface should be examined for striations, grooves, or major corrugations (called *mullion* by many American geologists). The trend and plunge of these linear features are indicative of the latest movement. The relative direction of movement may be indicated by feather joints (gash fractures), drag folds, or small steps on the face of the fault surface (Fig. 4–6). Because these details may not indicate the direction of net slip, they should be described or drawn to scale in the notes and re-evaluated later along with all stratigraphic evidence. The notes should also record the thickness and distribution of gouge, mylonite, breccia, veining, and alteration, as well as descriptions of foreign rock fragments that may give clues as to the direction and the amount of displacement.

Faults should be plotted as lines that are about three times as thick as contact lines or else as thin lines of a distinctive color (green and red are used commonly). Long dashes should be used for approximate locations, and the limits of "approximate" should be standardized for a given project. Question marks may be inserted between dashes when both the location and the existence of the fault are in question. Faults that have been observed at only one place and cannot be traced elsewhere should be plotted as a thick strike line with an arrow to indicate the dip. Various accessory symbols for faults are shown in the list of symbols (Appendix 4).

A question often arises as to how much movement qualifies a given fault for mapping. Space and time allowing, all faults that are observed or that are deduced with reasonable certainty should be plotted, for the attitudes and spacing of minor faults can be useful in interpreting large-scale deformation. For many projects, however, only those faults that offset contacts on the map are likely to be worth tracing out and plotting. The offset on the map should never be exaggerated to emphasize the position of a fault.

4–12. Methods of Approximate (Reconnaissance) Mapping

The term *reconnaissance* is applied to incomplete or approximate mapping that helps plan or reconnoiter more detailed and generally more local studies. Reconnaissance mapping can also enlarge the scope of local studies by providing a general geologic picture of the surrounding region. Some entire projects must utilize approximate methods, and they are sometimes called reconnaissance projects to indicate the scale and precision of the work. Mapping of this sort may be necessitated by limitations of time or funds, by the poor quality and small scale of base maps, or by heavy brush, forests, or deep soils that make it impractical to trace out all rock units and structures. Reconnaissance maps may therefore range from fairly complete geologic maps to mere sketches that block out only the largest rock units and structures of a region (Fig. 4–7).

Probably the most important step in reconnaissance work is to choose rock units that are large enough or distinctive enough to be mapped easily. This may require the grouping together of units that would otherwise be plotted separately, as sequences of units with

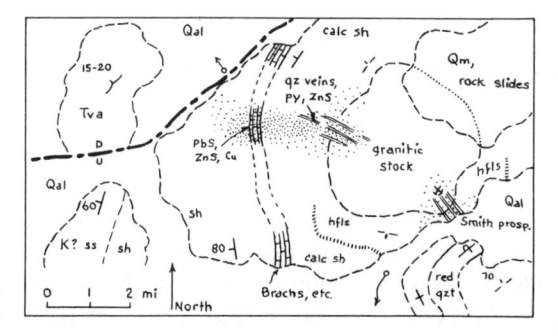

Fig. 4–7. Reconnaissance sketch map, showing the use of labels and brief notes.

gradational contacts. Unconformities are ideal contacts for reconnaissance mapping because they are generally sharp and meaningful. Faults may be plotted readily where they are expressed clearly in the topography, but only those faults that cause major offsets of rock units can be detected by stratigraphic mapping. Many faults must be sketched approximately or placed in the category of questioned or possible faults. Major folds will be shown by the mapping of rock units, but in metamorphic terrains it may be necessary to use small-scale folds and cleavage-bedding relations to detect major folding. Whether other small-scale features should be observed and plotted will depend on the purpose of the project. If, for example, one objective of a reconnaissance is to determine the approximate distribution of small-scale faults, a distinctive contact or key bed can be walked out and small offsets sketched in a qualitative way.

In contrast to exact studies, few contacts can be walked out completely in reconnaissance mapping. Instead, traverses are generally made along roads, streams, ridges, or trails that cross the strike, and contacts are plotted as they are crossed. The approximate trace of a given contact can often be seen from a ridge or road; thus it can be dashed in for a considerable distance. When there is doubt as to its continuity, question marks can be inserted in the contact line or descriptive queries can be lettered directly on the map. If contacts are completely concealed between traverses, the points must be joined by dashed lines that are drawn to conform to the bedding and the topography. This connecting of points should be done in the field, where changes in soil color, vegetation, or slope can be used to locate the trace of the contact. Field glasses will assist this work greatly, and aerial photographs will commonly show indications of contacts and faults that cannot be seen easily on the ground. Because of the scale of the work, structural attitudes must be averaged over a number of outcrops. A given strike and dip symbol may show the average attitude for an area of one quarter of a square mile or more.

4–13. Daily Routine of Field Work

To complete a map in a predetermined amount of time, it is often necessary to set up an exacting schedule and to follow it closely. Unforeseen factors usually work against a schedule. In an area of fairly complex geology, thorough mapping at a scale of 1:24,000 should average ½ to ¾ sq mi per day. One man will therefore require

3½ field months, working 6 days a week, to complete one 7½-minute quadrangle.

Ground must be covered thoroughly and efficiently, but not necessarily rapidly. The tendency to walk as fast as possible to the highest points in the area must be overcome, as must the feeling that mapping cannot be started until most outcrops have been seen and understood. One should use a vehicle when doing so will save time and should walk around brushy or dangerous ground where little if any useful data can be obtained. Study of aerial photographs the evening before each field day may greatly increase the efficiency of mapping.

The amount of notes taken will vary a great deal with the project. Regardless of the pace of the mapping, it is a good idea to pause every 20 or 30 minutes, think over what has been seen, and set down at least a brief note, even if no more than that the rocks are the same as those last described. Notes on minor lithologic variations, taken while catching one's breath, may prove very valuable at a later time. It is inefficient to have to return to outcrops to fill in items that were overlooked because the initial mapping was hurried. Often, several hours must be spent observing, collecting, and writing at critical outcrops.

Field work must be supplemented by work in the office or camp, and this is best done each evening. Locality numbers, structure symbols, and contacts *that are completed* should be inked in black waterproof ink. Faults may be inked in red or green to avoid making broad black lines that might obscure other features. Specimens may be unpacked, checked against the field sheets, and relabeled as necessary (Section 1–8). If space permits, they may be laid out in order of age or geographic distribution so that comparisons and cross checks can be made easily. Every few days, as parts of the area are completed, geologic features should be transferred from field sheets to the office map, which may be colored lightly to show the distribution of units. Another copy of the office map may be used to plot all specimen localities.

Cross sections should be prepared from time to time across various parts of the area because they provide a means of detecting structures that were missed in the field. They also permit a critical check of correlations of partially concealed rock units and structures. Using the cross sections, summaries of the field notes should be written to test the data and ideas developed at each stage of the work. Incomplete or puzzling relations can then be re-examined. The summaries and cross sections should be saved for use during the preparation of the final report.

References Cited

Billings, M. P., 1954, *Structural geology*, 2nd ed.: Englewood Cliffs, N.J., Prentice-Hall, 514 pp.

Greenly, Edward, and Howel Williams, 1930, *Methods in geological surveying:* London, Thomas Murby and Co., 420 pp.

Hodgson, R. A., 1957, *Precision altimeter survey procedures:* Los Angeles, American Paulin System, 59 pp.

Nevin, C. M., 1949, *Principles of structural geology*, 4th ed.: New York, John Wiley and Sons, 410 pp.

5

Mapping Geologic Features on Aerial Photographs

5–1. Kinds of Aerial Photographs and Their Uses

Three kinds of aerial photographs can be used by geologists. *Vertical photographs* are taken by a camera aimed vertically at the earth's surface. They are used widely as a base for geological mapping because they are similar to planimetric maps. *Low oblique photographs* are taken by a camera aimed at an angle to the vertical, but not so as to include the horizon. Aerial photographs showing the horizon are called *high oblique photographs*. Oblique photographs cover much larger areas than vertical photographs, but they are not generally used for geologic mapping because their scales change greatly from foreground to background. They provide overall views of large areas, however, and are therefore valuable in geomorphic and structural studies.

The great value of aerial photographs lies in their detailed picture of the earth's surface. The most detailed maps lack their exact rendering of trees, clearings, trails, gullies, and countless other small features that can be used to locate the positions of geologic features in the field. Moreover, adjoining vertical photographs overlap, making it possible to obtain a three-dimensional or *stereoscopic* image of the terrain covered by the overlap area. This image is so unmistakable that it can be used quickly to make approximate locations in hilly or mountainous country. Only two kinds of areas cannot be mapped readily on aerial photographs: areas of low relief (slopes less than about 5°) where the ground is covered monotonously with grass or brush, and areas where there is a heavy forest cover.

Besides serving as a base for plotting geologic features in the field, aerial photographs can be used before the field season to gain an idea of the geologic features and accessibility of an area and, after the field season, to aid in compiling geologic and geographic data on maps. Photographs should be examined each evening to plan the next day's mapping, even though the mapping itself may be done on a topographic map. This will locate routes that develop the geologic features effectively and give access through areas of heavy brush, cliffs, deadfall, or thick woods.

In some areas, aerial photographs show outcrops, rock units, and structures so clearly that contacts and other features can be drawn on the photographs after comparatively little field work. The term *photogeology* applies to geologic studies based entirely or almost entirely on examinations of aerial photographs in the office. Photogeologic methods should be used in the planning stage and also for finding and tracing structures that are obscure on the ground. These methods are described by Eardley (1941), Smith (1943), Lueder (1959), the American Society of Photogrammetry (1960), and Ray (1960).

Another field related to geology is that of *photogrammetry*, which includes methods of making maps and various measurements from aerial photographs. This chapter and Chapter 9 explain briefly how aerial photographs differ from maps and how a planimetric map can be made from them. Books such as those by Swanson (1949), Moffitt (1959), and the American Society of Photogrammetry (1952) describe photogrammetric methods in detail.

5–2. How Vertical Aerial Photographs Are Made and Indexed

Cameras used for most vertical aerial photography are mounted in such a way that they point downward, despite moderate tilting of the aircraft. The cameras have motor-driven, timed film winders and shutters that expose successive frames on a roll of film at a rate that gives about 60 percent overlap (the *end lap*) from one photograph to the next (Fig. 5–1). The photographs are numbered consecutively in the order in which taken. The aircraft is flown on as straight a *flight line* as possible, and the photographs taken along it comprise a a *flight strip*. If flying is executed perfectly, flight strips are straight;

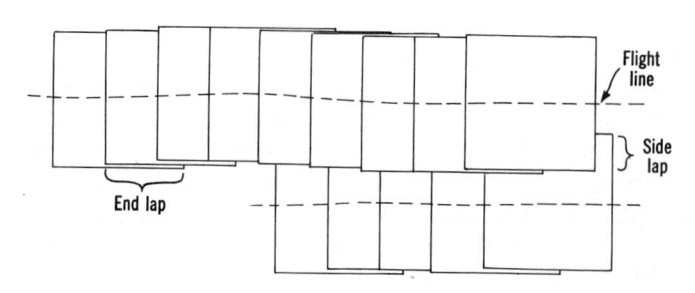

Fig. 5–1. Parts of two flight strips of aerial photographs, superimposed to show characteristic overlaps.

however, they are usually somewhat irregular because of lateral drift of the aircraft.

When one line has been flown to the boundaries of the area, another is started parallel to it and spaced so as to lap over the adjoining strip. This *side lap* is typically about 30 percent of each strip.

When all of the area has been photographed, the contact prints are laid out so that features in the overlap areas are superimposed. This layout is photographed and printed at a reduced scale to make a *photo index* of the area. A north arrow, a bar scale, the dates of photography, the number and geographic name of the project, and the approximate average scale of the photographs are all listed on the index. In addition, the photograph numbers can be read from the index.

5–3. Differences between Vertical Aerial Photographs and Maps

A precisely vertical photograph of a level plain is as accurate in its scale and directional relations as the best map. Most aerial photographs depart somewhat from this ideal because the camera may have been tilted from the vertical, or the terrain may not be level. The effect of an exaggerated *tilt* is shown graphically in Fig. 5–2. The rectangular grid shows that the scale varies across the photograph and the directional relations between such points as A and B are distorted. Fortunately, the amount of tilt is usually so slight its effects are negligible, although severely tilted photographs may be taken now and then in areas where relief is high and air conditions unsettled. Methods of detecting tilt and correcting approximately for its effects are described in Chapter 9.

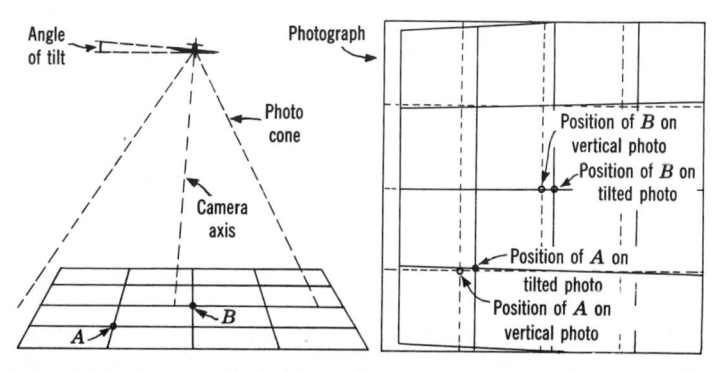

Fig. 5–2. Tilt of the aircraft (left) produces a distortion of the rectangular grid, as shown in the aerial photograph on the right.

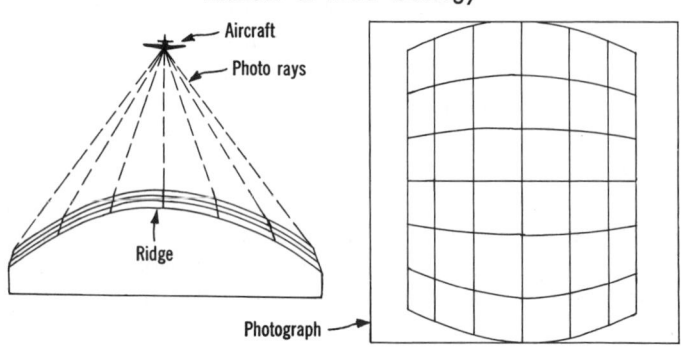

Fig. 5–3. Distortion caused by relief. The rectangular grid on the high ridge (left) appears in a photograph as shown on the right.

Distortions due to relief, which are far more common and more extreme than those due to tilt, are caused in part by variations in scale. The scale for each part of a photograph varies directly with the distance between the camera and the corresponding points on the ground. Figure 5–3 shows how a rectangular grid superimposed on a high rounded ridge is distorted in a vertical aerial photograph. Although

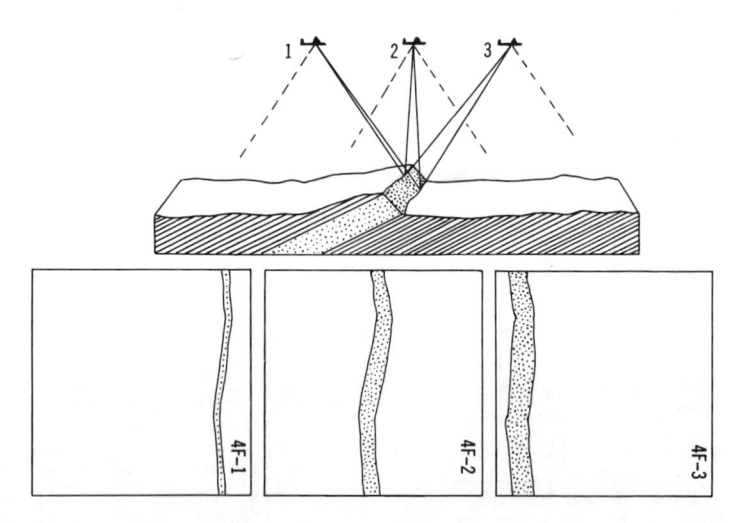

Fig. 5–4. Effect of the angular relation between slopes and photo rays. The outcrop widths of the sandstone unit are markedly different on three successive photographs (bottom) because of the position of the aircraft and the facing of the slope. The true map width of the unit is shown in photograph 2.

the relief in this figure is greater than average, the geometric relations are not exaggerated. Even moderate distortion of this kind must be considered when plotting geologic features on aerial photographs.

In addition to the distortions caused by variations in scale, further distortions are caused by the angular relations between photo rays and the slope of the ground. In Fig. 5–3, for example, this has resulted in a crowding together of the north-south lines on each side of the ridge. Because this effect is especially bothersome in the outer parts of a photograph, only the central part is used for mapping (Fig. 5–4).

5–4. Obtaining a Stereoscopic Image from Aerial Photographs

A stereoscopic image is one in which depth appears unmistakably real. It can be obtained from two aerial photographs because the camera recorded two separate views of one object. Two photographs taken in this way are called *stereo pairs*, and each flight strip of a photographed area consists of a series of such pairs. The stereoscopic image that can be obtained from one pair includes only the area of overlap (about 60 percent) of any one picture; however, by using one consecutive pair after another, an entire area can be examined stereoscopically.

A stereoscopic image can be obtained readily from two overlapping aerial photographs by using an instrument called a *stereoscope*. The most simple, least expensive, and handiest stereoscope to carry is the *pocket stereoscope*, which consists of two magnifying lenses set in a folding metal frame (Fig. 5–5). When this instrument is set correctly over the two photographs of a stereo pair, one eye looks at one photograph, the other eye at the second photograph. A clear stereoscopic image can then be seen. Unless this is done correctly, however,

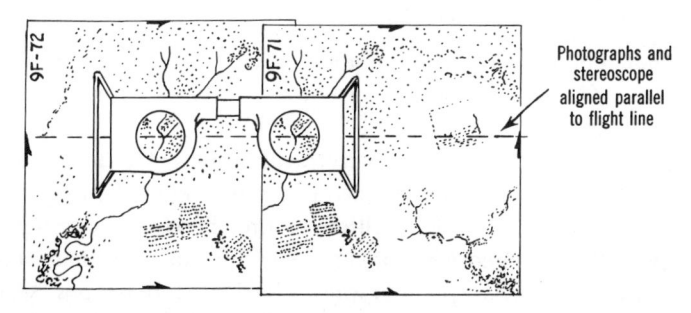

Fig. 5–5. Position of pocket stereoscope relative to two photographs of a stereo pair.

it can be tiring and perhaps damaging to the eyes, consequently the following instructions should be followed.

1. Set the adjustable sleeve of the stereoscope so that the centers of the lenses are the same distance apart as the pupils of the eyes.

2. Place the two photographs of a stereo pair on a smooth level surface, arranged so they are in the same consecutive order and orientation as in their flight strip (Fig. 5–5).

3. Pick a distinctive feature that lies in the overlap area near the flight line (this imaginary line passes through the centers of all photographs in a flight strip).

4. Place the photographs over one another so that their images coincide; then draw the pictures apart *in the direction of the flight line* until the two images selected are separated by about the same distance as the centers of the lenses.

5. Place the stereoscope so that one image is under the center of one lens, and the axis of the stereoscope coincides with the flight line. Carefully adjust the other photograph, moving it only in the direction of the flight line until the other image of the feature is under the second lens.

6. Look in the stereoscope; if a three-dimensional image is not seen at once, make the two parts of the double image coincide by moving the photographs, or perhaps the stereoscope, along the flight line direction.

7. If the three-dimensional image still is not seen, or if it is seen only by straining the eyes, the photographs or the stereoscope may require a slight turn from the supposed flight line. This is especially true in areas of high relief, where only a small part of the overlap area can be seen stereoscopically at one time.

8. In order to see the part of the overlap area concealed beneath one photograph, bend the overlying picture up and to one side, being careful not to fold or crease it.

Another type of stereoscope in general use is the *mirror stereoscope* (Fig. 5–6). It gives a view of an entire overlap area at one setting; however, it is too large to carry in the field.

Magnifying stereoscopes exaggerate the apparent relief of the terrain so that ordinary hills look impossibly steep and moderately dipping beds appear to dip steeply. Although one becomes accustomed to this aspect of the stereoscopic image, it is difficult to estimate dips of planar structures from photographs. Pocket stereoscopes seem to

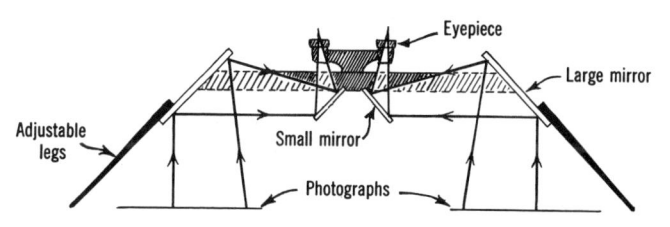

Mapping Geologic Features on Aerial Photographs 79

Fig. 5–6. Mirror stereoscope, with sets of heavy lines showing image cone from photographs to eyepieces.

exaggerate the relief about two or three times, and larger mirror models about two times.

Seeing a stereoscopic image without a stereoscope. If the eyes look in parallel directions (as when looking at a distant object), an overlap area can be examined stereoscopically without using a stereoscope. Perhaps the easiest way to do this is to look for several moments at a distant object, then insert the photographs into the line of view, holding them in position for stereoscopic viewing as described above. The three-dimensional image will be blurred at first, but after some moments of staring at the pictures, it will tend to become sharp. Another way to learn to see a stereoscopic image without a stereoscope is to practice with simple images like those of Fig. 5–7. By staring hard at the two dots for several moments and then relaxing the eyes as though looking at nothing in particular, the eyes will tend to assume their parallel position of rest, and the dots will drift together, finally merging into one image. This image will be blurred at first, but should come to focus shortly. The other figures should appear as three-dimensional forms. To apply this method to aerial photographs, it is necessary to choose points on the pictures that contrast with their surroundings.

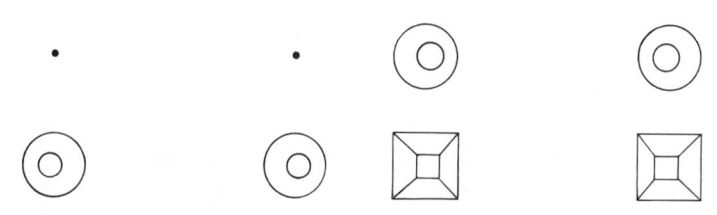

Fig. 5–7. Stereo pairs of simple images.

These methods should not be damaging to normal eyes, but a stereoscope must be used for systematic work. Some stereo viewers look cross-eyed when they force their eyes to make the images merge. The relief is then reversed (pseudoscopic), just as it is when two photographs of a stereo pair are reversed from their normal consecutive position in a flight strip.

5-5. Materials for Mapping on Aerial Photographs

Prints used for geologic mapping should be double weight and have a semimatte surface. Flight lines should be ordered complete so that stereoscopic images can be obtained for the entire area. The prints must be carried in a map folder large enough to protect them adequately (Section 1-3). It is advisable to carry a pocket stereoscope in the field. Only a very soft pencil (for example, 3B) and soft eraser should be used on photographs because the emulsion breaks and peels quite readily, especially if slightly damp. If the air is very dry, a harder pencil can be used. A fine needle can be used to prick a small hole through the print where location marks must be especially precise and permanent. The location can then be labeled on the back of the print. The needle may be mounted in a small wooden handle, with a cork to cover the pointed end. Many colored pencils are too hard to use on photographs; their marks are almost invisible, and they tend to cut the emulsion. Soft wax-base pencils make clear marks, but these rub off easily. Water-soluble pencils are undesirable because their marks spread and dye the emulsion deeply if the photograph is moistened. If a good deal of color work is required, it is best to use transparent overlays, preferably plastic drawing sheets that are frosted on one side so they will take pencil and ink marks. Tracing paper may also be used, but it is typically less transparent, less waterproof, and likely to crack in dry climates.

5-6. Determining Scales of Aerial Photographs

Three scales may be required for work with aerial photographs: (1) the average scale of the photographs of one or more flight strips, (2) the average scale of a single photograph, and (3) the scale in a particular part of a single photograph. The first two differ because the distance between the aircraft and ground vary during photography; the scale difference between adjacent flight strips is commonly large enough to be obvious. The second two differ because of the effect of relief on scale (Section 5-3).

Mapping Geologic Features on Aerial Photographs 81

The *approximate* average scale is recorded on the photo indexes of the area. It is also given in descriptions of the photography and is sometimes printed on the backs of the photographs. In some older coverage, only the distance of the aircraft above the ground and the focal length of the lens are given. The scale (expressed as a fraction) can be computed by dividing the focal length of the lens, in feet, by the distance of the aircraft above the ground, also in feet. Scales obtained in these ways may be 10 or 15 percent in error.

If a more accurate scale is required for any one flight strip, it is necessary to measure the distance between two points on the assembled photographs and to compare it with a distance scaled from a map of the same area. The photographs are first laid out in an overlapping strip with their detail carefully superposed from one picture to the next. As this is done, each is fixed in place with paper drafting tape. A distance is then measured between two points that are as far apart as possible on the strip, and the distance between the same two points is also measured on a map. The photograph scale, expressed as a fraction, can then be found from the following relation:

$$\text{photo scale} = \text{map scale} \times \frac{\text{distance scaled on photos}}{\text{distance scaled on map}}$$

Average scales are used when photographs are compiled into maps, but the scale of a given area on one photograph must be determined if geologic features are to be plotted by pacing methods. A large-scale map can be used to determine scales within single photographs by the method just described. When no such map is available, the scales can sometimes be determined from fields and roads laid out on sections or fractions of sections (full miles, halves of miles, or quarters of miles) of the Bureau of Land Management surveys. Lacking these features, scales can be determined by measuring ground-distances between points and relating them to distances scaled from the photographs. The points should be at about the same elevation, for the scale within any one photograph changes with elevation. A method of computing scale differences between more elevated and less elevated parts of a photograph is described in Section 5–9.

When the scale of one photograph has been determined, the scales of overlapping photographs may be obtained by scaling the distance between two points in the overlap area and using the ratio of distances to compute the scale from one photograph to another. Here again, the points should be at about the same elevation.

5-7. Locating Outcrops on Photographs by Inspection

Because aerial photographs show detail with great clarity, outcrops can be located most accurately and easily by comparing photograph images with features on the ground. The most useful details are the patterns of gullies or other minor valleys and the dark dots that mark trees and large bushes. A large and unmistakable feature is found first—for example, a large tree in a clearing of distinctive shape or the intersection of two gullies. The outcrop is then located exactly by tracing out the patterns of smaller trees or drainage features. Where bushes or small trees are closely grown, minor trails and small clearings can be used. Approximate locations can often be made quickly with a pocket stereoscope, and the stereoscope can be used to make exact locations at ridge crests or at distinctive breaks in slope. On most prints, a magnifying stereoscope or hand lens will disclose features that are otherwise too small to be seen. The date of the photography must be noted, however, for photograph patterns of deciduous vegetation change with the season, and many features of the culture may change with time.

5-8. Drawing North Arrows on Aerial Photographs

When there is not enough detail to locate an outcrop by inspection, compass and pacing methods can be used, but first a north arrow must be drawn on each print. The arrow can be drawn by (1) using a compass to determine the bearing between two points that can be located accurately on the photograph, (2) drawing a line through the two points, and (3) laying off a true north line with a protractor. The points should lie in the central part of the picture and be at about the same elevation; furthermore, they should be at least an inch apart. The entire procedure must be done as precisely as possible since many measurements will be based on the north arrow.

North arrows can sometimes be drawn on aerial photographs before the field season. An arrow can be based directly on roads and land boundaries laid out on the north-south, east-west plan of the Bureau of Land Management surveys. Some section lines may be off alignment by several degrees, but this will usually be obvious when several can be seen on one print. They should be checked where possible against accurate county land plats or other detailed property maps. An accurate map can also be used to transfer a north arrow to a photograph; however, its scale must either be close to that of the photograph or the bearing must be based on an extensive linear feature, as a principal road. Detailed route-survey maps of roads, aqueducts, or natural water courses provide excellent sources of accurate bearing lines; they can generally be obtained from state or county surveyors.

5-9. Locating Outcrops on Photographs by Compass and Pacing Methods

After a north arrow has been drawn on a photograph and the scale *for that photograph* has been determined, points may be located on it by most of the compass and pacing methods described heretofore (Section 4-2). These methods can be used without hesitation in areas of low relief unless the photograph has appreciable tilt distortion. In areas of moderate or high relief, compass and pacing methods will give only approximate results unless the distortions caused by relief are taken into account. In Figs. 5-3 and 5-4, for example, bearings read between many points on a photograph will not be the same as those read in the field. Furthermore, because of the change in scale with elevation, distances paced at one elevation will not correspond to those at another. It is important to recall that the effects shown in the figures are not exaggerated, and that they diminish to insignificance only as the relief and steepness of slopes diminish to nil.

Much of the distortion caused by relief can be avoided by using the central part of each photograph for plotting. Bearings used to locate outcrops should be taken between points lying at about the same elevation, for the positions of such points will be map-true *with relation to each other.* Bearings taken between points at different elevations will be correct *if the bearing line is radial with respect to the center of the photograph.* This is because points lying along lines that radiate from the center of an untilted photograph lie in vertical planes that include the optic axis of the camera and are radial with respect to the camera lens (Section 9-6). Bearing lines within 5 or 10° of radial will give usable intersections unless the points are at greatly different elevations.

A single bearing and a paced distance can be used to locate a point if the scale for that particular part of the photograph is known. When the scale for any one elevation within a photograph has been determined (Section 5-6), the scale at any other elevation can be computed by using the focal length of the camera lens. The change in scale, in feet on the ground per inch on the photograph, can be found by the following formula:

$$\text{change in scale} = \frac{1}{12} \frac{\text{difference in elevation (in ft)}}{\text{focal length of lens (in ft)}}$$

For photographs taken with a 6-inch lens (most large-scale photographs), this would give a scale change of 167 ft per inch for each elevation change of 1,000 ft. If the scale of a photograph has been determined to be 1 in. = 1,500 ft (1:18,000) at a certain elevation, the scale on a ridge lying 1,000 ft higher would be 1 in. = 1,333 ft, and the scale in a valley lying 1,000 ft lower would be 1 in. = 1,667 ft. For distances measured over sloping ground, the average elevation of the paced course should be used. These differences in scale are appreciable; thus differences in elevation must be estimated carefully or be determined with a barometer.

5–10. Geologic Mapping on Aerial Photographs

Though methods of locating points may be somewhat different, the general scheme of geologic mapping with aerial photographs is similar to that used with maps. In many cases the work will take less time because photographs may show actual images of contacts and faults, areas of outcrops, or trails across brushy or otherwise difficult terrain. Photographs are therefore valuable for projects where time is at a premium and detailed examinations of all features cannot be made. Time allowing, however, contacts and faults should be drawn a little at a time as they are traced on the ground. As this is done, the photograph should be examined frequently for hints of faults and variations in lithology. In many cases vague linear marks and slight differences in tone can be matched with vegetation, soil, or other textural changes on the ground. These features might never have been examined had they not been detected as the mapping developed. This is one reason why it is worthwhile to examine photographs with a magnifying stereoscope each evening before doing mapping.

All geologic features should be plotted exactly as they occur on the image of the print being used; they should not be corrected for distortions caused by relief. In Fig. 5–4, for example, the contacts are plotted correctly on each print, even though their most nearly map-true image appears in only one photograph. The necessary corrections are made when the geologic features are transferred to a map or when the photographs are compiled to make a map.

Suggestions for marking photographs are given in Section 5–5. The method of pricking small needle holes through prints is especially useful if many note locations must be made. If ground is being covered rapidly and only brief notes are being taken, as in reconnaissance mapping, the notes may be recorded on the backs of the prints—where they will never be separated from the features to which they refer.

If the north arrow of a photograph is not reliable, the strike or trend of each structure symbol should be recorded on the back of the photograph or in the notes. Attitudes can then be plotted correctly when compiled on a map.

Some geologists find it easier to remember the layout of a large number of photographs if they are renumbered in a simple digit sequence. The new number should be inked permanently on the front and back of the photographs, but not so as to obscure the original number. If a field party is using duplicate prints, one set should be marked with a capital letter or similar device. Perhaps the simplest system for note numbers is to start with 1 on each print. Each page of field notes must then show the photograph number, and specimens should be labeled with a complete photograph number. A typical specimen number would be 3F-16-5, meaning location number 5 on photograph 16 of flight number 3F.

Photographs must be kept dry because the emulsion is softened and easily rubbed off when even slightly moistened. Furthermore, photographs will warp severely if unevenly dampened, as by contact with perspiring hands. If photographs become wet, they should be interlayered with blotting paper and dried overnight under weights.

Because soft pencil marks on photographs rub off easily, location numbers and structure symbols should be inked each night, and contacts and faults should be inked as soon as their positions are as well known as they are likely to be. Because of the dark gray tones of many photographs, colored inks may be used to advantage to show up various kinds of structural features; there may also be an advantage in using a colored ink for note numbers. Waterproof colored inks that are stains rather than pigment-bearing fluids do not show up well on dark areas of photographs and should therefore be mixed with a small amount of postcard paint.

5–11. Transferring Geologic Features from Photographs to a Map

When areas larger than a few square miles are being mapped on aerial photographs, geologic features should be transferred to an office map as the work progresses. This is done to check structural continuity (or critical discontinuities) over areas too large to be seen on one print. It also insures that all necessary ground will be covered. The reasons the photographs themselves cannot be assembled to give a clear overall picture are: (1) each is likely to have only a small mapped area on it, (2) the side lap and end lap of adjoining prints cover most of this area, (3) ink work may be almost invisible on the darker areas of photographs, and (4) patterns of vegetation and culture may obscure the geologic patterns. Ideally, the compilation is made on a topographic base that has a scale close to that of the photographs (as the 7½-minute 1:24,000 topographic map series). Recent 1:62,500 quadrangle maps are sufficiently detailed for most field compilations, and they are far easier to use if enlarged to the average scale of the photographs. Modern planimetric maps (as the 1:24,000 sheets of the U.S. Forest Service and the U.S. Bureau of Land Management) also provide excellent bases for compiling photograph data. Where no suitable base map is available, a planimetric map can be constructed from the aerial photographs, as described in Chapter 9.

Transferring can be started by using a stereoscope to plot geologic features that lie on or very near peaks, saddles, valleys, breaks in slope, roads, trails, or any other features that can be matched exactly from photograph to map. If these points are close together, contacts and faults can often be drawn between them by observing their position in the stereoscopic image. A pair of proportional dividers should be used if the difference in scale between map and photograph is considerable. For areas of low relief, the dividers need be set only once for each photograph, but in areas of moderate to high relief, changes of scale with changes in elevation within each photograph must be taken into account. If proportional dividers are not available,

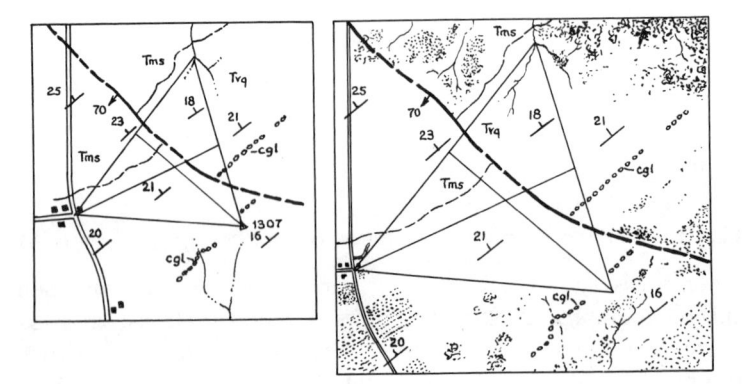

Fig. 5–8. Using a system of triangles and accessory lines to transfer geologic features from a photograph (right) to a map.

series of triangles may be drawn between points that have been located accurately on both the photograph and the map. Features are then transferred on the basis of proportional distances from corners, bisectrices, or other elements of the triangles (Fig. 5–8).

Generally, the map made at this stage need not be as accurate as the one that will be made when the field work is completed. Nonetheless, details should be added in at least a qualitative way, for it is important to see the general grain and character of geologic features as mapping develops them.

References Cited

American Society of Photogrammetry, 1952, *Manual of photogrammetry*, 2nd ed.: Washington, D.C., 876 pp.

———, 1960, *Manual of photographic interpretation:* Washington, D.C., 868 pp.

Eardley, A. J., 1942, *Aerial photographs: their use and interpretation:* New York, Harper and Brothers, 203 pp.

Lueder, D. R., 1959, *Aerial photographic interpretation; principles and applications:* New York, McGraw-Hill Book Co., 462 pp.

Moffitt, F. H., 1959, *Photogrammetry:* Scranton, Pa., International Textbook Co., 455 pp.

Ray, R. G., 1960, *Aerial photographs in geologic interpretation and mapping:* U.S. Geological Survey Professional Paper 373, 230 pp.

Smith, H. T. U., 1943, *Aerial photographs and their applications:* New York, D. Appleton-Century Co., 372 pp.

Swanson, L. W., 1949, *Photogrammetry* (Part 2 of the *Topographic Manual*): U.S. Coast and Geodetic Survey, Special Publication 249, 520 pp.

6
The Alidade and Plane Table

6–1. General Value of the Alidade and Plane Table

When no suitable base map or aerial photographs are available, the geologist must often make a complete map himself. Although a compass and a tape may be adequate for mapping very small areas, an alidade and a plane table are generally required, especially if the map must show both topographic and geologic features. The alidade and plane table can be adapted to many kinds of field projects, and they can be used for rapid sketching of features as well as for precise mapping. These instruments give a great advantage in that the map is drawn directly on the plane table as instrumentation proceeds. This insures that features will be plotted as they are examined in the field and will therefore be shown as naturally and completely as possible. Moreover, plane table methods generally are more rapid than other precise surveying methods. Using a transit, for example, the map must be compiled in the office from notes covering not only the topography, drainage, and culture but also all geologic features as well. The advantages of the alidade and plane table are, in fact, so great that they have become traditional to geologic field work.

6–2. The Peep-Sight Alidade and the Principles of Alidade Surveying

The peep-sight alidade illustrates clearly the principles of alidade surveying and thus will be considered before the telescopic alidade. It consists of a heavy ruler with folding sights mounted at each end that give a line of sight parallel to its straightedge (Fig. 6–1A). The Brunton compass can be converted to a peep-sight alidade by inserting it in a large celluloid protractor made especially for this purpose. The alidade is used with a small plane table board, called a *traverse board*, which is about 15 in. square and has four thumb screws for fastening a map or piece of drawing paper to it. The board fits over the flanged head of the *traverse tripod*, upon which it can be rotated freely. The board can be locked in position by tightening the knurled knob under the head.

The peep-sight alidade is a valuable instrument in its own right. It is particularly useful for geologic mapping where there are only a few map points that can be identified exactly. The board is set up

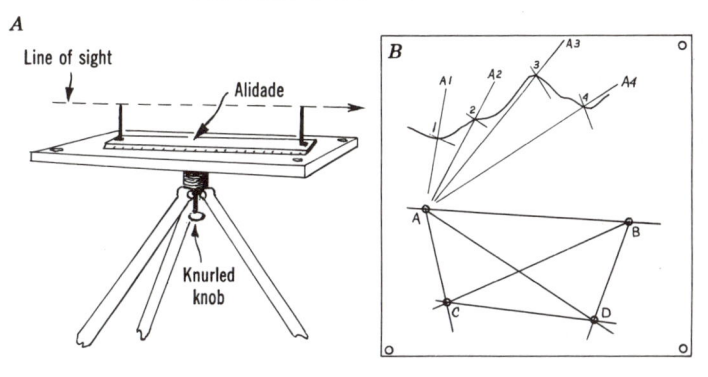

Fig. 6–1. (A) The peep-sight alidade and traverse board. (B) Map mounted on traverse board, showing points located with a peep-sight alidade.

in the field over a point located on the map. Then it is leveled by placing an open Brunton compass on it and adjusting the tripod legs until the bull's eye bubble is centered. If there is no magnetic disturbance, the board can be oriented by laying the compass parallel to a north line on the map and rotating the board until the compass needle points to 0. The map can be oriented more confidently, however, by laying the straightedge of the alidade through the point occupied and through a second point that can be seen from where the board is set up. The board is then rotated horizontally until the second point is centered in the sights of the alidade. When the knurled screw is tightened so that the board is rigid, the map will be oriented parallel with the surrounding terrain.

New points can now be located on the map by either of two procedures: (1) by drawing a ray from the point occupied to the point to be located and pacing or taping the distance to that point or (2) by intersection. The first procedure is as follows:

1. Move the alidade on the board until the point to be located is centered in the sights and the point occupied lies next to the edge of the alidade ruler.
2. Draw a thin pencil line from the point occupied along the straightedge toward the point sighted.
3. Pace or tape the distance to the point sighted.
4. Scale the distance on the pencil line and mark the point.

The procedure of locating points by intersection is illustrated in Fig. 6–1B, which shows a simplified map mounted on a traverse board.

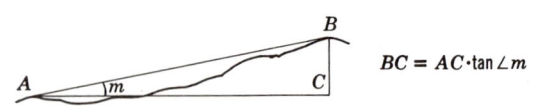

Fig. 6–2. Computation of difference in elevation from vertical angle (m) and horizontal distance.

$$BC = AC \cdot \tan \angle m$$

The two points A and B mark two intervisible points that have been located on the map and can be occupied with the traverse board and tripod. The board is set up over A and oriented by sighting on B with the alidade, as described above. The problem is to map a nearby contact, which has been marked by bright cloth strips at a number of points. To do this, the alidade is sighted on each cloth marker in turn, a pencil ray is drawn toward each, and the rays are carefully numbered in a clockwise direction ($A1$, $A2$, etc.). The board is then moved to B, oriented by sighting on A, and each of the flagged points is again sighted. As each pencil ray is drawn from B, the intersection with the corresponding ray from A marks the position of that point on the map (1, 2, etc.). With the contact located at these points, its details can be filled in quickly by removing the board from the tripod and walking along the contact. Nor is it necessary for all features to be flagged before they can be located by intersection; any points that are distinctive enough to be identified unmistakably from two instrument positions can be located by this method. It must be noted, however, that the locations will be unreliable if the angles of intersection are less than 20 or 30°.

Triangulation from a base line. A base map is typically used for mapping geologic features with the peep-sight alidade; however, it is also possible to construct a map from a blank sheet of paper with this instrument. Such a map can be started by measuring the distance between two points on the ground and plotting this distance on the paper, using the scale chosen for the survey. In Fig. 6–1B, for example, points A and B might have been plotted in this way. This plotted line is called a _base line_, and the survey is extended by measurements from its two ends. In the figure, instrument stations such as C and D can be located by intersection from A and B, and checked by sights from C or D. This general procedure of locating stations, called _triangulation_, is a basic method of starting a map; it will be considered more thoroughly in Chapter 7. Elevations can be carried to the points by measuring a vertical angle with the Brunton clinometer and computing the difference in elevation from this angle and the distance scaled from the map (Fig. 6–2). The map is completed by moving the board to the various intersected stations and mapping the geologic features around them by either method described heretofore.

6–3. The Telescopic Alidade

Compared to the peep-sight alidade, the *telescopic alidade* has a number of important refinements and accessories. These include a focusing telescope for taking long sights, a striding level and vernier scale for measuring vertical angles, a magnetic needle for determining magnetic bearings, stadia hairs set in the eyepiece for determining distances read on a stadia rod, and, generally, a Beaman arc for rapid conversion of stadia readings to true horizontal and vertical distances. The descriptions of the various parts and manipulations that follow are keyed by numbers to Fig. 6–3, a drawing of a typical telescopic alidade.

The *telescope* (1) is held in a yoke-like stand that is attached rigidly to the *blade* (3) by the *pedestal* (4). The only possible movement of the telescope relative to the blade is on the horizontal axis, and therefore the line of sight of the telescope is always in a plane perpendicular to the plane of the blade and parallel to its length. When the *axis clamp* screw (6) is loosened, the telescope can be pointed freely up or down as much as about 30° in each direction; when the clamp screw is tightened, the telescope can be moved slowly by turn-

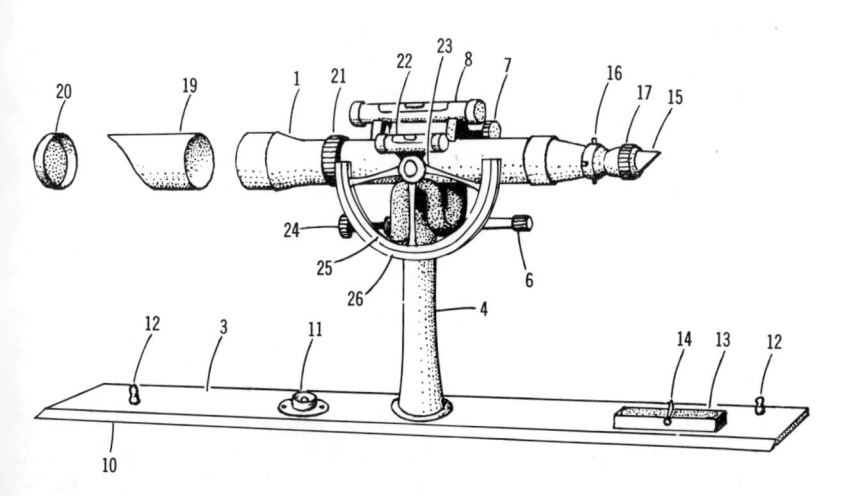

Fig. 6–3. The telescopic alidade.

ing the *tangent screw* (7). The *striding level* (8) lies loosely on bearing surfaces that bring its axis exactly parallel to the line of sight of the telescope. The striding level is readily removed by either a screw or spring clamp, and it is left mounted on the telescope only during use.

The blade has a straight beveled edge (10), called the *fiducial edge*, which is parallel to the sighting plane of the telescope. The *bull's eye level* (11) (or, in some models, two tube levels set at right angles to each other) is used to level the plane table when it is being set up. The two knobs (12) provide a means of moving the alidade on the plane table without touching the telescope or other moving parts. The *compass* box (13) lies parallel to the fiducial edge, so the instrument can be aligned to magnetic north with the compass needle. The needle must be lifted off its jewel bearing by the *lifting lever* (14) when the compass is not in use.

The telescope may be fitted with either an ordinary eyepiece or a prism eyepiece (15), which turns the line of sight up at a right angle and permits the instrumentman to look downward to sight through the telescope. *Cross hairs* and *stadia hairs* are mounted just in front of the eyepiece (16), and they appear as in Fig. 6–4 when they have been properly focused by turning the knurled ring (17) of the eyepiece. The telescope is focused by turning a knob on its right side (this is hidden in the drawing). A sunshade (19) may be slipped over the end of the telescope, and this is replaced by a protective lens cap (20) when the instrument is not in use. When the knurled *retaining ring* (21) is released, the telescope can be rotated on its sighting axis through 180°; during ordinary use the telescope should be turned firmly against its stop (so that the focusing knob is on the right side) and then secured by the ring.

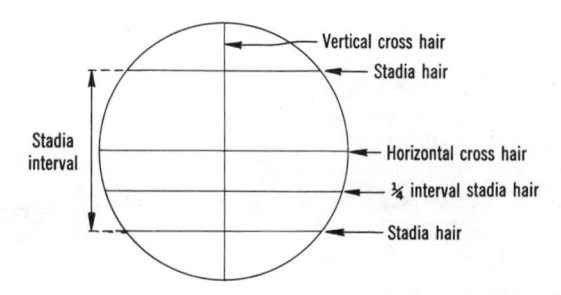

Fig. 6–4. Cross hairs and stadia hairs, as seen through the eyepiece.

The Alidade and Plane Table

On the left side of the alidade is a vertical angle arc and accessories used for reading vertical angles and stadia correction factors. It includes the vernier level (22), a frame (23), a tangent screw (24) that moves the vernier scale and index lines (25), and a calibrated arc (26) used in stadia and vertical angle measurements. The arc is calibrated with both a vertical angle scale and, in most alidades, the useful Beaman arc scale. Some Beaman arcs (as in the figure) are read from the side, some from the rear of the instrument, and others through an optical eyepiece.

Each alidade has a strong wooden carrying case in which it is kept during transport and in which it should be placed when not in use.

The alidade shown in the figure is commonly called a *high-standard* or *topographic* model. The *explorer's* alidade is 2 to 3 in. shorter than the model illustrated, and its stand is attached directly to the blade, making it a low and compact instrument. Its carrying case measures about $3 \times 3 \times 10$ in. as compared to $3 \times 8 \times 12$ in. for the standard model, and this makes it handier to transport. Its telescope, however, can be elevated and depressed only about 8° from level, and its blade may be too short for some sights. Otherwise its construction and operation are in every way the same as for the standard alidade. A third type of alidade has a blade 20 to 30 in. long and is used where long rays must be drawn for control surveys or for exacting large-scale mapping on large plane table boards. The *self-indexing alidade* is an instrument that automatically adjusts for minor tilting of the plane table and that has optical devices to increase the accuracy and speed of readings. Other alidades developed recently have rapid-reading optical devices and are more compact than the explorer's alidade. Detailed descriptions of these new instruments can be obtained readily from manufacturers. In principle, they do not differ from ordinary alidades.

6–4. Adjusting the Alidade in the Field

The adjustments needed most commonly in the field are described below.

Focusing cross hairs. The cross hairs can be focused sharply by turning the knurled ring of the eyepiece. So that no disturbing image will appear in the field of view, the telescope should be pointed at the sky, or at an opened notebook held about 6 in. in front of the instrument.

Correcting the parallax effect. A sighted point may appear to shift in relation to the cross hairs when the eye is moved slightly. This is the parallax effect, caused by imperfect focusing of the instrument. It can be corrected by carefully focusing and refocusing until the image of the point sighted remains exactly at the cross hair when the eye is moved.

Magnetic needle. When an instrument is first used in a given area, its magnetic needle may dip enough to rub against the compass box. This can be corrected by opening the box and sliding the small coil of wire or metal clip along the needle until the needle is balanced. Before the compass is reassembled, the needle bearing should be cleaned, but never oiled.

Striding level. It is occasionally necessary to bring the two ends of the striding level to the same height above the telescope, and this can be done by the so-called *reversal method*. The level is placed on the telescope and the bubble is centered by moving the tangent screw; then the level is reversed end for end. If the bubble moves from center, the capstan screws at one end are turned until the bubble moves halfway back toward the center of the tube. The bubble is then centered again with the tangent screw, and the procedure is repeated until the bubble remains centered when the tube is reversed.

Vernier level. The vernier level should be adjusted frequently when it is used for Beaman stadia work. This can be done by leveling the telescope with the striding level (adjusted by the reversal method), and turning the vernier tangent screw (24 in Fig. 6–3) until the Beaman index is exactly at 50 on the V scale. The vernier level bubble is then centered by turning its adjusting screws.

6–5. Major Adjustments of the Alidade

The alidade should be cleaned and adjusted in an instrument shop before the field season. If it is dropped or jolted strongly in the field, it should be repaired professionally. Where this is not possible, adjustments can be made according to the instructions in the manufacturer's booklet pertaining to the instrument. Whether damage is apparent or not, the following two adjustments should be checked before attempting to use the instrument.

Set of cross hairs. The cross hairs should be set so the vertical cross hair is at right angles to the horizontal axis of the telescope. Before this is tested, the telescope must be rotated firmly against its stop, and its retaining ring tightened. The alidade is then placed on a solid level surface, aligned on a distinct point, and adjusted for parallax. The point should remain exactly on the vertical cross hair when the telescope is tilted slowly up and down with the tangent screw. If it does not, it is necessary to loosen the four capstan screws of the cross hair mount and rotate them by very light taps in the desired direction. After checking and resetting as necessary, the following adjustment should be made.

Line of collimation. The line of collimation is the axial line of the telescope. The cross hairs must coincide with this line. Their coincidence can be tested by (1) placing the alidade on a smooth solid surface, (2) releasing the knurled telescope ring (number 21 in Fig. 6–3), (3) sighting on a well-defined point, (4) adjusting for the parallax effect, and (5) rotating the telescope through 180°. If the cross hairs move off the point, they should be brought halfway back by screwing the appropriate pair of capstan screws on the cross-hair mount. They should then be tested again and readjusted until the cross hairs remain on the point.

6–6. Care of the Alidade in the Field

Instruments should be cleaned and adjusted before the field season. They will generally stay in adjustment if the following precautions are taken:

1. Lift the alidade by the pedestal, never the telescope.
2. Place the alidade in its box when not in use; never leave it on the plane table unattended, for a gust of wind can easily slide it off the board.
3. Remove the striding level before putting the alidade in its box.
4. Turn up the shield of the striding level when removing it from the alidade.
5. Lock the magnetic needle off its bearing when it is not being used.
6. Clean lenses with a soft camel's hair brush or rub them (only when absolutely necessary) with a clean piece of lens tissue or a clean cotton cloth. Never use silk to clean lenses.
7. Do not oil an alidade while in the field because oil must be applied sparingly and expertly. When the instrument is being cleaned and adjusted for use in an area that is likely to be dry and dusty, specify that it be oiled appropriately.

The alidade should always be transported in its case. In a vehicle, the case should ride in a cushioned place, where there will be no danger of its falling off a seat if the vehicle stops suddenly.

6–7. The Plane Table and Tripod

The plane table is a well-made drawing board that has a brass *base plate* by which it can be attached to the tripod. Eight screw fasteners are used to fasten a plane table sheet to its upper surface. Most boards measure 18 × 24 in. or 24 × 31 in. The smaller is preferred for most work because it is transported easily, but the larger is useful in surveying control networks or in very large-scale mapping, where there is an advantage in fitting a large area on one plane table sheet. The wood used in the boards should be both light and strong, and it must be seasoned and joined so that the board will not warp. Occasionally, however, new boards warp appreciably when the weather is hot and dry. Most boards will shrink so much in a dry climate that open spaces will form at each join. This need cause no particular concern because a plane table sheet is strong enough to lie rigidly over the gaps.

Most tripods used with plane tables have an ingenious head, called the *Johnson head*, which permits the board to be tilted into a level position. The board can be partially tightened in this position and then rotated into suitable orientation, where it can be completely tightened. Figure 6–5 shows a diagrammatic view of this head. The board is screwed down over the plane table screw (1) until it is seated firmly against the head's upper surface (2). When both screws (3 and 4) are loosened (by the usual counterclockwise turn), the head moves freely on both its central axis (5) and its cupped bearing surfaces (6). When the upper screw (3) is set tightly, the cupped bearing is locked, and the board moves only on the central axis. When the lower screw is set tightly, the central axis is also locked, and the board cannot be moved.

There are two kinds of tripods, those with rigid split legs and those

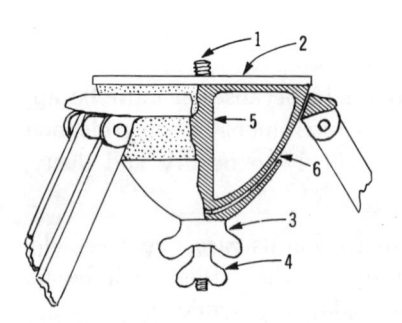

Fig. 6–5. The Johnson tripod head, with right side cut away to show bearings.

with telescoping legs. The telescoping type is heavier but is more compact in transport and permits more rapid setups on irregular ground. There is the possibility, however, of forgetting to clamp its legs firmly in place at each setup—not only a matter of accuracy but also of dropping and damaging an alidade.

On both kinds of tripods, the wing nuts that hold the legs to the head are adjustable, and they should be set just tightly enough to hold a leg out at 45°, unsupported. A protective screw cap or a heavy leather or canvas case should be placed over the Johnson head when it is not in use. This is done primarily to protect the brass plane table screw, which is soft enough to be easily and seriously damaged. When being transported, the tripod should be wrapped so the Johnson head will be kept free from dust and grit.

Before accepting a plane table and tripod for a season's field work, the following should be checked:

1. The plane table screw must be an *exact* mate to the base plate of the board. At least four different screw sizes exist, some of them nearly enough alike so that mismates are not obvious until several weeks' use has worn threads to the point of slipping. To guard against damage, the fit of unknown units must be tested with care.

2. The condition of the Johnson head bearings can be tested by setting the outfit up, tightening only the upper screw (number 3 in Fig. 6–5), and determining whether the edge of the board can be moved more than a quarter of an inch up and down. Excessive play will not permit rapid, accurate surveying.

3. The board must be reasonably smooth and unwarped, and this can be tested by laying a steel straightedge at various angles across its surface.

6–8. Plane Table Sheets and Their Preparation

Paper plane table sheets consist of two pieces of heavy drawing paper joined with an insert of linen. This reduces uneven shrinkage or swelling and keeps a dry sheet from buckling when its dimensions change. The dimensional changes can be determined when the plane table sheet is prepared, as follows:

1. Select a smooth, unbuckled sheet, fit it over the plane table board, and find the positions of the screw holes by punching through the paper with a needle or a small sharp nail.

2. Enlarge the holes to ¼ in. across with a sharp knife or paper

punch, and then screw the sheet to the board loosely, allowing freedom for shrinkage or expansion.

3. Using a sharp 9H pencil and a steel straightedge, draw two lines across the sheet, one parallel to each set of edges.

4. Make two fine needle holes on each line, 12 to 18 in. apart; scale these distances carefully and record their lengths near the border of the sheet.

5. Set the board and sheet out to season under conditions as similar as possible to those of the area to be mapped, and check the lengths of the lines (using the same scale) to determine when the paper has reached a reasonably consistent regimen of daily change. A record of the progress of change should be kept, and the final lengths of the lines recorded permanently near the edge of the sheet so that the map scale can be corrected accurately at any time.

Seasoning the sheets in this way will usually take a few days. It is important that the thumb screws are not set tightly during the seasoning, and that they are loosened each night during the period of actual mapping; otherwise the map may be strained and distorted.

To keep the sheet clean, a piece of heavy wrapping paper may be fastened over it before field work is begun and pieces can then be cut out as the work progresses. The paper cover will help protect the sheet from moisture, but as soon as rain begins to fall, the canvas case for the board should be laid over the sheet because dampened sheets buckle severely. If showers are expected, a piece of plastic or oilcloth should be carried to cover the board.

For many projects, there is an advantage in using sheets that are waterproof and have a smoother and more durable surface than paper sheets. Sheets of tough plastic are excellent for plane table work when their surfaces have been prepared properly. The Stabilene Scribe Coat sheets of the Keuffel and Esser Co., for example, have a colored surface layer, which is cut away by a stylus in order to make lines and points (Section 4–1). This scribing technique is now used almost universally by topographers of the U.S. Geological Survey. One of its advantages is that a map can be printed photographically on the surface layer so that existing topography or planimetry can be used as a base for more detailed work, including geologic mapping.

Plane table sheets with aluminum rather than linen inserts have been used to reduce the effects of moisture; excellent sheets may also be made by enameling a sheet of aluminum or masonite and then grinding the enamel smooth with a very fine abrasive. The enamel should be colored because the glare from a white or metallic sheet makes surveying difficult.

6-9. Setting Up and Orienting the Plane Table

Instructions for setting up the plane table are:

1. Set the tripod up approximately over the station occupied, loosen both screws of the Johnson head, and move the upper part of the head until it lies evenly upright with reference to the cup bearing (Fig. 6-6A). Next, tighten both screws.

2. Screw the plane table down onto the Johnson head until it fits tightly against the top of the head.

3. Pick the tripod and board up and turn them so that the map is oriented with the terrain; then place them down in such a position that the stake or other station marker lies approximately under the corresponding point on the map. For very large-scale mapping, this can be checked exactly by holding a pebble under the point on the map and seeing if it will fall onto the stake (Fig. 6-6B).

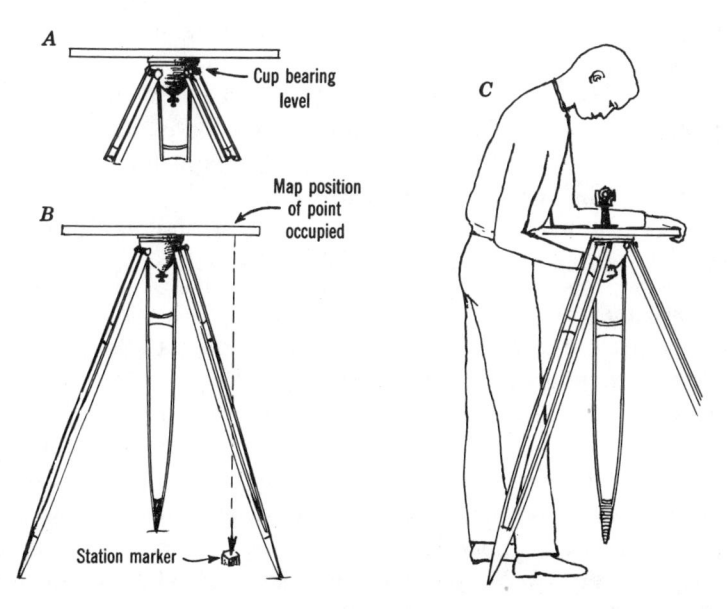

Fig. 6-6. Steps in setting up the plane table.

4. Adjust the position or length of the tripod legs until the board looks level and the legs are inclined at least 30° to the vertical. Two legs should be downhill and one uphill.

5. Force the tripod legs into the ground with the heel, or brace them with stones until they are set firmly. Tighten the wing nuts on each tripod leg.

6. Place the alidade (without striding level) in the center of the board and parallel to its length.

7. Place the left arm over the blade, grip the far edge of the board with the left hand, and press the board against the waist as the left arm pins the blade to the table (Fig. 6–6C). With the alidade thus held safely, loosen the screws of the Johnson head with the right hand and carefully tilt the board until the blade level bubble is centered. Tighten the upper Johnson head screw firmly.

If the plane table has been set up at the first station of a new survey, with no points marked on the sheet, it must first be oriented suitably relative to the area that will be covered. After it has been clamped in this position, the initial station is marked on it and a magnetic north arrow is drawn on the sheet by using the magnetic needle of the alidade. The arrow should be labeled with the station number and the date, as magnetic declinations at other stations may be somewhat different.

If the map has already been started, so that two or more stations are plotted on the sheet, the board can be oriented by using the magnetic arrow drawn at the nearest station. If this method is used, the reliability of the magnetic north arrow should be considered. In general, it is preferable to orient the board by sighting on another point, as by the instructions that follow.

1. Place the alidade so that the fiducial edge just bisects the pinpoint that marks the station occupied and also the pinpoint of another well-located station whose signal (generally a flag on a vertical pole) can be seen. The points on the board should be at least 4 in. apart. A reading glass or hand lens should be used to check this setting.

2. Turn the board slowly and carefully until the signal is bisected by the vertical cross hair. Focus and refocus the telescope to make sure that parallax has not affected the setting.

3. Tighten the lower screw of the Johnson head while watching the signal through the telescope. Give the board a budge or two to make sure it will not be disoriented readily.

The Alidade and Plane Table

4. Align the alidade on other signals to check the orientation.
5. Draw a magnetic north line on the sheet.

6–10. Measuring Vertical Angles with the Alidade

An elevation datum is carried through most areas by vertical angles, which must therefore be measured precisely. Although the alidade and plane table are not ideally suited for measuring vertical angles, they are adequate if manipulated with care. The procedure is:

1. With the plane table set up and oriented as described in the last section, measure the vertical distance from the station marker to the horizontal axis of the alidade, using a tape or stadia rod. Record this as the H.I. (height of instrument) in the notebook (Fig. 6–7). If a signal flag or cairn marks the point sighted, record its height (H.F.) above the ground at the station sighted. These values must be used when computing the difference in elevation.

2. Regardless of where the station occupied and the station sighted occur on the sheet, place the alidade near the center of the board to improve the stability of the setup.

3. Loosen the axis clamp screw (no. 6 in Fig. 6–3) and bring the signal into view in the eyepiece by moving the telescope manually. Tighten the axis clamp screw and use the tangent screw (no. 7 in Fig. 6–3) to bring the horizontal cross hair (*not* a stadia hair) onto the top of the station signal. The final movement of the tangent

Station occupied	Station sighted	Level readings	Reading to signal	Vertical angles
K	CC	27° 04' 00"	30° 06' 30"	
H.I.= 4.1	H.F. = 6.0	27° 04' 20"	27° 04' 10"	+3° 02' 20"
		23° 10' 00"	26° 12' 20"	
		23° 10' 20"	23° 10' 10"	+3° 02' 10"

Fig. 6–7. Notebook page for vertical angle measurements, showing both the entries and the computations of angles averaged from two sets of readings.

screw should be made with a forward (clockwise) turn because the return movement is activated by a spring that occasionally sticks. From this point on the board should not be touched, nor should anyone walk around it until the entire measurement is completed.

4. If the reading is being taken on a rod rather than on a permanent station signal, read and record the point where the horizontal cross hair cuts the rod.

5. Using a reading glass, read the vertical angle vernier to the nearest minute and, if possible, estimate the nearest half or quarter minute. Record the reading in the notebook.

6. Loosen the axis clamp screw and level the telescope approximately; then tighten the clamp screw and level the telescope exactly by using the tangent screw.

7. Read the vernier and record the reading; then reverse the striding level, center the bubble again with the tangent screw, and take another reading. Any lack of adjustment of the striding level is compensated for by using the average of these two readings (Fig. 6–7).

8. To detect both reading and operational errors, repeat steps 3 through 7 after shifting the vernier a few degrees by turning the vernier tangent screw (no. 24 in Fig. 6–3). Record the second set of vernier readings as in Fig. 6–7.

9. Subtract each of the sets of readings, and if the results do not agree within acceptable limits, repeat the procedure.

It will be noted that the "0" point of the vertical angle scale has arbitrarily been given the number 30 on a scale that reads from 0 to 60° (Fig. 6–10). This has been done to eliminate errors caused by reading and recording + and − angles close to zero. The second reading—with telescope leveled—is always substracted from the first reading; thus inclined sights always have a + sign and depressed sights a − sign.

The difference in elevation is computed as described in Section 7–10.

6–11. The Stadia Method and the Beaman Arc

Most of the horizontal and vertical distances that must be determined to make a map with an alidade are measured by the *stadia method*. The principle of this method is simple. The two stadia hairs of the eyepiece are mounted just such a distance apart that they intercept 1 ft on a graduated stadia rod held 100 ft away. Therefore, to find the distance in feet from the alidade to the point where the

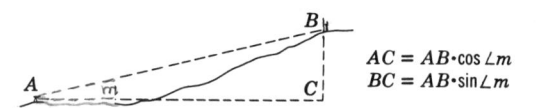

$$AC = AB \cdot \cos \angle m$$
$$BC = AB \cdot \sin \angle m$$

Fig. 6–8. Computations of horizontal distance and difference in elevation from stadia slope distance.

rod is held, the amount of the stadia rod included between the hairs —called the *stadia intercept*—is multiplied by 100. Stadia rods are commonly marked so the stadia intercept can be read directly to the nearest tenth of a foot, while the nearest hundredth of a foot can be estimated accurately if the rod is less than 300 ft away (this varies with the instrument, the rod, and the lighting). These readings therefore give a distance that is within a foot or two of being correct.

Except for level sights, the distance from alidade to rod is a *slope distance*, which must be converted to a horizontal distance before being plotted on a map. Moreover, if a contour map is being made, it is also necessary to determine the difference in elevation between the instrument and the point where the rod is held. These corrections can be made by measuring the vertical angle to the rod and then computing both the horizontal and vertical distances trigonometrically (Fig. 6–8). The second computation, however, gives the difference in elevation from the axis of the instrument to the point where the cross hair cuts the rod. This difference must be corrected so as to obtain the difference in elevation between the point occupied and

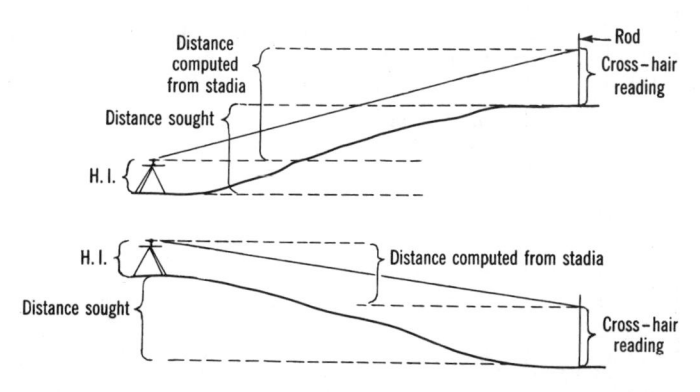

Fig. 6–9. Correcting vertical distances for the height of instrument (H.I.) and the cross hair reading.

the point where the rod is held (Fig. 6–9). This is done by giving the H.I. a + sign, the cross-hair reading a − sign, and adding them algebraically to the difference first computed (*BC* in Fig. 6–8). If, for example, the H.I. is 4.1 ft, the cross-hair reading is 8.3 ft, and the difference in elevation computed from the vertical angle is −34.5 ft, the difference in elevation between the station occupied and the point on which the rod is held is −34.5 + 4.1 − 8.3, or −38.7 ft.

Stadia reduction tables or a stadia slide rule can also be used to convert the stadia slope distance to a horizontal distance and a difference in elevation; however, these values can be obtained more quickly with a Beaman arc. As shown in Fig. 6–10, the Beaman arc consists of two graduated scales scribed on the same arc as the vertical angle scale but read by a separate index. The arc shown in the figure is an *edge-mounted arc*, read from the rear of the instrument, while the arc shown in Fig. 6–3 is a *side-mounted arc*. The scale graduations

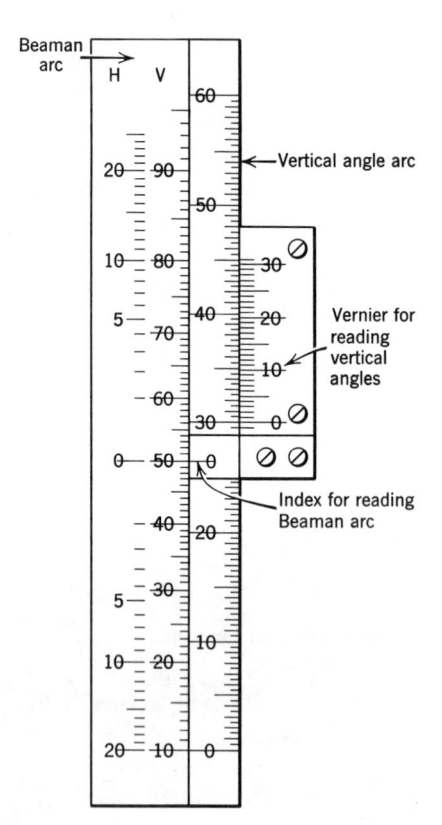

Fig. 6–10. The Beaman and vertical angle arcs, as they appear when the line of sight is level.

are the same on each and are spaced to give proportionately increasing correction factors as the telescope is tilted to steeper and steeper sights. To correct the stadia distance to a horizontal (map) distance, the horizontal or H scale is read (although the reading is made by visually projecting the index across the V scale, this can be done precisely enough for the measurements needed). The H reading is then multiplied by the stadia intercept, and the product is subtracted from the stadia distance. If, for example, the H reading is 2 and the stadia intercept has been determined to be 2.50, the horizontal correction is 2×2.50 or 5 ft, and the correct horizontal distance is $250 - 5$ or 245 ft.

With regard to the vertical correction, it must be noted that the V scale reads 50 when the telescope is level (Fig. 6–10). This is to eliminate errors caused by transposing signs when sights are close to level. The number 50 is subtracted from the V reading to obtain the vertical correction factor, which will therefore have a $+$ sign for inclined sights and a $-$ sign for depressed sights. The sense of sign must be carried throughout the computations. The vertical correction factor is multiplied by the stadia intercept to give the difference in elevation between the instrument and the point where the cross hair intersects the rod. If, for example, the V reading is 36, the correction factor would be $36 - 50$ or -14; if the stadia intercept were 2.50, this would give a vertical correction of -14×2.50 or -35 ft.

This number must then be modified to give the difference in elevation between the instrument station and the point where the rod is held on the ground, as described earlier (Fig. 6–9). The resulting number is added algebraically to the elevation of the instrument station to give the elevation at the point where the rod is held.

6–12. Stadia Procedure with the Beaman Arc

Stadia surveying involves so many repetitious operations that it is generally necessary to adopt a reliable procedure and to use it consistently. Once the principles of the procedure are understood, however, it is often possible to save time or to improve working precision by improvising. The procedures suggested in this section are designed for mapping at scales larger than 1:12,000 (1 in. = 1,000 ft), where stadia distances will generally be less than 1,000 ft. The arrangement of the notebook (Fig. 6–11) can be modified if desired. The Topographic Branch of the U.S. Geological Survey, for example, uses a much more abbreviated form, and various other printed books can

Instrument station	Rod point	Stadia intercept	Beaman V	Beaman H	H x stad. intercept	Horiz. distance	V factor
ΔM (Elev.=372.1)	17	3.21	52	0	0	321	+2
H.I. = 4.1	18	2.52	10	20	50	202	-40
Elev. inst.= 376.2	19	1.89	50	0	0	189	0

Fig. 6–11. Two (facing) pages of notebook, set up for stadia surveying. The column headings can be abbreviated much more than those shown.

be purchased. The lower half or third of each page should be ruled into compartments for calculations. For mapping at intermediate scales, the notebook should be arranged so that readings can be recorded for the methods described in Sections 6–14 and 6–15. Computations of elevations will be easier if the H.I. is added to the elevation of the station occupied, to give the *elevation of the instrument*.

With the plane table set up and oriented, the stadia measurements can be made as follows:

1. Align the telescope on the rod by sighting over the top of the striding level while alternately glancing down to make sure the fiducial edge is just alongside the pinhole that marks the station occupied (Fig. 6–12A). If this is done carefully, the rod will be in the field of view of the telescope, and only a slight movement will be necessary to bring the vertical cross hair onto the rod.

2. Check to be sure the station point is just a pencil line's width from the fiducial edge; then draw the pencil ray with a sharp 8H or 9H pencil. Do not draw the ray through the station point because this would soon spoil the point on the map.

3. Tighten the axis clamp screw and turn the tangent screw until the lower stadia hair is exactly on a whole foot mark of the rod; then *count* the feet and tenths and estimate the hundredths up to the second stadia hair (Fig. 6–12B). Counting prevents errors caused by substracting readings at the upper and lower stadia hairs. To guard against parallax, refocus carefully and repeat the operation, for this

V factor X stad. int.	Cross hair reading	Net diff. in elev.	Elev. at point	Remarks
+6.4	-3.2	+3.2	379.4	Fence at crest of ridge
-100.8	-4.9	-105.7	270.5	Gully confluence ~100' n. of road
0	-10.3	-10.3	365.9	Ss otcp on road E of pt 18

will be the most important reading taken. Record the intercept at once. If both stadia hairs cannot be brought onto the rod simultaneously, read a half stadia intercept (Fig. 6–12B) and multiply it by 2 before recording it.

4. Level the telescope by means of the striding level, and, using a reading glass, bring the Beaman scale index exactly to 50 on the V scale by turning the *vernier* tangent screw (the one on the *left* side of the alidade). The reason for this step is that the board tilts slightly each time the alidade is moved to a new sight, and the Beaman index must therefore be readjusted for each stadia shot. If the instrument has a vernier level, it may be used instead. After this adjustment, it is important not to touch the plane table until the reading is completed.

5. With the telescope still level, check to see if the horizontal cross hair is on the rod or close enough above or below it to estimate the difference to the nearest foot. Where this is the case, time and ac-

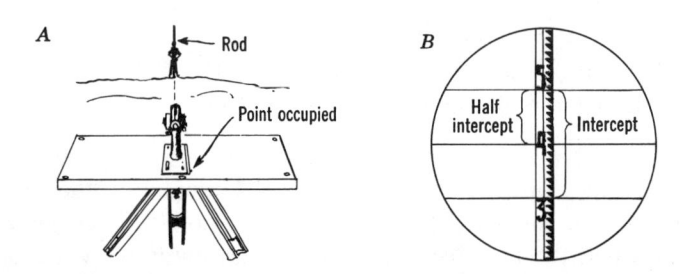

Fig. 6–12. (A) Aligning the alidade on the rod by sighting over the telescope. (B) Appearance of the rod in the telescope.

curacy are gained by a level shot; H is 0, V is 50, and it is necessary only to record the distance between the horizontal cross hair and the base of the rod.

6. If the horizontal cross hair is not on or near the rod when the telescope is level, unclamp the telescope and tilt it until the cross hair falls somewhere near the center of the rod. Clamp the telescope again and turn the tangent screw (the one on the *right* side of the instrument) until the *V scale graduation nearest the Beaman index is brought exactly to the index mark*. This is done because only whole numbers can be read precisely on the V scale; furthermore, whole numbers simplify computations. A reading glass or hand lens should be used to make this setting exact. Read the V and H scales, and record these numbers.

7. Look through the telescope to determine where the horizontal cross hair cuts the rod. Record this number—always with a minus sign—as the *cross-hair reading*. Even for precise work, this number need be read only to the nearest tenth of a foot.

8. Signal the rodman on to another point; then begin computing at once.

9. Multiply the stadia intercept by the H reading and record the product to the nearest tenth of a foot in the column marked $H \times Stad.$ *Intercept*. Subtract this number from the stadia distance (stadia intercept $\times$ 100) to get the true horizontal distance (Section 6-11). Scale the distance on the pencil ray; prick a small hole with a needle to mark the point, and letter the point number by it.

10. Subtract 50 from the V reading, record the resulting *V factor*, and then multiply it by the stadia intercept. Do the multiplication in a numbered box on the lower part of the notebook page where it can be checked.

11. Algebraically add the cross-hair reading (which always has a — sign) to the product just determined. Record the result as the *net difference in elevation*.

12. Algebraically add the net difference to the *elevation of the instrument* to get the elevation of the point sighted. Write the elevation on the map next to the point.

The difference in elevation can also be determined by the *stepping method* (Section 6-15). This procedure will save time and improve accuracy when the line of sight is nearly horizontal, as when one of the stadia hairs falls on the rod or near it. The number of steps should be recorded in the column labeled V *factor*.

Computations should be carried out point by point as instrumentation progresses. Where points are closely spaced and geologic features fairly simple, the instrumentman will need a recorder to keep up with the rodman. The instrumentman should glance up occasionally while computations are in progress to see where the rodman is; otherwise it may be difficult to find him when he is at a new point.

6–13. The Stadia Interval Factor and the Stadia Constant

The stadia distance corresponding to a stadia intercept of 1 ft is not precisely 100 ft in all alidades. This difference is likely to be so small (less than 1 ft in 100 ft) that it can be ignored in all but the most detailed mapping. To make certain it is small, however, the actual stadia interval factor should be determined. This is done by taping a distance of 100 ft and determining the exact intercept by using a Philadelphia rod or a scale that can be read accurately to a hundredth or thousandth of a foot.

For external-focus alidades, the point of origin for a stadia measurement is not at the axis (pedestal) of the instrument but generally about a foot in front of it. This short distance, called the *stadia constant*, is inscribed in the instrument box by the manufacturer. It may also be determined by the method described by Bouchard and Moffitt (1959, p. 370). It is too small to be considered a serious factor in most mapping; however, it must be taken into account when determining the stadia interval factor.

6–14. The Gradienter Screw

The telescope tangent screw of many alidades includes a cylinder calibrated into 100 divisions and constructed so that one complete turn moves the telescope through exactly one stadia interval. These calibrated tangent screws, called *gradienter* or *Stebinger screws*, can be used to determine horizontal distances when the rod is held so far away that a half stadia interval cannot be read directly, or when so much of the rod is concealed in brush or trees that a half intercept cannot be read.

The gradienter must be adjusted to give reliable results, especially where sights are long. Adjustment should be specified when the instrument is cleaned before the field season. The screw can be checked in the field by testing the calibration against distances that have been taped accurately. Errors in adjustment smaller than one

At	Point sighted	Cross hair readings	Grad. readings	Stadia distance	H rding	Horiz. dist.
$\triangle C$	9	2) 12 1) - 7 5	0.43 0.23 0.20	$\frac{5}{0.20} \times 100 = 2500$	1	2475

Fig. 6–13. Computing horizontal distance from gradienter screw readings.

division of the cylinder may be neglected because the rod cannot be read with a corresponding precision at long distances.

The procedure with an adjusted instrument is as follows:

1. Place the cross hair on the lowest visible division of the rod by turning the tangent screw in a clockwise direction (the screw may not move smoothly in the other direction).

2. Read and record the cross-hair reading and the gradienter reading (Fig. 6–13).

3. Using the tangent screw, elevate the telescope until the cross hair is on another full division (preferably the top) of the rod; then read and record the rod and gradienter cylinder for this position.

4. Compute the stadia distance by dividing the difference between the rod readings by the difference between the gradienter readings, and multiplying the result by 100, as shown in Fig. 6–13.

The lower reading is always made first because *all movements must be made by a forward motion of the screw.* The parallax effect must also be corrected at the lower setting, for a moderate error will cause considerable inaccuracy. Mistakes can be detected by releasing the telescope clamp screw, depressing the telescope slightly by hand, and then repeating the readings on a different part of the gradienter screw.

The stadia distance can be corrected approximately to a horizontal distance by using the Beaman H factor, or, more precisely, by taking a vertical angle reading to a given point on the rod and computing with the cosine of this angle (Fig. 6–8).

6–15. Differences in Elevation by the Stepping Method

The vertical angle arcs of alidades can be read only to the nearest 1 or ½ minute. It may be desirable to use more precise methods to measure differences in elevation, and the *stepping method* is useful

The Alidade and Plane Table

where lines of sight are inclined at low to moderate angles. The correspondence of this method to the Beaman arc method can be understood by noting that the Beaman index moves across one full division on the V scale when the telescope is moved through one full stadia intercept on the rod. The reader can demonstrate for himself that the telescope can be moved through a stadia intercept more accurately by observing the stadia hairs than by observing the V scale.

The method may be carried out as follows:

1. Level the telescope with the striding level.
2. Examine the telescopic view, and find a distinct point or line to key the position of the upper stadia hair (if the rod is above the level of the instrument) or the lower stadia hair (if it is below the instrument) (Fig. 6–14).
3. Move the telescope with the tangent screw until the other stadia hair lies on the point chosen.
4. Repeat steps 2 and 3 after choosing another suitable point or line, and continue with further steps until the horizontal cross hair is on the rod.
5. Record the cross-hair reading and also the number of steps used in moving the telescope to this position.
6. Read and record the stadia intercept (typically a half intercept will be used because of the distance to the rod).
7. Compute the difference in elevation between the alidade and the foot of the rod by multiplying the stadia intercept by the number of steps and then subtracting the cross-hair reading.
8. Compute the elevation of the point sighted by adding or subtracting the difference in elevation to the elevation of the instrument (elevation of instrument station + H.I.).

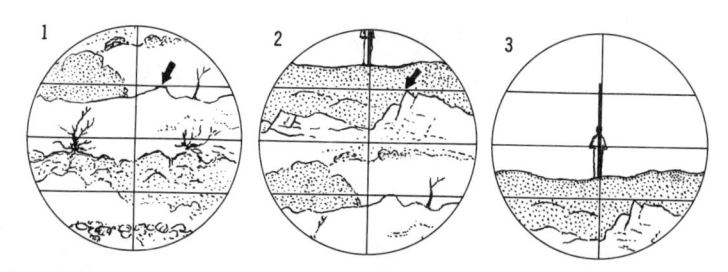

Fig. 6–14. Determining difference in elevation by three successive steps of the upper stadia hair. The arrows indicate points used for the steps.

6–16. Using the Stadia Rod

Stadia rods are 10 to 15 ft long, are made of spruce or some other suitably strong light wood, and are generally hinged for ease of transport. They are marked in feet and in tenths or whole multiples of tenths of feet. A variety of marking patterns are used, each designed to give optimum readability for certain working conditions and distances. The clarity of the pattern and the brightness of the colors have a considerable effect on the accuracy of plane table work, and therefore the colored surfaces must be handled carefully, washed occasionally with soap and water, and touched up with paint when chipped or scratched. Weathering in the sun tends to turn white paint yellow and to make red paint dull, so the rod should be kept indoors when not in use. Rods should be carried carefully in the field, for a damaged hinge may be difficult or impossible to repair.

The rod must be held plumb (perpendicular to a level surface) while stadia readings are being made. On a windless day it can be balanced between the finger tips. On a windy day it should be checked by holding a Brunton compass (with clinometer set at $90°$) against it. The rod should be lighted by the sun if possible, and this can often be achieved by turning it 10 or $20°$ away from the line of sight.

The rodman should carry a notebook in which to record descriptions of the features that make each point useful. The points must be numbered in the same consecutive sequence as those in the instrumentman's notebook, and the tally should be checked from time to time. The details of the rodman's procedures in geological mapping are described in Chapter 8.

References Cited

Bouchard, Harry, and F. H. Moffitt, 1959, *Surveying:* Scranton, Pa., International Textbook Co., 664 pp.

7

Control for Geologic Maps

7–1. General Nature of Control Surveys

Accurately located *control points* serve as a basis for surveying the topographic and geologic features that make up a complete map. They are generally distributed over the area in what is called a *control network*, which is planned and surveyed during the early phases of a mapping project. The survey should be sufficiently precise, within predetermined useful limits, to satisfy the following specifications: (1) the horizontal scale will be constant over the entire map; (2) all elevations and contours will refer to one vertical datum (generally mean sea level); and (3) all features will be oriented to true north. If possible, the survey should include points of known geodetic position (as government triangulation stations) because lines of latitude and longitude can then be drawn on the map, and the map can be related to any other points on the earth's surface.

There are many kinds of control systems; each can be surveyed with almost any degree of precision. Generally, the precision of the survey should be about three times that of the methods used for mapping geologic and topographic features. For local, moderately precise mapping at large scales, a base line can be taped and expanded into a small triangulation network with a peep-sight alidade, as described in Section 6–2. For a topographic map of a small area, it may be sufficient to lay out instrument stations by making a stadia traverse with an alidade and plane table. Maps made from aerial photographs can also be compiled on the basis of moderately accurate control data, as county plats of property lines or old Land Office corners. When a geological survey will extend over several seasons, however, or when maps must cover large areas with consistent accuracy, it is necessary to use more precise methods. In these cases, a transit should generally be used to measure angles between control stations, while key distances should be measured with a steel tape. As presented in this chapter, these procedures require only ordinary surveying equipment. Many additional methods and improvisations are possible, and the reader is urged to consult the sources listed at the end of the chapter. When a precise modern instrument is available, as a compact theodolite, specific instructions can usually be obtained from manufacturers.

7–2. General Plan of a Triangulation Survey

In a triangulation survey, points are located by measuring the angles formed by the lines of sight between the points. The word triangulation originates from the fact that the lines are treated as the sides of a series of joined triangles. This has great practical value because triangles give a simple basis for trigonometric checks of accuracy as well as for computations of distances between points. The length of one side of one triangle must be determined before the distances can be computed. This side, called the *base line*, may be measured on the ground or computed from existing geodetic data. An example of simple triangulation is given in Section 6–2. A more precise triangulation survey can be executed as follows:

1. Determine the distribution of existing triangulation and level-line data by inquiries to federal and state agencies.
2. Plan the survey on the basis of time, funds, equipment, and the purpose of the project.
3. Reconnoiter the area to select and mark control points and to locate existing triangulation stations and bench marks.
4. Measure the base line (if required).
5. Level from a nearby bench mark to bring an exact elevation to the network, preferably at the base line.
6. Determine the azimuth of the base line, as by a sight on Polaris.
7. Measure the interior angles of triangles and the vertical angles between triangulation stations.
8. Compute the lengths of triangle sides, the elevations of stations, and the horizontal coordinates of stations.

7–3. Using Existing Control Data for Surveys

Regardless of whether a control survey will be made by triangulation or traverse methods, a good deal of time can be saved if it is based on existing survey data rather than on a new base line. In many areas, for example, there are two or more established triangulation stations that can be plotted on a plane table sheet and then used to intersect any number of other points. Search for control data must be started long before the field season begins since it may take several months to sound out all sources and to compute distances from

Control for Geologic Maps

geodetic data. Control data may be obtained from the following agencies (regional offices should be contacted if their addresses are known):

Geological Survey, Department of the Interior, Washington 25, D.C.
Coast and Geodetic Survey, Department of Commerce, Washington 25, D.C.
Bureau of Reclamation, Department of the Interior, Washington 25, D.C.
Bureau of Land Management, Department of the Interior, Washington 25, D.C.
Forest Service, Department of Agriculture, Washington 25, D.C.
Soil Conservation Service, Department of Agriculture, Beltsville, Md.
Mississippi River Commission, P.O. Box 80, Vicksburg, Miss.
Lake Survey, U.S. Army District, 603 Federal Building, Detroit 26, Mich.
Office of the Chief of Engineers, Corps of Engineers, Building T-7, Washington 25, D.C.
Various State geodetic and land survey offices, State highway departments, State natural or water resources bureaus, and State geological surveys.

The following data may be available from these agencies: geographic positions (precise latitudes and longitudes) of stations, azimuths and distances between various pairs of stations, elevations and descriptions of stations, and State plane coordinates of stations. Precise level-data that may be obtained consist of elevations and descriptions of bench marks as well as maps showing the lines of leveling stations. All inquiries must be specific regarding the boundaries of the area concerned, and an index map showing latitude and longitude lines should be submitted if possible.

When only latitudes and longitudes of stations can be obtained, the distance and azimuth between two stations can be computed by *inverse position* methods, which are based on the equations that define the average or idealized spheroidal shape of the earth. The equations and methods are given in U.S. Coast and Geodetic Survey Special Publication 8 (1933) and in U.S. Geological Survey Bulletin 650 (Gannett, 1918, pp. 292–293).

Nongeodetic surveys can also be used to control geologic mapping. The best sources are the detailed highway route surveys made by state and county surveyors, railroad survey and route data, and accurate plats of property lines and corners (typically held by county surveyors and assessors). The surveyor's blueprint maps are generally not accurate enough to permit direct scaling of distances. Instead, the survey coordinates should be used to compute distances and bearings.

7-4. Selecting Stations for a Triangulation Network

The base line and the triangulation stations can be selected by first making a stereoscopic study of aerial photographs and then reconnoitering the possibilities in the field. The optimum spacing of control stations depends in part on the scale and nature of the final map. If the map will be compiled from aerial photographs or from several plane table sheets, there should be at least two or three stations on each sheet or each photograph. If map or photograph scales are between 1:12,000 and 1:24,000, the sides of the triangles will therefore be about 1 to 3 mi long.

The size and arrangement of the triangles and the base line also affect the precision that can be obtained in a given amount of time. The probability for errors increases as measurements are carried farther and farther from the base line. The examples that follow illustrate four general arrangements. Networks like that shown in Fig. 7-1A are used where the base line must be short and visibility is limited

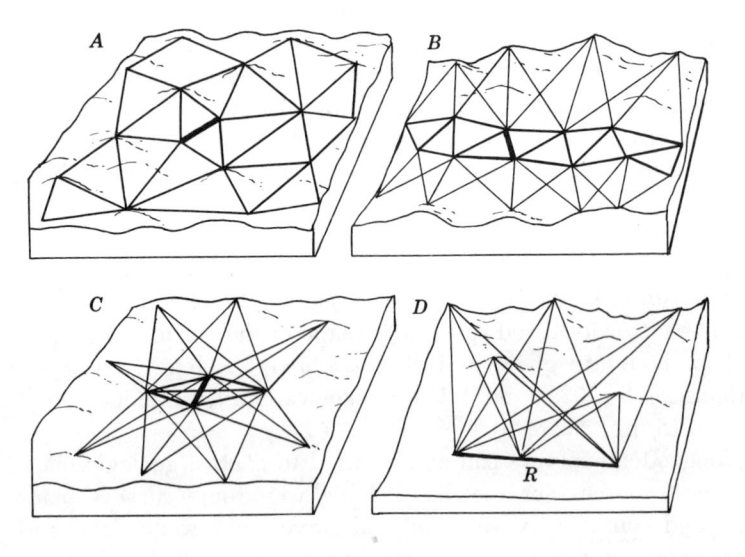

Fig. 7-1. Four types of triangulation networks. The heaviest line in each is the base line.

to relatively short sights. Each triangle must be closed as accurately as possible, and therefore all stations are occupied with a transit. Although this is time consuming, the network can be expanded later in any direction without loss of accuracy.

The arrangement shown in Fig. 7–1B is particularly suitable for mapping a high range bordered by hilly ground. The stations within the heavy-lined triangles are occupied with an instrument, and the outlying stations are located by intersection. The intersected locations are not reliable in that they are not checked within single triangles. Much time can be saved by this scheme, however, and the mapping can be expanded at each end of the chain of occupied triangles.

In Fig. 7–1C, a short base line has been expanded into an exact quadrilateral whose four corner stations serve as a base to intersect the other stations with three rays. This is an excellent scheme for mapping an equidimensional area rapidly and fairly accurately. Quadrilaterals may be used to improve the accuracy of other networks; for example, the chain of triangles of Fig. 7–1B could have been expanded as a chain of quadrilaterals.

Finally, there are cases where a long and accurate base line can be used to intersect all stations, typically where an open linear valley is bordered by increasingly elevated hills or mountains (Fig. 7–1D). If an accessory point (R in the figure) is located on the base line by taping, the intersections can be checked with a third ray. This scheme is efficient in that it takes far less time to tape a long base line than to occupy a number of high-lying trangulation stations. The network cannot, however, be expanded from the outlying points until they are occupied and their constituent triangles closed precisely.

The accuracy of a survey will be improved by placing the base line near the center of the network. Generally, angles should be larger than 15°, but this may be altered to fit the precision required by the survey. Small angles may be adequate if they are measured by repetition and 7- or 8-place log tables are used in computations.

Each station must be visible from enough adjoining stations so that it can be surveyed accurately, and it must be situated so that it can be reached in a reasonable time and occupied safely with an instrument. Triangulation stations used in long-distance geodetic measurements are commonly placed on mountain tops, but there is no need to do this for local control surveys. Stations near the centers of valleys (often near roads) are ideal in many cases, and low hills, spurs, or saddles may afford adequate visibility over rough terrain.

After the probable station sites have been selected on aerial photographs or by reconnaissance with field glasses, each site must be visited, checked for suitability, and marked with a signal. This selection must be made with care because the stations should not be moved once instrumentation starts. Signals of adjoining stations must be visible *at instrument height.* The signal should be as completely in the open as possible so that accessory instrument stations can be located from it easily. There must be ample space for the instrumentman to move safely around the instrument, especially if the station will be occupied with a plane table.

Sketch map of network. As the stations are selected and marked in the field, a sketch map should be made of the network by reading angles to adjoining stations with a compass. This map is valuable for estimating the strength of the network and for planning and executing the instrumentation. Notes regarding visibility and routes of access should be made directly on it; it may also be used to make a reconnaissance map of the main geologic features.

7–5. Signals for Triangulation Stations

Each station should be marked by a firmly set wooden stake or some other permanent mark. A signal is then set up over the station as a sighting point for instrument work. This is done when the station is first visited in order that each succeeding station can be checked for intervisibility. The signal shown in Fig. 7–2 is light in weight, can

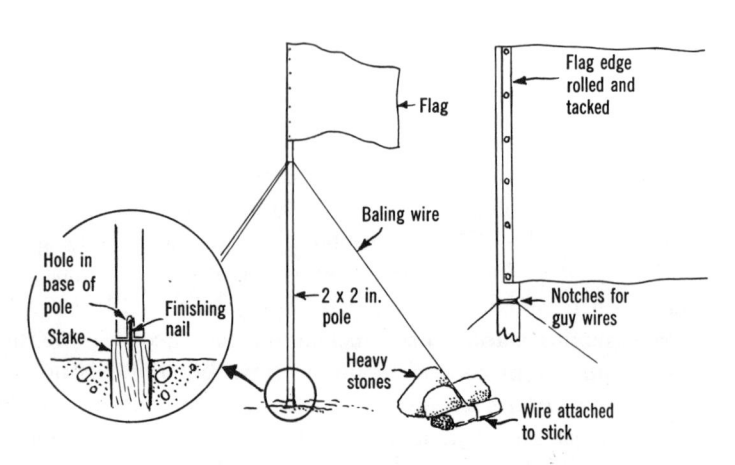

Fig. 7–2. Signal for triangulation stations.

Control for Geologic Maps

be plumbed exactly over the station hub, and can be taken down and set up easily during instrumentation. The flag color is chosen to contrast with background colors. In most areas, a brightly colored flag on a pole 6 to 8 ft high can be seen with an instrument for distances of at least 5 mi. There will be an advantage in later computations if all poles are the same height. In addition to the materials shown in Fig. 7-2, the following will be needed: a small axe or hatchet, pliers with wire cutters, a Brunton compass for plumbing the pole, materials for making a sketch map, and, if possible, a pair of field glasses for observing other stations.

7-6. Measuring the Base Line

The base line should be measured with somewhat greater precision than that required of the control work as a whole. The measurement is generally made with a steel tape, but it may be made by pacing or stadia methods where their precision is adequate. Precise measurements can be made quickly with a theodolite and subtense Invar bar. The taping procedure itself may be modified to yield various degrees of precision. In any case, both the precision and speed of measurement can be improved by selecting a base line that lies along level or evenly sloping ground, with end signals visible from all points along the line. Furthermore, the length and orientation of the base line relative to adjacent triangles must be considered (Section 7-4).

Taping with the tape held level. Distances can be taped most rapidly by holding the tape level. Measurements made in this way across rough or variably sloping ground need not be in error by more than 1 part in 2,000; if the ground is even and slopes gently, the method can be at least twice that precise. The general procedure is to accumulate full lengths of the tape where possible and to break tape (use part of a length) where the slope is steep. The following items are required: steel tape, 2 plumb bobs, hand level or Brunton compass, taping pins, tape thermometer, and 2 notebooks and pencils. A spring scale for measuring pull on the tape should be used in relatively precise work, but the standard pull of 10 lb can be approximated with practice. The instructions that follow cover the procedure step by step.

1. Party examines markings at ends of tape to insure correct readings (some tapes are graduated beyond the 0 and 100 ft points).

2. Head tapeman carries 0 end of tape ahead (downhill if possible) to that point where the tape must be held at chest height in order to be level. Rear tapeman sits over station and lines head tapeman in

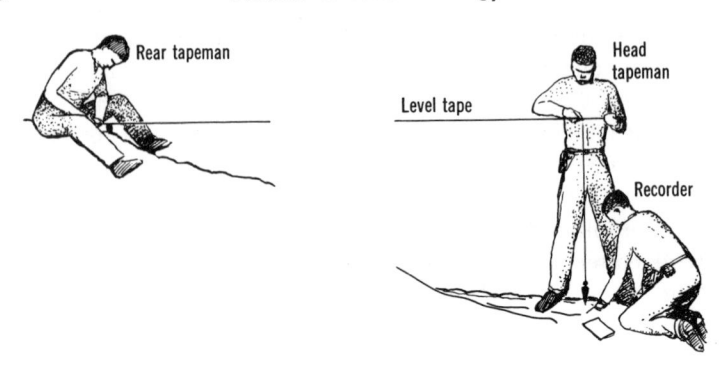

Fig. 7–3. Taping with tape held level.

by sighting over him to far station signal. Recorder stands near head tapeman and, using hand level, instructs him to raise or lower tape within a few inches of level.

3. Rear tapeman holds 100-foot mark or other whole foot mark at the station point, while head tapeman holds plumb bob string against 0 mark of tape, with plumb bob point just above ground (Figure 7–3).

4. Recorder clears ground around point of plumb bob, and head tapeman stills plumb bob by dipping it gently to the ground.

5. Rear tapeman calls out the whole foot he is holding, and recorder drives in a taping pin at the point of the plumb bob.

6. After the distance of the measure is recorded, the last step is repeated as a check.

7. Rear tapeman examines marks on both sides of mark used to be sure he read it correctly, then he records the distance.

8. The tape is moved ahead, and when the next measure is completed, the rear tapeman pulls out the pin and carries it with him an additional tally of the number of measures taken.

The line should be measured at least twice to detect gross mistakes and to ascertain the approximate precision of the measurement. Temperatures are recorded alongside the appropriate measures. If maximum precision is desired, 2×2 in. wooden stakes may be set at each point instead of chaining pins, and the recorder marks the exact measure on each stake with a pencil. The stakes permit an easy remeasurement of the line.

The measurement must be corrected for temperature and for the sag of the tape. The temperature correction is based on a coefficient of linear expansion of 0.00000645 per degree F. If the temperature change during taping was slow and totaled only about 5°, the correction can be made as an average for the line. It is computed by finding the difference between the standard temperature (68°) and the average temperature during taping, multiplying this difference by 0.00000645, and then multiplying the product by the length of the line. If temperature changes were rapid and large, the correction is made by separate sections.

The correction for sag is likely to be between 0.08 and 0.16 ft per full tape length. A fairly exact correction factor can be found by laying the tape on a smooth floor, marking its length under a pull of 10 lb, and then marking its length again when holding it off the surface with a pull of 10 lb. The correction can also be calculated approximately from the general formula

$$\text{sag correction} = \frac{w^2 l^3}{24 P^2}$$

where w is the weight of the tape in pounds per foot, l is the distance between supports, and P is the pull applied in pounds.

Taping on slope. If maximum precision is needed, as when a short base line will be expanded into a large quadrilateral, the line should be taped by setting heavy stakes on slope at each 100-foot mark, taping each measure on slope, and then correcting each measure to a horizontal distance.

In order to obtain suitable precision, the tape should be sent to the U.S. Bureau of Standards at Washington, D.C., where it can be standardized for a moderate fee. The bureau should be instructed to determine its length when fully supported and also when supported at both ends, using a 10-pound pull. The tape will be marked as standardized only if the graduations are etched or engraved in the steel ribbon, not on separate soldered patches.

Gullied or boulder-strewn ground should be avoided because the exact position of each stake cannot be determined at the outset. The stakes, preferably of 2×4 in. lumber, should be set by placing a transit over one end of the base line and lining in a taping party to each 100-foot mark. This is done so that an exact 100-foot measure falls somewhere on each stake. The exact measures are then taken by a taping party, and they should be scratched precisely on tin can strips nailed to the tops of the stakes. A transit or engineer's level and Philadelphia rod can be used to determine the difference in elevation between each pair of stakes. Each slope measure (the standardized length with tape supported at two ends) is then corrected for temperature, as described for taping with tape held level. The exact horizontal distance of each measure is found by the relation

$$\text{horizontal distance} = \text{corrected tape length} - \frac{h^2}{200}$$

where h is the difference in elevation for that measure.

7-7. Triangulation with the Transit

Triangulation with the transit consists of occupying the control stations and measuring the various interior angles of the network as well as the vertical angles to adjacent stations. To do this efficiently, the network should first be studied and the angles listed in the notebook in the order they will be measured. Generally, only the angles between adjacent stations are measured; otherwise the computations become too involved. All angles within quadrilaterals are measured, however, because both the triangles and the quadrilateral afford checks of the sums of angles in each figure. Wherever there is a choice between lines of sight to outlying stations, as in the system of Fig. 7-1C, the sights used are those that intersect most nearly to 90° at that station.

The work can be done by one man, but a second man can help carry gear and assist at the instrument. In addition to the transit and the plumb bob, the equipment must include a reading glass ($2\times$ to $3\times$), notebook and pencil, and tape for measuring height of instrument; a pair of field glasses is useful for finding signals.

The height of the instrument and the height of the flag should be recorded before the measurement is made. The instrument is set up firmly and leveled with care, for if the plates are not level, steep sights to distant stations will be appreciably in error. The various horizontal angles should be measured by repetition, as described in surveying texts. A reading should also be taken on magnetic north at each station to check the conformity of the magnetic declination within the area; this data will be very useful in later mapping (Sections 8-11 and 8-12).

Measuring vertical angles with the transit. Vertical angles can be measured more precisely with the transit than with the alidade, but the procedures are essentially the same (see Section 6-10). The angle is always found as the difference between the vernier reading to the station (the top of the flag) and the reading with the telescope leveled. The + or − sense of the small correction from 0° must be noted with care. To eliminate errors caused by a poorly adjusted telescope level, one reading is taken in the normal position and one with the telescope reversed. The cross hairs and objective must be focused carefully to eliminate parallax. It cannot be emphasized enough that vertical angle readings *must be free of mistakes*, because the method can introduce errors that are in themselves appreciable. Observations should be made when the air temperature gradient between the instrument and signal is as small as possible, as on cloudy, windless days. Refraction by heterogeneous air masses can be compensated for precisely by observing vertical angles simultaneously in both directions with two instruments. If this cannot be done, the effects of irregular refraction can be compensated for approximately by making a reading in the reverse direction when conditions are the same as when the first reading was made.

For relatively low-angle sights, the stepping method (Section 6–15) may give greater precision than a vertical angle measurement.

7–8. Surveying an Elevation to the Control Network

An elevation above mean sea level must be surveyed to the base line or some other well-located station of the control network. This survey should be started at a bench mark, a highway or railroad route point, or an established triangulation station whose elevation is known precisely. The elevations and descriptions of these points must be obtained well before the field season (Section 7–3). Local surveyors, forest rangers, or ranchers can often help locate points in the field.

It will generally be adequate to make the measurement by vertical angle readings because all other elevations in the network will be determined in this way. If greater precision is needed, a line of levels may be run between the two points, as by the method called *differential leveling*, described in all surveying texts. Differences in elevation can also be determined by controlled altimeter measurements (Bouchard and Moffitt, 1959; Hodgson, 1957).

7–9. Observation on Polaris at Elongation

An accurate bearing must be brought to the network, and this is usually done by observing Polaris at elongation. The position of Polaris is somewhat eccentric to the earth's axis of rotation so that the star appears to move in a small circle around the celestial pole (Fig. 7–4). Polaris is said to be at elongation when it appears to move vertically, as along the vertical cross hair of a transit. The precise direction of true north can be obtained by reading the bearing of the

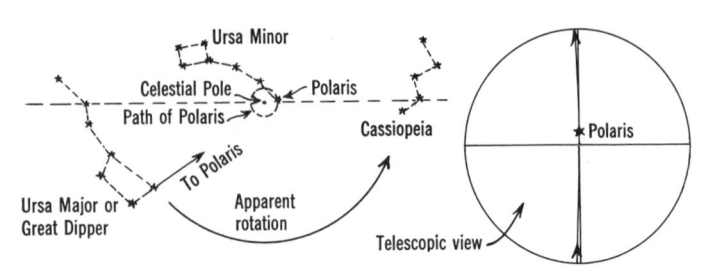

Fig. 7-4. Polaris at eastern elongation, showing relations of principal constellations on the left and view through noninverting telescope on the right.

star at elongation from tables in a current *solar ephemeris,* such as the booklets prepared annually by the W. and L. E. Gurley Company and the Keuffel and Esser Company (supplied free on request).

The position of eastern elongation can be observed before midnight from midsummer to early fall, and the exact time can be determined from tables in the ephemeris. If the transit is not equipped with a mirror reflector to illuminate the cross hairs, a flashlight can be pointed toward the objective and somewhat to one side of the line of sight. The observation is made as follows:

1. Set a transit up over a control station (preferably one end of the base line) and level it precisely by means of the telescope level. Practice with illumination.

2. Sight on Polaris 20 or 30 minutes before elongation and then depress the telescope to line in approximate north point on the ground, about 200 ft away.

3. Drive in two stakes to bracket this point, and nail a board on top of them.

4. Adjust the telescope and observe the star until it appears to move exactly along the vertical cross hair, then depress the telescope to the board and have an assistant drive a 2- or 3-in. nail vertically in the board in the line of sight.

5. Elevate the telescope to make sure the star is still on the cross hair; if it is, rotate the telescope 180° horizontally, reverse it, and sight again on the star.

6. Depress the telescope again and line in another nail on the board (it will coincide with the first if the instrument is in perfect adjustment).

7. Next morning, set the transit up over the same point, align it on the midpoint between the two nails, and measure the horizontal angle to an adjacent control signal.

8. Determine the bearing of the line of sight to the nails by using the tables in the ephemeris.

7-10. Computations and Adjustments of Triangulation Data

Computations of triangulation data should be made in the following order: (1) adjustments of triangles and quadrilaterals by summation of angles, (2) computations of horizontal distances of occupied triangles by law of sines, (3) adjustments of unoccupied stations, (4) computations of elevations, (5) corrections of elevations for curvature and refraction, (6) computation of bearings of triangle sides, and (7) computation of coordinates. The computations should be planned so that each sheet can be checked by a second independent trial. The second set should be run at once, item by item, after the first; otherwise an error would necessitate redoing all subsequent computations. Because few field parties will have a computing machine, the forms illustrated here are based on logarithms. Seven-place tables should be used for angles measured to the nearest 5 seconds; less detailed tables require time-consuming interpolations. The sheets should be stored systematically in a looseleaf binder as they are completed.

Adjustments of figures. The sum of the interior angles of each triangle should be 180° and the sum of the angles of each quadrilateral 360°. The allowable difference will depend on the precision of the survey; if it is greater than 30 seconds, the angles must generally be remeasured. The residual error is distributed equally to the angles of each triangle. If quadrilaterals have been measured as well, errors can be distributed more accurately because each quadrilateral angle is the sum of the adjacent angles of two triangles. These adjustments may be made by trial and error; however, if the network is large and the results must be unusually precise, they can be made as described by Reynolds (1928) and Bouchard and Moffitt (1959).

When all angles have been adjusted so that triangles sum to 180°, the resulting angles should be listed in the order in which they will be used.

Computations of occupied triangles. Triangles whose three corners have been occupied are calculated first, starting with those that include the base line. The basis of the calculation is the law of sines;

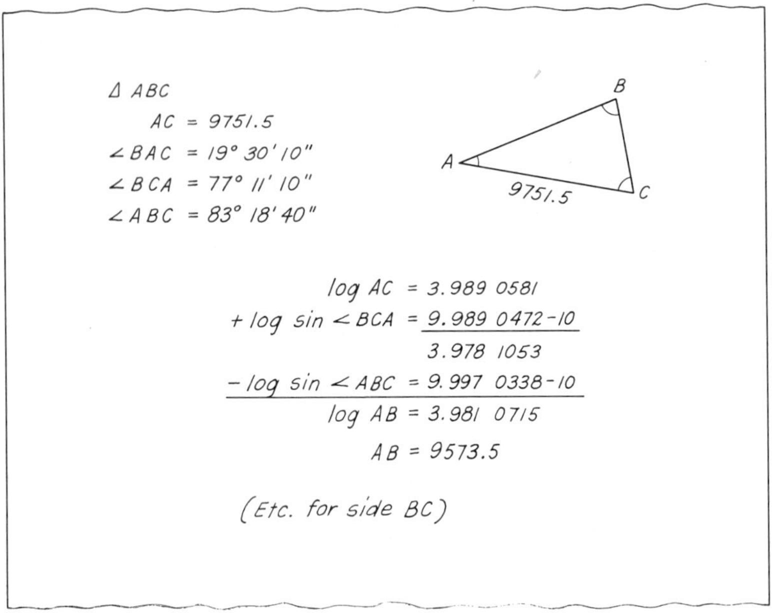

```
△ ABC
    AC = 9751.5
  ∠BAC = 19° 30' 10"
  ∠BCA = 77° 11' 10"
  ∠ABC = 83° 18' 40"
```

              log AC = 3.989 0581
        + log sin ∠BCA = 9.989 0472-10
                         3.978 1053
        - log sin ∠ABC = 9.997 0338-10
              log AB = 3.981 0715
                  AB = 9573.5

            (Etc. for side BC)

Fig. 7–5. Computation of horizontal triangles.

using the notation shown for the triangle at the top of Fig. 7–5, the unknown sides are solved by the following equations:

$$AB = AC \frac{\sin \angle ACB}{\sin \angle ABC} \quad \text{and} \quad BC = AC \frac{\sin \angle BAC}{\sin \angle ABC}.$$

Computations with logarithms may be set up as shown in the figure, a separate sheet of paper being used for each triangle.

Adjustments of unoccupied stations. The angle at an outlying unoccupied corner of a triangle can be found only as the difference between 180° and the sum of the two measured angles, and therefore the accuracy of the measurements cannot be checked by summation. Instead, adjustments are made after completing the sine computations of the sides of the triangles. In Fig. 7–6, for example, the outlying point M can be adjusted by averaging the two solutions of side MB in the triangles AMB and BMC, and the outlying point N can be adjusted by averaging the two solutions of side NC in triangles BNC and CND. If the distance MN is needed (for example, to be plotted on a plane table sheet) it can be found by computing with the law of cosines. Using the notation shown in the figure, the equation is:

$$MN = MB^2 + NB^2 - 2(MB \cdot NB) \cos \angle MBN$$

This computation can be set up for logarithms in a way similar to the sine computation.

Computations of elevations. Elevation computations are based on the law of tangents for right triangles, as shown in Fig. 7–7, a typical computation sheet. The + or − sign of the angle is carried throughout the computation. The height of instrument is always given a + sign, the height of the signal a − sign, and the net difference in elevation is found by adding algebraically. As shown in the example, the vertical angle used is an average of the angles read from each station, but the sign used has the sense of carrying the elevation forward in the network. The sketch made in the corner of the computation sheet insures that this will be done correctly. Elevations of a station must be computed in at least two separate vertical triangles, and the results are averaged as shown in the example.

Corrections for curvature and refraction. The earth's surface curves nearly 0.7 ft per mile, and thus appreciable corrections are needed where level or inclined sights are made over long distances. Light traveling from a signal to the instrument is also refracted downward as it passes from less dense to more dense layers of the atmosphere, affecting readings in the opposite sense to the earth's curvature, but to a much smaller degree. Refraction varies with atmospheric conditions, but on windless days an average correction can be used. The combined correction for curvature and refraction, in feet, is found by multiplying the square of the distance, in miles, by 0.574 (Gannet, 1918, p. 335). This number is always given a + sign and is added

Fig. 7–6. Adjustments of outlying (unoccupied) stations.

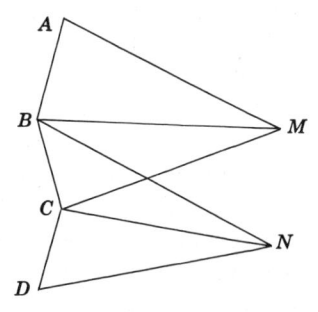

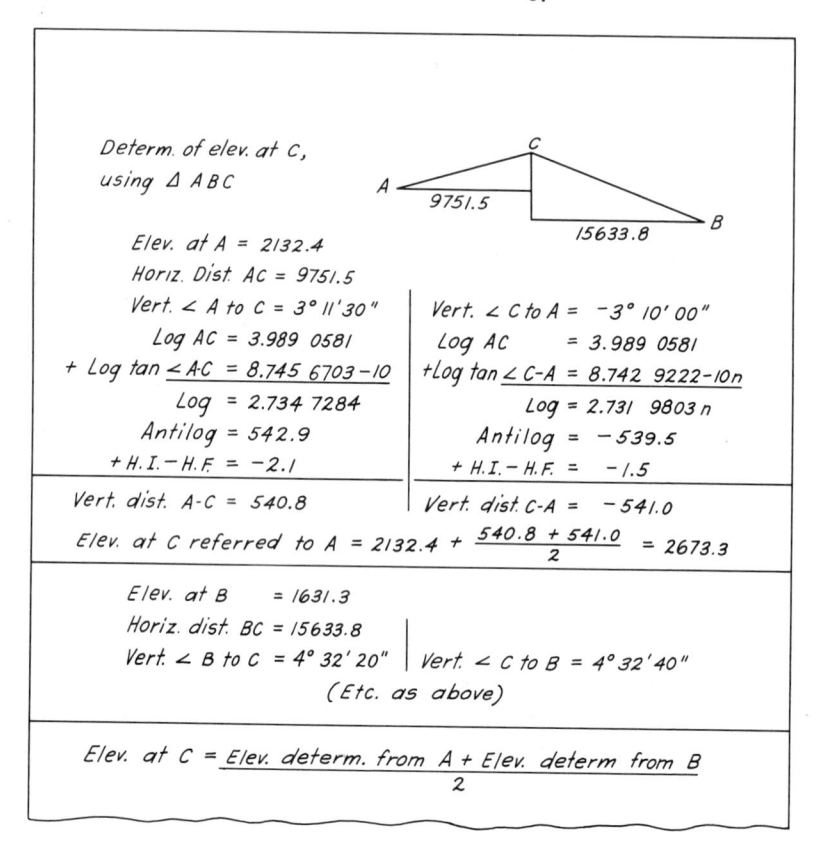

```
Determ. of elev. at C,
using △ ABC
                              A _____ C
                                    9751.5  |
  Elev. at A = 2132.4                       |_____ B
  Horiz. Dist. AC = 9751.5            15633.8
```

Vert. ∠ A to C = 3° 11' 30"	Vert. ∠ C to A = −3° 10' 00"
Log AC = 3.989 0581	Log AC = 3.989 0581
+ Log tan ∠AC = 8.745 6703 −10	+Log tan ∠C-A = 8.742 9222−10n
Log = 2.734 7284	Log = 2.731 9803 n
Antilog = 542.9	Antilog = −539.5
+ H.I.− H.F. = −2.1	+ H.I.− H.F. = −1.5
Vert. dist. A-C = 540.8	Vert. dist. C-A = −541.0
Elev. at C referred to A = 2132.4 + $\frac{540.8 + 541.0}{2}$ = 2673.3	

```
    Elev. at B     = 1631.3
    Horiz. dist. BC = 15633.8
    Vert. ∠ B to C = 4° 32' 20"   |  Vert. ∠ C to B = 4° 32' 40"
              (Etc. as above)
```

Elev. at C = $\frac{\text{Elev. determ. from } A + \text{Elev. determ. from } B}{2}$

Fig. 7–7. Computation of vertical angles. The small letter *n* indicates logarithms of negative numbers.

algebraically to the *difference in elevation* as measured by a given sight between two stations (Fig. 7–8). Study of the figure will show that if sights are made in both directions, they can be averaged directly to give the correct result. The corrections are generally made, however, to detect errors made in one sight or the other.

Computations of bearings. Bearings of triangle sides are computed before coordinates for control stations. This is done as shown in Fig. 7–9, starting with the side used for the observation of Polaris and proceeding systematically through the network, one triangle at a time. The computation is made counterclockwise around each triangle, and the final back-bearing serves to check the arithmetic.

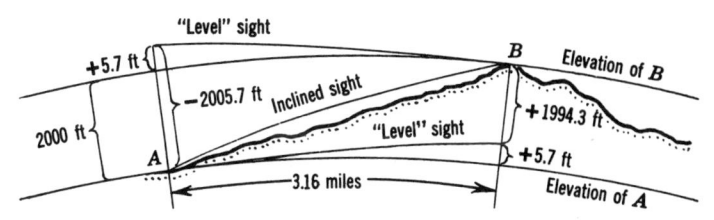

Fig. 7–8. Relations resulting from curvature and refraction of lines of sight between two stations. The vertical scale of the drawing is exaggerated greatly, but the numbers are approximately representative.

Computations of coordinates. Coordinates provide a simple means of plotting control stations on a map base, and inter-station distances and bearings can be computed easily from them. The coordinates considered here are *plane coordinates*, plotted on the basis of a rectangular grid. The coordinate axes are oriented *N-S* and *E-W*, and the coordinate (grid) interval is selected on the basis of the scale of the

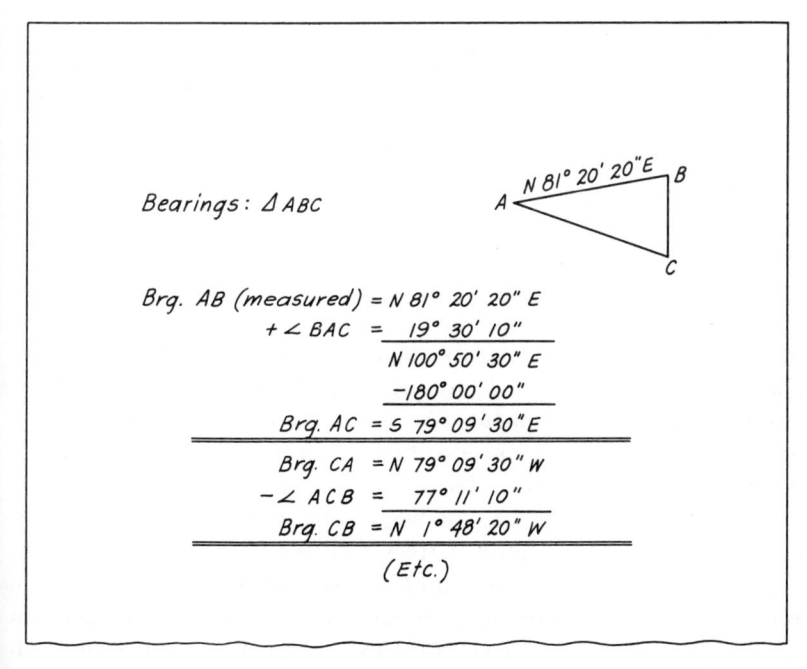

Fig. 7–9. Computation of bearings.

$\frac{\Delta}{Side}$	Data	Departure Comp.	Latitude comp.
AB	Brg. = Distance = Coords. A = 1000, 1000	log sin Brg. =_____(1) +log Dist. =_____(2) (1) + (2) =_____(3) Departure = (antilog of 3)	log. cos. Brg. =_____(4) +log. Dist. =_____(5) (4)+(5) =_____(6) Latitude = (antilog of 6)
	Coords. of B = 1000 + Lat., 1000 + Dept.		

Fig. 7–10. Computation of coordinates.

map, generally at even multiples of 100 ft. The most southwesterly station is arbitrarily given large enough coordinates (as 1,000, 1,000) so that all coordinates within the area will have + signs. The computations are started at this station and are made as shown in the computation sheet of Fig. 7–10. The sign for each latitude and departure is determined by inspection of the bearing; westerly bearings have − departures; easterly bearings have + departures; northerly bearings have + latitudes; and southerly bearings have − latitudes. The computations are first made around the entire perimeter of the system, closing on the initial station to provide a check of the arithmetic. Computations of the remaining stations are made in lines across the network so that an arithmetical check can be made by closing on a perimeter station.

State plane coordinate systems. Control data can also be computed and plotted on the basis of national or state-wide coordinate systems. The local state plane coordinate system should be used because these coordinates can be computed from the latitudes and departures considered above. All that is needed to start the computations is to know the state plane coordinates of one of the network stations. If one of the stations is a government triangulation station, the coordinates can generally be obtained on request, or they can be converted from geodetic coordinates by using tables published by the Coast and Geodetic Survey. A brief description of these coordinate systems is given in most modern surveying texts. Complete descriptions and tables can be purchased for a nominal charge (see lists in *Publications of the*

Coast and Geodetic Survey, which can be obtained from the Director, U.S. Coast and Geodetic Survey, Washington 25, D.C.).

7–11. Rectangular and Polyconic Grids

Control points computed to plane coordinates can be plotted on a rectangular coordinate grid that can be constructed with a long steel straightedge, an accurate scale, and a beam compass of suitable length. An important preliminary step is to practice drawing straight, thin lines. This can be done by (1) using a hard, sharp pencil, (2) holding it at right angles to the straightedge and tilted slightly in the direction in which the line is being drawn, and (3) drawing the line in one steady and calculated stroke. It is also advisable to practice with the beam compass until an arc can be drawn without enlarging or distorting the needle hole at the point held.

The instructions that follow are based on Fig. 7–11.

1. Draw a vertical line near the center of the sheet and construct a perpendicular at *E* by first striking arcs *EA* and *EB*, then arcs *AC, BC, AD,* and *BD*.

2. Using the scale, set the beam compass exactly at a whole number of grid intervals and strike arcs 1, 2, 3, 4, 5, and 6 from points *C, E,* and *D*.

3. Lay the straightedge next to these arcs and draw east-west lines that are tangent to them.

4. Now set the beam compass for some whole number of grid intervals (for example, to give the approximate proportions shown in

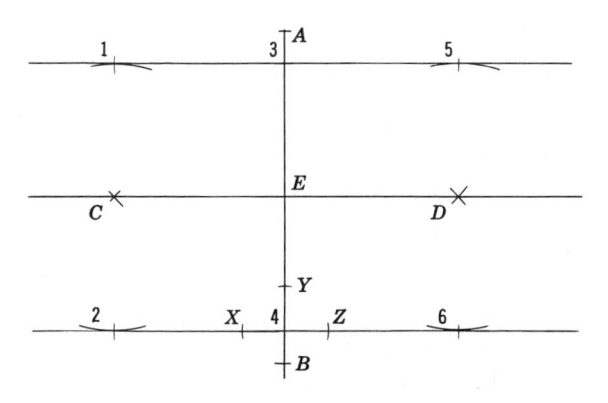

Fig. 7–11. Constructing a rectangular coordinate grid.

the figure) and strike arcs to 1, 2, 5, and 6 by holding at points 3 and 4.

5. Check the diagonal distances 1–6 and 2–5 with the scale; if they are equal, continue with the next step; if not, erase the work and start again.

6. Prick all intersected points with a needle and reading glass.

7. Set the beam compass for a single grid interval, and strike arcs from all pricked points to intersect all lines (as at X, Y, and Z in the figure).

8. Check these arcs with the scale to make sure the compass did not slip, then prick them and continue to expand the grid until it is completed. Draw the grid lines with a straightedge.

The scale of the sheet must remain reasonably constant during the plotting, and the following instructions will help in this regard: (1) season all materials in advance, (2) use one master scale or meter bar only, (3) keep materials at the same temperature and humidity, and (4) work carefully *but steadily* from the beginning to the end of the construction.

Control points can be plotted on the grid by the procedure given in Section 8–5.

Polyconic grids. Polyconic grids are customarily used as a base for geodetic coordinate data, especially where a map covers a large area. They are constructed in the same general way as rectangular grids, except that tables are needed to determine the linear distances subtended by various arcs of latitude and longitude. The tables of U.S. Geological Survey Bulletin 809 (Birdseye, 1929) are used to make the construction in feet and inches, whereas those in U.S. Coast and Geodetic Survey Special Publication 5 (1935) are used to make it in meters. Instructions accompany each of these sets of tables.

7–12. Triangulation with the Alidade and Plane Table

The alidade and plane table are less suited for triangulation surveys than the transit; however, *wind and weather allowing*, plane table work is faster and is precise enough for many projects. Horizontal locations are made by intersecting lines of sight on the plane table sheet, while differences in elevation are based on vertical angle measurements.

Top-quality waterproof plane table sheets should be used; if linen-filled sheets must be used, they should be seasoned thoroughly (Section 6–8). The base line should be plotted on the sheet with a steel straightedge and the end points pricked with small needle holes. If geodetic or other established control will be used instead of a base line, it must be plotted on the sheet by means of coordinate lines (Sections 7–11 and 8–5). The notebook should be set up ahead of time and the order of sights from each station listed.

Starting with the base line, each station is occupied and rays are drawn to adjacent stations and to those outlying stations that would otherwise not be intersected with three pencil rays. The pencil lines should be made the full length of the fiducial edge so that the board can be oriented accurately by backsights. Each pencil ray must be labeled clearly. Vertical angles can be measured (Section 6–10), or the stepping method can be used (Section 6–15). Special attention must be given to the level and stability of the plane table; if the wind causes it to flutter, for example, the instrumentman must wait for a lull. If the telescope is the type that can be released and rotated in an axial sense, as the one shown in Fig. 6–3, the vertical angle may be repeated with the telescope in reversed position. An average of the two readings will compensate for incorrect alignment of the cross hairs with the line of collimation.

Precision. Distances can be scaled along thin pencil lines with considerable precision, for a needle hole need be no more than 0.003 in. across (at a scale of 1 in. = 1,000 ft, this is 3 ft on the ground). The pencil ray itself, however, cannot be drawn along the fiducial with more than half this precision. The suggestions for drawing straight lines given in Section 7–11 may be helpful in this regard. Plotting errors can, with care, be held to less than 0.01 in. Because the greatest probable plotting error is at right angles to a pencil ray, it is essential in plane table triangulation that intersection angles be greater than 30°.

7–13. Surveying Control by Transit Traverses

Where trees or other obstructions make it impossible to locate stations by triangulation methods, tape and transit traverses can be used to establish control. The traverses must be run in closed circuits or must be closed on points that can be located at least as precisely as the points of the traverse. Because of the precision required, the courses should be reconnoitered so they can be directed along open, evenly sloping stretches. Distances may be taped with the tape held level, and a tape 200 or 300 ft long can be used efficiently in country with little relief.

The first leg of the traverse should be turned from a line of known bearing or azimuth. If it must be started from an isolated triangulation station, the bearing of the first leg can be determined by an observation on Polaris. This accurate bearing will not be essential until the traverse data are computed to coordinates; however, it is useful to compute bearings of lines as the traverse proceeds. Vertical control should be run at the same time by reading vertical angles or by stepping vertical distances between stations.

Traverses that close on their starting points may be made efficiently by measuring the interior angles of the circuits, while traverses that start at one known point and end at another are best run as deflection-angle traverses. The procedures for both types of traverses are described fully in most surveying texts (as Bouchard and Moffitt, 1959, ch. 8). They can be executed by two men, but at least one other man is helpful for taping and recording. Computations, which should be kept up-to-date in camp, will generally include checks on bearings computed in the field, correction of taped distances (Section 7–6), computations of latitudes and departures, computations of plane coordinates of stations, and computations of differences in elevation from vertical angles and corrected distances. For closed traverses, it will be helpful in the computations of latitudes and departures if any small closing error in the bearings is first distributed equally among the angles of the traverse. The latitudes and departures should then close evenly.

References Cited

Birdseye, C. H., 1929, *Formulas and tables for the construction of polyconic projections:* U.S. Geological Survey, Bulletin 809, 126 pp.

Bouchard, Harry, and F. H. Moffitt, 1959, *Surveying:* Scranton, Pa., International Textbook Co., 664 pp.

Gannett, S. S., 1918, *Geographic tables and formulas:* U.S. Geological Survey, Bulletin 650, 424 pp.

Hodgson, R. A., 1957, *Precision altimeter survey procedures:* Los Angeles, American Paulin System, 59 pp.

Reynolds, W. F., 1928, *Manual of triangulation computation and adjustment:* U.S. Coast and Geodetic Survey, Special Publication 138, 242 pp.

U.S. Coast and Geodetic Survey, 1935, *Tables for a polyconic projection of maps and lengths of terrestrial arcs of meridian and parallels based upon Clark's reference spheroid of 1896:* 6th ed., Special Publication 8, 101 pp.

———, 1933, *Formulas and tables for the computation of geodetic positions:* 7th ed., Special Publication 8, 101 pp.

8
Geologic Mapping with the Alidade and Plane Table

8–1. Plane Table Projects and Their Appropriate Scales

The alidade and plane table are esssential for most projects requiring an accurate geologic and topographic map with a scale larger than 1:6,000 (1 in. = 500 ft). Few topographic base maps with such scales exist, and large-scale aerial photographs do not give the exact vertical control needed for many structural and quantitative studies. Accurate topographic and geologic maps are especially valuable in resolving such problems as offsets on faults, continuity of beds or ore bodies in complex folds, stratigraphic details in intertonguing sequences, and relations of various alterations to faults or intrusive bodies. Maps used in these studies typically cover 1 to 25 sq mi at scales of 1 in. = 200 ft to 1 in. = 500 ft, and the procedures described in this chapter are appropriate for such projects. Mapping at scales larger than 1 in. = 200 ft may be necessary where measurements and sampling must be particularly precise, as in some economic studies; this type of project is considered in Chapter 10.

Geologic and topographic mapping at a scale of 1 in. = 1,000 ft can be done readily by methods described in this chapter; however, a controlled aerial photograph survey should also be considered because it takes less time than a plane table survey (Chapter 9). Complete topographic and geologic mapping at scales smaller than 1 in. = 1,000 ft is not usually undertaken by geologists because it is more efficient to map on aerial photographs or to have a base prepared by an engineering or aerial survey company. It may be necessary, however, to map certain limited features or structures at small scales. In unsurveyed regions, for example, petroleum exploration may require accurate positions and elevations on a given unit over a large area. Planning the control for such a survey requires ingenuity and judgment, and plane table procedures must often be improvised to meet particular needs. Methods that will be especially useful are the gradienter screw and stepping methods (Sections 6–14 and 6–15) and the resection and three-point methods (Sections 8–10 and 8–11). The suggestions given by Low (1952, chs. 4, 6) will also be helpful.

8–2. Planning a Plane Table Project

Most large-scale mapping projects proceed through a series of steps, as those indicated in the following suggestions:

1. Order and prepare materials and control data before the field season.
2. Lay out a local control system and at the same time reconnoiter geologic features.
3. Survey the primary control and make control computations.
4. Plot primary control on plane table sheets that will be used for stadia mapping.
5. Walk over the area to make geologic sketch maps, flag contacts and faults, and select instrument stations for stadia work.
6. Survey instrument stations or secondary control points.
7. Map geologic features, topography, drainage, and culture by stadia methods.
8. Complete the map by checking geologic features, making non-instrumental additions, and walking cross-section lines.

Suggestions for preparing and ordering materials and data are given in Chapters 1, 6, and 7, and Appendix 1. Considerations of plane table sheets, instruments, and established control are particularly important. Aerial photographs are always valuable to a plane table project.

The steps listed above can be done by two men, unless a base line is to be measured precisely. A three-man party might consist of an experienced geologist, a second man who knows surveying methods as well as some geology, and an assistant. The geologist would be the key man in step 5, above, and he would generally be rodman during the stadia mapping. The second man would help plan the control work and would be responsible for its execution. If he could do steps 3, 4, and 6 while the geologist did step 5, a good deal of time could be saved. The helper would work with the surveyor and would be a recorder either at the instrument or rod during the stadia mapping.

For larger parties, it is unwise to switch major jobs during the control work. To give continuity, the examination and stadia mapping of any one part of the area (as a plane table sheet) should be guided by one geologist.

The time required for a plane table project depends on many factors, especially the complexity of the geology, the openness of the terrain, and the weather. Assuming the scale chosen is such that geologic features can be plotted easily, an area covered by four sheets will, under average conditions, require about 20 six-day weeks for the three-man party described above. This assumes that the control and stadia surveying are done precisely enough so stadia survey points are within about 0.02 in. of their true positions.

Choosing the scale. The scale for the plane table mapping should be considered in a general way before the field season, but the final choice requires a geologic reconnaissance. In particular, it must be determined *how large the smallest units or features are that must be mapped to scale.* The ideal scale is just large enough so these features can be plotted easily and accurately. General working precision of the alidade and plane table are such that errors on the map can be held to less than 0.02 in. (Section 7–12). If, for example, measurements made from the map can be in error by no more than 2 or 3 ft on the ground, the map scale should be 1 in. = 200 ft. At this scale, features 20 ft across (0.1 in. on the map) can be plotted easily to scale. Many economic and detailed geologic studies require this sort of detail. If mapping errors may be as large as 5 to 10 ft on the ground, and features no larger than 40 or 50 ft across need be shown to scale, mapping should be done at scales of 1 in. = 400 ft or 500 ft.

8–3. Reconnoitering the Control System and Geologic Features

The first step in the field is to reconnoiter the area in order to get a general view of the geologic features, plan the scale of the work, and lay out the control network or plan the control traverses. Aspects of planning control surveys and selecting stations are given in Chapter 7. A compass sketch map of the control system should be used as a base for a geologic sketch map (Section 7–4). The scale of this map will typically be one-fourth or one-fifth that of the final sheets, and its accuracy need be adequate for planning only. More detailed sketch maps and sections showing areas of special interest can be made in field notes. In many cases, aerial photographs make an ideal base on which to sketch features that will later be mapped in detail because some structures and rock units can be recognized on them at once.

8–4. Choosing the Layout of Plane Table Sheets

After the control system has been surveyed and computed (Chapter 7), the plane table sheets are prepared for stadia mapping. The first step is to determine the optimum layout of plane table sheets relative to the area to be mapped. Each sheet must (1) include at least two and preferably three or four intervisible control points, (2) have none of these points closer than 2 in. from its edges, (3) overlap at least 2 in. with each adjacent sheet, and (4) cover the area so that no stadia points need be plotted closer than 1 in. to its edges. The layout can be selected as follows:

1. Plot the control points on one piece of paper, at about one-fourth the scale to be used in mapping (nearest-$\frac{1}{2}°$-accuracy is adequate).

2. Cut several rectangles from tracing paper, each representing a plane table sheet at the same scale as that used in step 1.

3. Arrange and rearrange these rectangles over the control plot until an optimum layout is obtained, then prick the corners through to the plot and draw the positions of the sheets on the plot.

This procedure is simple if the ground is reasonably open, but if appreciable parts of the area cannot be seen from certain nearby control stations, the sheets must be arranged to cover these blind spots. The scale of the work and the minimum spacing for stadia instrument stations should be considered (Section 8–6).

8–5. Plotting Primary Control on Plane Table Sheets

After the plane table sheets have been prepared, the primary control points can be plotted on them by using coordinates or by intersecting with a beam compass. To plot them by coordinates, the coordinate grid should first be constructed with moderate precision on the small-scale layout described in Section 8–4. The approximate positions of two coordinate lines are then found by transferring distances from the small-scale plot, as distances MN and MP in Fig. 8–1. Start-

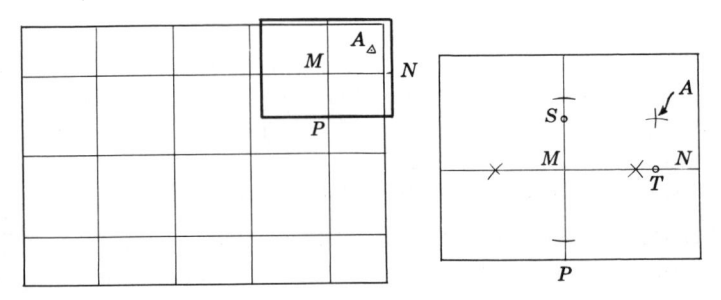

Fig. 8–1. Transferring coordinate grid from small-scale sheet (left) to plane table sheet (right) and locating a control point (A).

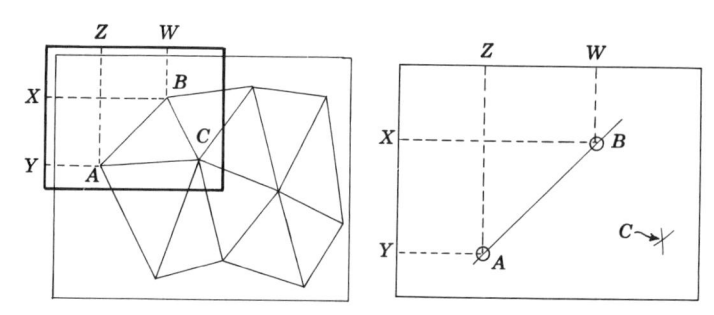

Fig. 8–2. Transferring control points A and B from small-scale sheet (left) to plane table sheet, and intersecting a third point, C.

ing with a meridian line such as *MP*, a perpendicular is constructed at *M* on the plane table sheet. A control point such as *A* is then plotted by (1) scaling its coordinates from *M* to *S* and *T*, (2) setting a beam compass for the distances *MS* and *MT*, and (3) striking arcs from *S* and *T* to intersect the point. When all points are located and checked, the coordinate grid is inked in very thin lines.

If coordinates were not computed, the points can be plotted by intersecting with a beam compass. In Fig. 8–2, for example, the approximate positions of points *A* and *B* are first located by transferring such distances as *WB*, *XB*, *YA*, and *ZA* from the small-scale plot to the plane table sheet. A thin pencil line is then drawn through the points on the sheet with a steel straightedge. A fine needle hole is pricked first for one station, and then the other station is pricked after carefully scaling the exact (surveyed) distance *AB*. A beam compass is next set to the exact distance *AC* and an arc is struck in the approximate area of *C*. When the same is done for *BC*, *C* is located by the intersection. The distances should be checked with a scale, and the point should be pricked with the help of a reading glass or hand lens. Finally, a north-south line can be drawn on the sheet by referring to the bearings of one of the lines.

8–6. Instrument Stations for Stadia Work

The primary control points cannot be spaced closely enough to provide instrument stations for all the stadia mapping, and additional stations must therefore be located. How this will be done depends on the scale of the project and the nature of the geologic features and terrain. If the scale is large, the geologic features complex, and the terrain reasonably accessible, instrument stations should be selected during a preliminary and rather detailed examination of the entire area, as described in Section 8–7. The new stations are marked with signals and surveyed by intersection from primary stations. These instrument stations are sometimes called the *secondary control* because they are located as a subsidiary network of points, prior to the stadia mapping. Where mapping scales are rather small, geologic features are simple, or the terrain is so rough that it is not efficient to work in any part of it more than once, instrument stations should be selected and surveyed as the mapping progresses. These stations must be located by resection, three-point methods, and traverses (Sections 8–10, 11, and 12).

Although there should be as few instrument stations as possible, they must be spaced closely enough so that stadia measurements will be adequately precise. The stadia intercept can usually be estimated to the nearest 0.02 ft at distances less than 600 ft. The precision of readings and Beaman corrections falls off rapidly for sights of more than 1,000 ft. Instrument stations should therefore be spaced about 2,000 ft apart for detailed mapping. This distance may be increased moderately where the map scale is small or where errors may be larger than 8 or 10 ft on the ground. The limiting distance should be tested by each party, because stadia accuracy depends not only on the distance to the rod but also on the qualities of the instrument, the rod, and the lighting.

Instrument stations can be selected systematically and quickly in open country of moderate relief, but obstacles often limit this selection. Each station must give a clear view of the ground to be mapped and must be visible from at least two primary stations (three primary stations if three-point methods will be used). Ideal stations are in open valleys or on the tops of low knolls from which there is a view up and down a valley as well as to the valley sides. Gently sloping ridges and saddles are often good locations. The tops of steep hills or ridges are typically unsuitable because the instrumentman will not be able to see their slopes.

A stake or solid rock can be used to mark each station. Stations that will be located by intersection must be marked with a distinctive signal, as a cairn or a small flag and pole similar to the one described in Section 7–5.

8–7. Examining Geologic Features and Flagging Contacts

In areas that must be mapped in detail, geologic features should be examined before stadia mapping is started. This can be done at the same time the instrument stations are selected and marked (Section 8–6). Sketch maps should be made of the more complex parts of the area. Points that will be used in the stadia mapping can be entered on these sketch maps and marked in the field by small strips of bright cloth or plastic ribbon. Examples of such points are sharp turns in contacts, intersections of contacts and faults, or places where critical structural details can be seen or important specimens collected. The flags may be hung on bushes, limbs, or rock outcrops where they can be seen easily by the rodman. The examination and flagging should give so clear a picture of the main geologic features that the stadia mapping will be more nearly a systematic coverage than a painstaking exploration. Otherwise, the instrumentman and recorder may stand idle for long periods while the rodman puzzles out each contact. It is not desirable, however, to flag all the points that will be used to make the final map. Additional critical points will be discovered during the mapping, and many points will have to be added to map nongeologic features as well as geologic details.

8–8. Choosing the Contour Interval

When the control survey and preliminary geological examination are completed, it should be possible to choose the best contour interval for the stadia work. The standard intervals for large-scale plane table maps are 5, 10, 20, 25, 40, and 50 ft. The interval should be chosen on the basis of the scale and purpose of the map, the steepness of slopes, and the probable spacing of stadia points. Commonly, too small an interval is chosen; thus more time is spent locating and drawing contours than is justified. An interval of 10 ft or less should be used only where gently sloping topographic forms must be mapped, or where unusually detailed vertical control is needed for mining and construction operations. An interval of 20 or 25 ft is the best choice for most large-scale mapping projects, even where spacing of stadia points will be such that 10-foot contours could be drawn. A contour interval of 40 or 50 ft is generally preferable for projects where the map scale approaches 1:6,000 (1 in. = 500 ft) and where slopes are steep.

8–9. Intersecting Instrument Stations

After the primary stations have been plotted on each plane table sheet (Section 8–5), each primary station is occupied and rays are drawn to all visible instrument station signals. The procedure is the same as that for primary triangulation (Section 7–12). Stations located by three-line intersections should be pricked carefully with a needle and labeled with ink. Computed elevations should be lettered in pencil beside the station, but they must not be inked until the instrument stations are occupied for stadia work and the vertical angle to each primary signal is checked by a reverse reading.

Instrument stations that can be seen from only one or two primary stations must be located or checked by accessory rays from other secondary stations. In some cases, a ray can be drawn to them by offsetting a short distance from a primary station. This is done by setting up at the primary station, drawing a ray toward a point from which the instrument station can be seen, measuring the distance by taping or stadia, and plotting the new accessory station. When set up at this new point, the plane table must be oriented carefully on the primary point used to locate it and *then checked by sighting on other points.*

8–10. Locating Instrument Stations by Resection

Instrument stations in outlying parts of an area can often be located more efficiently during the course of stadia mapping than during the intersection of the secondary control. This is particularly true where the map scale is small. By the method of *resection,* the location is made by drawing a ray toward the point occupied, as explained in the following example.

A rodman has completed all stadia work in a given direction from the instrument which is at S5 in Fig. 8–3A. He examines the ground beyond and chooses a new instrument station, S6, from which he can see at least one primary station, M, from instrument height. He marks the point with a rock or stake and plumbs the stadia rod on it. The instrumentman checks the orientation of the plane table by sighting on a primary signal; then he sights on the rod, draws a ray, and reads a vertical angle. When the plane table is moved to the new point, it is oriented by a backsight on station S5. The instrumentman then lays the front part of the fiducial edge against the needle hole marking the primary station M and sights on the signal of that station (Fig. 8–3B). Intersection is achieved by drawing a ray *back* along the fiducial edge to cross the ray drawn from S5 toward S6. A vertical angle is read,

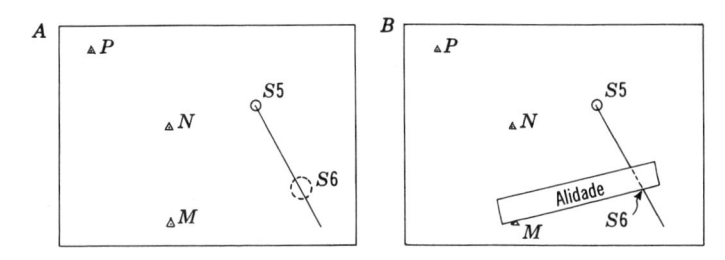

Geologic Mapping with the Alidade and Plane Table 143

Fig. 8-3. Locating a new instrument station (S6) from an established one (S5) by resection.

and then other primary signals are sighted and other vertical angles read to serve as checks. Stadia mapping may begin at S6 as soon as its elevation is computed from the vertical angle and the horizontal distance scaled from the map.

8-11. Three-Point Locations

When the plane table must be set up at a new station toward which no rays have been drawn previously, it is necessary to orient the board and intersect its position by *three-point methods*. These methods are especially useful for intermediate-scale and small-scale work or where terrain is so rough it is inefficient to set up all station signals before the stadia mapping. The control for these projects is typically a network of very large triangles, and therefore the primary signals should be placed so that they can be seen over large areas.

Tracing sheet method. This quick and almost foolproof method of three-point location requires several sheets of tracing paper and some drafting tape. The procedure is as follows:

1. Set up the plane table at the new station, in view of at least three, *and preferably four* signals that are plotted on the sheet.
2. Tape a piece of tracing paper anywhere on the sheet and prick a small needle hole in it.
3. Tighten the Johnson head, lay the fiducial edge against the needle hole, and sight each of the three or four signals through the telescope. Draw rays toward each signal from the needle hole, but stop each ray just short of the hole.
4. Remove the tape and shift the tracing paper until each of the pencil rays passes exactly through its corresponding plotted point on the plane table sheet (Fig. 8-4).

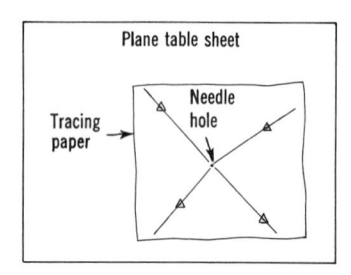

Fig. 8-4. Using a sheet of tracing paper to orient the plane table over a new station. The rays in the figure are on the tracing paper, whereas the triangles are control stations that can be seen through it.

5. Prick a hole in the plane table sheet by inserting the needle through the hole in the tracing paper; then remove the tracing paper.

6. Loosen the lower screw of the Johnson head, place the alidade on the sheet so that the fiducial bisects the new needle hole and the hole marking the farthest station, then orient the board by sighting on the signal of that station.

7. Tighten the lower screw of the Johnson head, and check the orientation by sighting on the signals of the other stations.

The only points that cannot be located in this way are those that lie on or very near the circle that passes through the three points sighted (the *great circle* of Fig. 8-5). These cases become obvious when step 4, above, is attempted, and either a fourth point must be used or the plane table must be moved.

Correcting a compass orientation. When tracing paper is not available, the board must first be oriented by the magnetic needle or by

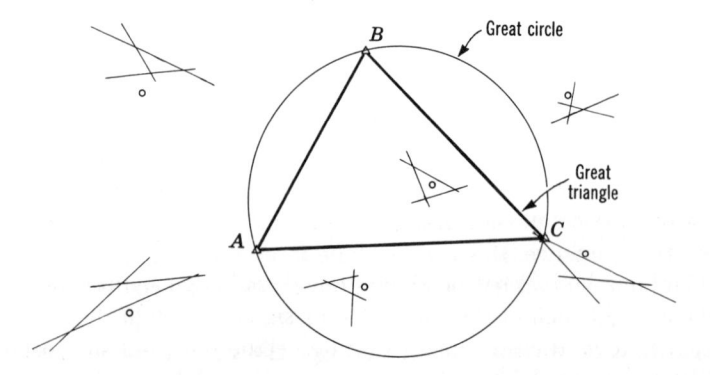

Fig. 8-5. Relation between control stations (*A*, *B*, and *C*), the great circle that passes through them, and the small triangles of error formed by three-line intersections on an imperfectly oriented plane table. In each case the small circle shows the correct position of the point sought. After Birdseye (1928, p. 203).

estimation. Three or more points are then sighted, and rays from each of them are drawn back along the fiducial edge. If the board is oriented correctly, these rays will intersect in a point, and the location is completed *unless* all the points lie on a circle, as noted above. If the rays form a small triangle, the orientation must be corrected, and this is done essentially by trial and error. The following rules, based on those given by Beaman in U.S. Geological Survey Bulletin 788 (Birdseye, 1928, p. 203) will help to determine the approximate position of the true point relative to the small triangle. The rules and the terms used are illustrated by Fig. 8–5.

1. If the small triangle is within the great triangle, the true position of the point is within the small triangle.

2. If the small triangle falls between the great triangle and the great circle, the true position of the point lies outside the small triangle, opposite the side formed by the ray from the middle station sighted.

3. If the small triangle lies outside the great circle, the point sought lies outside the small triangle and on the same side of the ray from the most distant station sighted as does the intersection of the other two rays.

The actual position of the point sought can be estimated by considering the distances to the three points sighted. The triangle was caused by *rotation* relative to these three stations; the true position of the point is therefore closest to the ray from the nearest station and farthest from the ray from the farthest station. Moreover, the distances between the true position of the point and the three rays are proportional to the distances from the point occupied to the three points sighted.

After a new point is marked, the alidade is placed alongside it and also alongside the point representing the farthest station. The plane table is reoriented by sighting on the farthest station, and rays are drawn back from each station, as before. The procedure is repeated until any remaining small triangle can be resolved into a point. With moderate experience, this can be done in about three trials; however, the final point should not be chosen too quickly (a needle hole is very small).

Regardless of whether this method or the tracing paper method is used, the reliability of the locations will depend largely on the trigonometric strength of the intersections. Points within the great triangle can be located most exactly, while those outside it become less and less reliable as intersections of rays become more acute. In Fig. 8–5, for example, locations southeast of station C are inherently weak, and some other primary station to the northeast should be sighted if possible.

8-12. Traversing with the Alidade and Plane Table

Instrument stations that cannot be intersected from other points must be located by traverses. In heavily wooded country all stations must often be surveyed in this way. The traverses may be started at primary triangulation points or (as is typically the case in wooded country) from the stations of primary transit-tape traverses (Fig. 8-6). The legs of the traverse should be measured with a tape where they must be quite precise, but less precise traverses can be run by stadia methods. Where the country is reasonably open and the going is not difficult, traverses are sometimes run only to locate instrument stations, and they are therefore closed before stadia mapping is started. Where the going is rough, it will generally be more efficient to map geologic features and topography as the traverse is run so that the circuit need be taken only once. When this is done, the traverse course should be reconnoitered thoroughly ahead of time and surveyed with extra care.

A stadia traverse is the same in principle as the traverses described in Chapter 3 and Section 7-13. The plane table is first set up over a point of known location and oriented by sighting on another known point, or, perhaps less reliably, by using a magnetic meridian (section 6-9). The rodman then selects the first forward station and his position is measured by the usual stadia methods (Section 6-11). The stepping method should be used where possible to determine the difference in elevation. The length of each leg should be checked by a backsight measurement when the instrument is moved ahead; otherwise mistakes would not be detected until the traverse was closed.

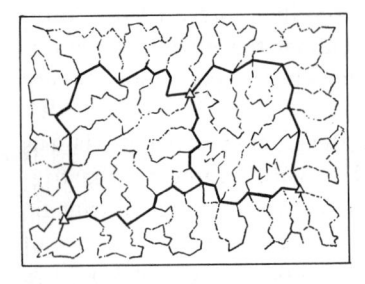

Fig. 8-6. Plane table sheet, showing primary traverses (heavy lines) and system of closed stadia traverses used to map the area.

Geologic Mapping with the Alidade and Plane Table

The legs of the traverse may also be measured by taping. In most cases, the taping need be no more precise than the plotting limitations of the plane table method. The procedure described in Section 7–6 may therefore be modified by (1) leveling the tape by eye, (2) carrying a measure to the ground by dropping a pebble rather than using a plumb bob, and (3) reading each measure to the nearest tenth of a foot. To prevent big mistakes, taping pins or comparable markers should be used to give the rear tapeman a check on the number of measures taken.

8–13. Mapping by Stadia Methods

Stadia mapping should be started in an area where the geologic and topographic features are relatively simple. It should then be expanded into surrounding areas without leaving gaps that would necessitate occupying stations a second time. This systematic coverage will be aided greatly by a preliminary examination and sketching of geologic features, as described in Section 8–7. When this system is used, the rodman is the key man of the party and he should direct the work; the instrumentman may need a recorder to keep up with the rodman if the scale is large. Where the scale is intermediate and geologic features relatively simple, the geologist should generally work at the instrument. To map efficiently, he may need more than one rodman. Finally, where large amounts of data and samples must be collected, it may be advantageous to have two geologists at the rod and one man at the instrument.

With the plane table set up at an instrument station, the rodman and instrumentman should decide on the layout of the first series of stadia points. This is particularly important in rough or partly wooded country where the rodman will commonly be out of sight while walking from one stadia point to the next. The course of points will generally form a loop, perhaps up one valley and down an adjacent ridge; the idea being to complete the points needed to map that small part of the area. When the loop is closed and the rodman returns to the instrument, he should draw the geologic and geographic features immediately. As he does this, the instrumentman should complete the calculations of all elevations. After these are penciled on the sheet, the instrumentman and rodman should together sketch the small segments of contour lines within the area mapped. With all the data plotted on the sheet, the next series of points can be planned. The number of points that should be located in a series varies with the scale and the

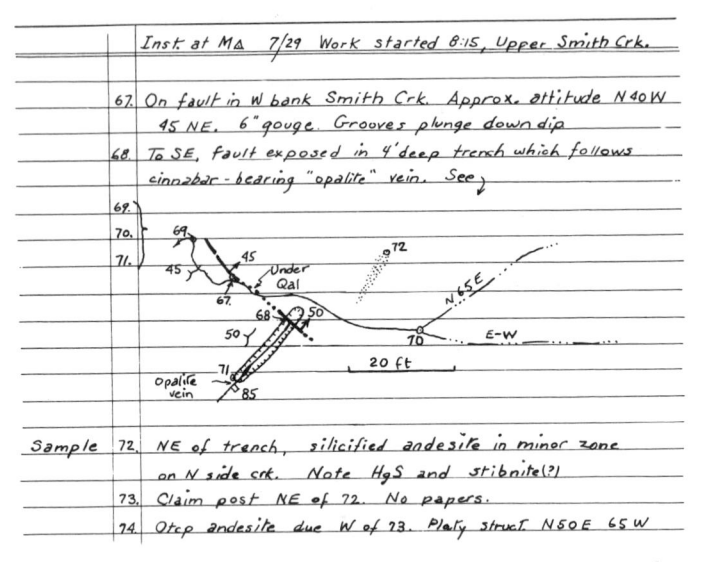

Fig. 8–7. Part of a page from a rodman's notebook, showing how a large-scale sketch map is used to record features that are based on several stadia points.

complexity of the geology, but between 12 and 25 will be right for most cases. If too many points are located before sketching is started, the principal value of the plane table method will be lost.

In addition to working the alidade, the instrumentman helps the rodman draw geographic features. His notebook and procedures are described in Chapter 6. The duties of the rodman-geologist are manifold. Not only must he select and describe the points that will be most useful in building up the map, but he must plumb the rod and carry it from one point to the next as swiftly as is safe.

The rodman-geologist should carry a notebook to record descriptions or sketches of the features on which he takes points. His notes should cover not only specific geologic items, such as contacts, faults, specimen locations, and small-scale structures, but also pertinent nongeologic data. In a mining district, for example, he should carefully delimit all workings, including minor exploratory pits and trenches, all property posts and cairns, and all dumps. Unless the layout of the geologic and geographic features is simple, he should make scaled sketches in the notebook that show positions of stadia points relative to nearby features (Fig. 8–7). The stadia points should be used as a framework from which minor features are located by pacing, by

Geologic Mapping with the Alidade and Plane Table

estimating, or by using the stadia rod as a scale. These sketches should be made at four or five times the scale of the plane table sheet so they can be used to draw details accurately on it when the rodman returns to the plane table.

Drawing contours. Contours are drawn by two methods—interpolation between points of known elevation and visual projection of level lines across the ground. Interpolation is used to find the spacing of a given number of contours on a constant slope. In Fig. 8–8, for example, the positions of the contours nearest the two stadia points are first estimated and marked on the map; in this case they are the 400- and 480-foot contours. The intervening contours are then spaced evenly between these two lines. Although the spacing can be computed and scaled precisely, as described in surveying texts, it is generally preferable to estimate the spacing of lines by examining the slope visually. Irregularities in the slope can thereby be shown on the map. When the spacing of contours has been determined at a number of places, the lines are connected by visually projecting their imaginary traces across the ground and sketching these traces on the map. This can be done more easily if all adjacent ridge lines, drainage lines, and other prominent features are drawn first.

If the geologic features are so large that many stadia points will be used for contours only, the rodman can use a hand level to find a given contour and then walk a level line to get several points on it. If this is done for every third or fourth contour, the intervening lines can be interpolated accurately.

Marking stadia points. It is worthwhile marking most stadia points if the map will be brought up to date from time to time or embellished with more detail. The points can be marked by writing the point number in large color-crayon figures on the outcrop or on a flat stone placed at the point. If the point was flagged before mapping, the number can be written on the flag. A cairn of a few stones or a stick held upright between stones can be used if there is no place to write the number.

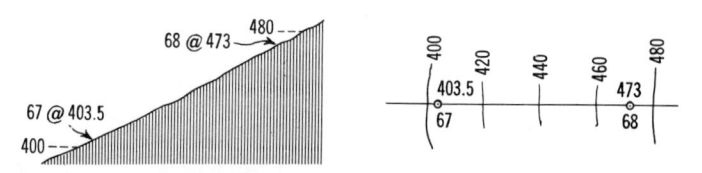

Fig. 8–8. Interpolating 20-foot contours between two stadia points; profile at left, map at right.

Most points marked in these ways can be found easily by anyone carrying the plane table sheet, even several years after the survey.

8–14. Using Aerial Photographs in Plane Table Mapping

Time can often be saved by transferring points or features from aerial photographs to plane table maps. This can be done by intersecting from three or more points that have been plotted on the plane table map and identified precisely on a photograph. The method is similar to the three-point methods described in Section 8–11. In Fig. 8–9, for example, a sheet of tracing paper has been placed over the photograph and rays have been drawn from the point to be transferred through any three points located on the plane table sheet (they need not be control points). The tracing sheet is then placed over the plane table map so that the three rays pass over the corresponding points on the map. The new point lies under the intersection of the rays, and it can be marked by pricking through the intersection with a needle. This method can be used precisely in areas of low relief or where all the points lie at similar elevations (Section 5–3). In areas of moderate to high relief, it is necessary to use the radial line methods described in Chapter 9.

In areas of low relief an aerial photograph can be enlarged to a suitable scale and used directly as a plane table sheet. If the photograph has appreciable tilt distortion, it can be rectified to an untilted image after three or more control points have been located on it (see Moffitt, 1959, ch. 10). The photograph is prepared for mapping by drawing a magnetic north line on it and plotting three or more control stations that are at similar elevations. Mapping can be started from any of the control points or from other points located by three-point methods (Section 8–11). Stadia or intersection methods can be used

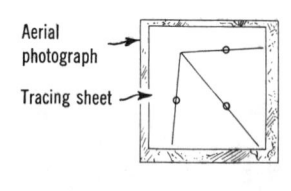

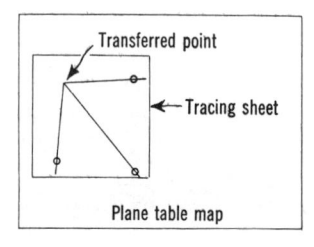

Fig. 8–9. Transferring a point from a photograph to a plane table map by intersecting on a sheet of tracing paper.

to locate points exactly as on a map; however, the change in scale with elevation must be taken into account (Section 5-9). The photograph image should be used wherever possible to delineate features directly; commonly, only a few stadia points will be needed to draw a given feature accurately. Contours can be drawn most easily by using a stereoscope and sketching directly on the photographs. Generally, fewer elevations will be needed than would be used in ordinary plane table mapping.

8-15. Method of Moving the Plane Table around the Rod

The procedure that follows may be preferred for mapping at intermediate scales. Its only special requirement is that the magnetic declination must be nearly constant over the area or the magnetic meridian must be determined for each control station.

1. The rod is placed on a station marked on the plane table sheet.
2. The geologist sets up the plane table at a feature he wishes to plot and then orients the board by the magnetic meridian.
3. To locate his position, he draws a ray toward the rod and makes the usual stadia readings and computations.
4. After plotting the point and sketching the features around it, he moves the table to another point.

An advantage of this method is that the plane table is directly over the outcrop, where relations can be drawn accurately and completely in one step. Moreover, the method is the only one that enables one man to make a stadia survey, for the rod can be replaced by a graduated pole that is guyed in place by three wires. For small-scale work, an isolated tree can be stripped and painted with one-foot graduations.

8-16. Work on the Plane Table Sheet in the Evening

After each day's mapping, the stadia points located during the day should be marked with waterproof ink by placing a small dot in the needle hole that marks the point. The point number should be lettered about $\frac{1}{10}$ in. high beside the point and facing so as to read from the south edge of the sheet. Elevations should be left in pencil. Structure symbols may be inked in from day to day, but other geologic features are best left in pencil until the final field check of the map, unless the features are simple and are located by many stadia points. If possible, contours should be left in pencil until the map is completed. Pencil work in completed areas can be protected against smudging by overlays of brown paper.

The instrumentman or recorder should also check through the stadia computations for arithmetical mistakes, while the rodman-geologist examines the notes and sketch maps of the preliminary geologic examination to determine what should be covered in the next day's mapping.

8–17. Vertical Cross Sections from Plane Table Maps

Almost any project requiring a plane table map will also require detailed vertical cross sections. Section lines should be chosen as the mapping approaches completion. They can then be walked out during the final stages of the mapping, when geologic and geographic details can be added easily (Section 8–18). The sections should be drawn in pencil before the party leaves the field.

The section lines should be chosen to intersect as many geologic features as possible at about right angles to their structural trend. The lines should be drawn with a sharp hard pencil and a steel straightedge.

8–18. Completing Plane Table Maps

After the stadia work on a plane table sheet has been completed and cross-section lines penciled on it, the sheet should be carried into the field to check the shapes and positions of the various features. Details overlooked in the stadia mapping should be added, especially along the cross-section lines. At this stage of the work, the value of the stadia point markers described in Section 8–13 becomes apparent. The markers permit exact recovery of most stadia points, so that data can be added accurately.

The map is generally carried on a plane table board, and the changes and additions are made by compass and pace (or tape) measurements from the stadia points and instrument stations. Some of the following are commonly added at this stage: structure symbols to complete coverage (especially those minor structures that were overlooked until the mapping was well along), thin or partially hidden key beds, lithologic variants with gradational boundaries (especially in igneous and metamorphic rocks), areas of hydrothermal alteration, and contacts of surficial materials. This is also the best time to make systematic collections of rocks, minerals, or fossils, particularly where there are specific problems to be solved.

After the field work is completed, the map should be inked, labeled, and perhaps colored. Geologic features should be inked first, either

Geologic Mapping with the Alidade and Plane Table 153

all in black, or by using green for faults and red for veins, ore bodies, or other features that require emphasis. The culture should then be inked in black, followed by the drainage in blue. Finally, contours may be inked in brown. The entire map should then be cleaned with a soft eraser, after which it may be colored in pale erasable tints. Symbols for rock units may be added in black ink if space permits.

Each plane table sheet should then be given an inked title, including project or geographic name, sheet number (or an index map showing its location), names of organization and party members, and dates of the survey. A north arrow and bar scale must be added, even if the map has a coordinate grid. Finally, an explanation should be drawn that includes the rock colors, symbols, structures, and culture symbols. If there is not room for all these items on the front of the map, they can be put on the back. Although they may seem unimportant in the field, the map would eventually become worthless without them.

Before packing gear to leave the field area, all notebooks, sketch maps, computations, aerial photographs, and other items bearing information on or about the survey should be accumulated, checked to see that they are suitably labeled, put in correct sequence, and packed for transport. Plane table maps must always be packed flat. The instruments and their carrying boxes should be dusted with a soft brush or cloth and then packed for transport.

References Cited

Birdseye, C. H., 1928, *Topographic instructions of the United States Geological Survey:* U.S. Geological Survey, Bulletin 788, 432 pp.

Low, J. W., 1952, *Plane table mapping:* New York, Harper and Brothers, 365 pp.

Moffitt, F. H., 1959, *Photogrammetry:* Scranton, Pa., International Textbook Co., 455 pp.

9

Making a Geologic Map from Aerial Photographs

9–1. General Value of Aerial Photograph Compilations

Geologic features plotted on aerial photographs are generally transferred to a base map (Section 5–11), but a map can also be compiled directly from the photographs. This method of making a geologic map is often superior to the plane table method, especially if the scale is small or intermediate. The control network can usually be planned and surveyed more quickly than that for a plane table project, and there is no need for time-consuming stadia measurements. Moreover, unless a base line must be taped, the entire photograph survey can be made by one man. The compilation of the final map is not nearly so simple as drafting a map from plane table sheets; however, this compilation does not take up any field time.

The plane table method is generally superior for mapping at large scales. Photographs with scales larger than 1:18,000 are usually not available, and it is difficult to compile a map from photographs enlarged more than two times. Moreover, the detailed relations shown on such maps are, in a sense, misleading, for the compilation can be no more accurate than the original photographs from which the enlargements were made.

The methods described in this chapter can be used to make a planimetric map from an overlapping set of aerial photographs. No special equipment is required; however, photogrammetric instruments and office equipment will aid the compilation considerably. Compiling contour maps from aerial photographs requires special instruments and involved procedures, as those described by the American Society of Photogrammetry (1952) and Moffitt (1959). The third dimension on the planimetric compilation can be shown locally by elevations measured during the control surveys. It can also be shown by a series of cross sections if section profiles are surveyed during the field season.

9–2. Preparations for an Aerial Photograph Project

The most important preparation for the field season is to determine that photographs of suitable scale and quality are available (Section

5-1). These photographs must have more than 50 percent end lap. Little side lap is necessary if the compilation can be controlled adequately; otherwise, 30 percent side lap is desirable. Ideally, the photographs should be tilted little if at all, but this cannot be determined until points are surveyed on the ground. In any case, the compilation compensates for tilted pictures if enough control points are surveyed.

Data on established ground control should be obtained well in advance, and the photographs that have established control points on them should be listed. It will also be useful to determine the average scale of the photographs and to make a thorough photogeologic study of them before the field season. The control network can often be planned in detail when this is done.

Equipment and supplies are listed in Appendix 1, and additional suggestions will be found in Chapters 5 and 7. The materials described in Section 9-6 should also be included in field gear because the photographs can often be prepared for the compilation while in the field.

9-3. Mapping Geologic Features on Aerial Photographs

Geologic mapping can be started at once in the field, typically long before the control survey. This great advantage is possible because the features plotted on a given photograph are based solely on the photographic image of that print. Geologic features can be plotted by the methods described in Sections 5-5 through 5-10. A numbering system should be devised at once so that specimens collected during reconnaissance can be plotted and labeled in sequence with all subsequent collections. Prints that will be used for compilation must be handled with particular care because their surfaces should be smooth and unmarred.

The unusually detailed images of some photographs may make it tempting to map far more detail than is justified by the project and the scale limitations of the final map. If the purpose of the project does not clearly delimit what should be mapped, the area should be reconnoitered thoroughly before beginning detailed mapping.

9-4. Ground Control for Photograph Compilations

The compilation must be based on control points that have been surveyed on the ground and marked precisely on the photographs. These points should be spaced closely enough so that three or four fall on each photograph. For photographs with scales of about 1:20,000, this would space control stations 1 to 3 mi apart, distances that can be surveyed precisely with a transit or adequately, in many cases, with the alidade and plane table.

Geologic mapping can be started without a controlled base, and therefore the control system can be planned and set up over a considerable period of time. The network can be planned tentatively by examining the photographs before the field season, and possible station sites can then be checked and marked with signals as geologic features are reconnoitered and mapped. Established control points, such as triangulation stations, bench marks, and Bureau of Land Management Survey corners, should be located as each part of the area is mapped. Control points should be marked precisely on the photographs when they are visited, preferably by pricking a small needle hole through the print and labeling it on the back.

If the control is surveyed by plane table triangulation, it is ideal to set up the control sheet at an even scale that is close to the average scale of the photographs. A scale of 1:18,000, or 1 in. = 1,500 ft, is commonly suitable. Where the control will be based on established triangulation stations, however, and where these stations will be plotted on the plane table sheet by geodetic coordinates, it may be preferable to construct the map at one of the standard scales used in constructing map grids, as 1:24,000. If large areas must be covered in a short time, the control can be intersected at a smaller scale than that of the photographs. Very large areas can be covered on one plane table sheet by using a large (24 × 31 in.) board. The following example illustrates how a large number of photograph points can be located rapidly if the basic framework of the control is strong.

1. Established triangulation stations, as Miner, Dudley, and Brisk in Fig. 9–1, are plotted on a plane table sheet; this may be done on the basis of their geodetic coordinates and a polyconic grid (Section

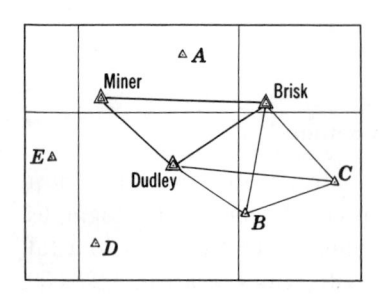

Fig. 9–1. Ground control based on established triangulation stations (double triangles).

Making a Geologic Map from Aerial Photographs 157

7–11), on the basis of their state plane coordinates and a rectangular grid (Section 7–10), or simply by intersecting with a beam compass set at the interstation distances (Section 8–5).

2. As the geologic mapping progresses, additional stations are located in the field and marked with signals (as A, B, C, D, and E in the figure).

3. These supplementary stations are intersected from the established triangulation stations and, where necessary, by accessory sights (C, for example, requires a ray from B for a strong location).

4. As each of the stations is occupied, additional points are selected by comparing the photographs with the terrain. These points must be sharp and identifiable from at least three instrument stations. They will commonly be sharp peaks, forks of streams, road intersections, houses, lone trees, and sharp indentations on the perimeters of irregular clearings or patches of brush. Enough points should be selected so that each photograph will include three or four.

5. The survey is completed by drawing rays to each of these points and measuring vertical angles. Each ray is labeled clearly with the number used to identify it on a photograph.

If established triangulation points are not available, the survey may be started with a base line, as described in Chapter 7. For many projects, time can be saved if a base line as long as 1 or 2 mi is taped with moderate precision.

9–5. Surveying Cross-Section Lines for Photograph Compilations

Section-line surveys are made to measure accurate topographic profiles and to make thorough observations and checks of the geologic features along them. If outcrops permit, these data should be about twice as detailed and precise as that plotted throughout the area. Cross sections are especially important in aerial photograph projects, for the compiled planimetric map will not show the vertical dimension of geologic features. The exceptions are areas where there is little or no relief or where rock units and structures are nearly vertical. The section lines must be chosen before the end of the field season because the section profiles must be surveyed in the field. If their approximate positions can be selected before the control survey, the section lines can be controlled by this survey.

If slopes are even for considerable distances and geologic features reasonably simple, a profile can be based on a few accurate locations

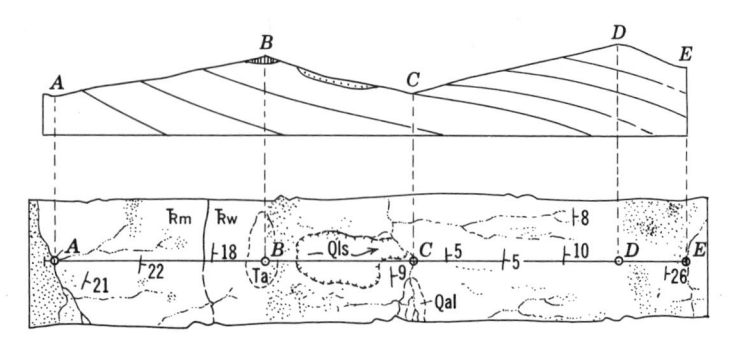

Fig. 9–2. Simple cross section (top) and strip from an aerial photograph (bottom), showing lettered points marked at principal breaks in slope.

at breaks in slope. In Fig. 9–2, for example, the elevations and positions of five key points on the profile were determined by intersection during the control survey for the map. After the main geologic features had been mapped, the section lines were walked out and structural details were added.

When the final map is compiled from data such as these, it may be found that the key points do not lie quite on the final section line (because of distortion within the photograph image). They should be close enough to it, however, so that elevations along the line can be estimated closely by stereoscopic examination of the photographs.

More involved section-line surveys must be made if slopes are uneven and geologic features so complex that elevations and positions are required at each contact or change in dip. These accessory elevations can commonly be measured by hand level traverses along the section line (Section 2–6) or by using a barometer (Section 4–4). The data may be plotted on an acetate sheet placed over the photograph. If relief is so moderate that distances can be scaled directly from the

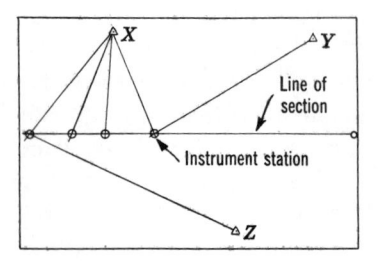

Fig. 9–3. Surveying a profile on a plane table sheet by resecting positions along the line of section; X, Y, and Z are control stations.

photographs, the cross section can be plotted directly in the field (preferably at about twice the scale of the photographs).

The alidade and plane table should be used when an especially thorough survey is needed. If signals are set up at each end of the section line, instrument stations along the line can be located by (1) setting the plane table up on the line, (2) orienting it by sighting to a signal on either end of the line, and (3) resecting the point by sighting on any other signal (Fig. 9–3). Three-point methods and accessory traverses can be used where the view along the section line is obstructed. Details between plane table stations can be added by stadia methods, and notes and geologic features can be plotted directly on the plane table sheet. Each instrument station should be marked on a photograph when it is occupied.

9–6. Radial Line Compilation

The *radial line method* is used for photograph compilations when special photogrammetric equipment is not available. The method can be used easily and accurately if the photographs are of good quality, if the relief is moderate, and if several ground control points are included on each print. When relief is considerable and an occasional photograph is severely tilted, the compilation is more laborious and less accurate unless many control points have been surveyed.

The radial line method is based on the principle of intersection: a point can be located by drawing rays toward it from two or more previously located points. Figure 9–4 illustrates the application of this principle to aerial photographs. If overlapping vertical photographs are taken at A and B, as shown in the perspective views, any one point on the ground, as C, will lie *in the vertical plane that includes the*

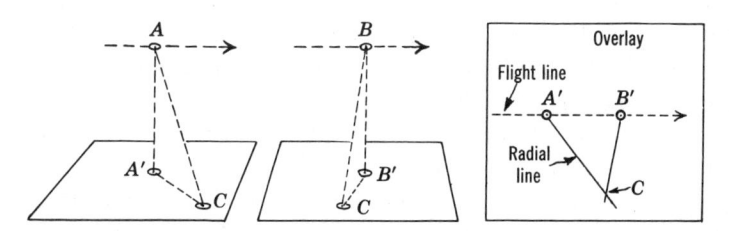

Fig. 9–4. Two overlapping photographs, showing the vertical planes that include the camera, the centers of the prints, and the images of a point, C. In the drawing on the right, the position of C has been intersected by placing a transparent overlay over each of the prints and drawing the radial rays $A'C$ and $B'C$.

camera and the point at the center of that photograph (points A' and B'). It must follow that C will lie on a line (a *radial line*) that passes from the center of each print out through the image of C. This will be true regardless of the relative elevation of points A', B', and C. If the center points of two overlapping photographs can be marked accurately on a transparent overlay, any point such as C can be located by placing the overlay on the photographs and drawing the two rays $A'C$ and $B'C$. This corrects photograph distances that would otherwise be modified by the effects of relief. The location of centers on the overlay and the orientation of each print are achieved in part by the use of ground control and in part by the compilation methods described in the remaining sections of this chapter. Briefly, the steps in the compilation are as follows:

1. The center points and the points to be intersected are marked on the prints.
2. The ground control points are plotted on a transparent overlay at a scale close to the average scale of the prints.
3. The center points of the photographs are located on the overlay by resecting from the plotted control points.
4. The photograph points are intersected on the overlay by drawing radial lines from photograph centers.
5. Geologic features, drainage, and culture are penciled on the overlay by referring to the locations of photograph centers, intersected points, and control points.

Materials for the compilation. In addition to the instruments needed for plotting the control points on the overlay (Section 9–8), the following materials are needed: several bright-colored opaque inks, drop-circle compass, pencils ranging from 3B to 6H, file for pointing pencils, 10-inch triangle, soft eraser, transparent sheet for overlay, tracing paper, black waterproof ink, and crow quill or other fine pens. The opaque inks can be prepared by mixing brightest red, green, or orange transparent inks with a tempera paint of the same color. They must be mixed thinly enough so that they will flow readily from the point of the compass, and they must also be opaque enough when dry to show clearly against either light or dark areas of a photograph.

The transparent sheet must be large enough for the whole map and must be as strong and transparent as possible. Frosted plastic drawing stock and top quality water-resistant linen are suitable. The tracing paper should be exceedingly transparent.

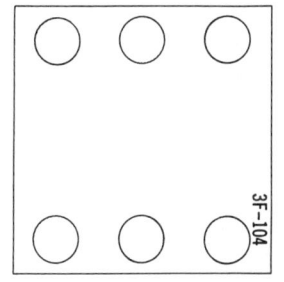

Fig. 9–5. General locations for the six pass points.

9–7. Marking and Transferring Photograph Points

Three kinds of points must be marked on the photographs before the compilation is started: (1) the *control points* located by the ground survey, (2) the *center points* of the prints (as A' and B' in Fig. 9–4), and (3) various *pass points*, which are intersected by the radial line method and are selected on the basis of their usefulness in compiling the map. Pass points are typically placed at stream intersections, sharp turns of contacts or other critical geologic features, or on features of the culture. In areas of high relief, almost every feature that must be plotted accurately on the map should be located by a pass point. In areas of low relief, only six pass points are needed on each photograph, and they should be located within the six circles shown in Fig. 9–5. These six points must fall within the areas of side lap if photographs do not include two or three ground control points.

The various points can be marked on each photograph as follows:

1. Locate the center point on each print by laying a straightedge against each set of *collimation marks* and drawing a small cross with a sharp 3B pencil (Fig. 9–6A).

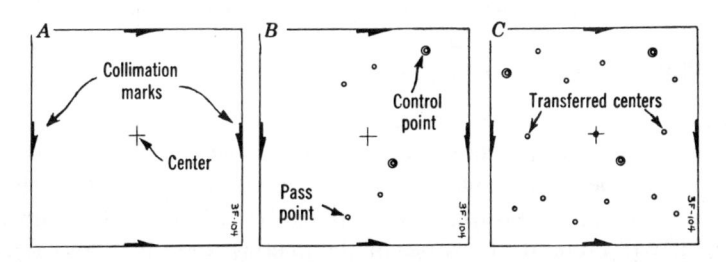

Fig. 9–6. Steps in marking and transferring centers and other points on prints.

2. Fill a drop-circle compass with opaque colored ink and set it to draw circles approximately 0.2 in. in diameter.

3. Select pass points within the central one third of each print, set the compass carefully on each point, and draw the circles (Fig. 9–6B). Draw double circles at the needle holes that mark control points.

4. Using an ink of a different color, draw a circle at each center point.

The points must now be transferred to all photographs on which they fall. This can often be done by identifying the marked image on each adjacent photograph, placing the point of the compass exactly on it, and drawing a circle of the same size and color as that drawn around the original point. If it is not certain that the transfers can be made accurately in this way, the following steps can be followed:

1. Starting at one end of a flight line, obtain a stereoscopic image of the first stereo pair, using a magnifying stereoscope. The colored circles as well as the needle holes made by the compass will appear to be superimposed on the photographic image, although they are drawn on only one photograph.

2. Looking into the stereoscope, place the point of the compass onto the *unmarked* photograph, exactly where the needle hole seems to lie. Look away for a moment to rest the eyes and then back again to be sure the compass is placed correctly. Draw a circle of the same size and color as that used on the marked print.

3. Put the compass to one side and examine the stereo image. If the transfer was made correctly, the two circles will merge and will appear to lie *exactly at the same level as the needle hole in the print*.

4. If the circle appears to float above the ground, quickly and gently wipe away the inked circle with a damp cloth and relocate it.

The last two steps should also be used to check circles transferred by inspection. When all transfers have been completed and checked, lay out the photographs to make sure that all points have been transferred. It is inconvenient to draw circles after the compilation has been started.

9–8. Plotting Control on the Overlay

An overlay is prepared for the radial line compilation by plotting the ground control points on a transparent sheet. The scale of this overlay must be close to the scale of the photographs if tracing methods are to be used to construct the map (Section 9–11). If projection methods will be used, the scales of the overlay and photographs need not be the same. The average scale of the photographs is determined

Making a Geologic Map from Aerial Photographs

by laying them out and measuring some overall distance on them, as explained in Section 5–6. The distances determined by the control survey can be used for this. A scale of 1:18,000 is useful for many plots because it is close to the scale of most photographs and provides an even scale of 1 in. = 1,500 ft for constructing the control plot.

The control points can be plotted by either the coordinate or the intersection methods described in Sections 7–11 and 8–5. They can also be traced from the plane table sheets used to intersect the control, provided the scale is suitable and the sheets have not been distorted since the field season. Each point should be pricked with a fine needle hole, which is then filled with black ink and marked with a small triangle and the point number or letter.

9–9. Compiling Points from Controlled Photographs

The radial line method can be used quickly and precisely to compile a map from photographs that include three or more intersected control points. The photographs are first marked with radial rays that pass from the center out through the control points (Fig. 9–7A). This can be done either with a transparent triangle and a sharp 3B pencil, or, if the prints are so dark that the pencil lines cannot be seen through the overlay, with a ruling pen and partly opaque yellow ink.

The compilation should be started at one end of a flight line where relief is moderate. The steps are as follows:

1. Fasten the first photograph to a smooth surface with drafting tape and orient the overlay so each control point on the overlay lies exactly over the corresponding radial line on the print (Fig. 9–7B). This will orient the photograph correctly unless the center falls on the circle that passes through the control points used in the location (see Section 8–11).

2. Using a sharp hard pencil and a triangle, draw a small cross over the center of the photograph, then draw a line through the center of the cross and over the circle marking the center transferred from the next photograph (Fig. 9–7C).

3. With the overlay still in place, draw radial lines across each pass point of the underlying photograph (Fig. 9–7D).

4. Tape down the next photograph and shift the overlay until the control points lie exactly over the radial lines drawn on the print, just as in step 1. Draw a center cross and radial lines as in steps 2 and 3 (Fig. 9–7E). Unless one of the photographs is severely tilted, the sec-

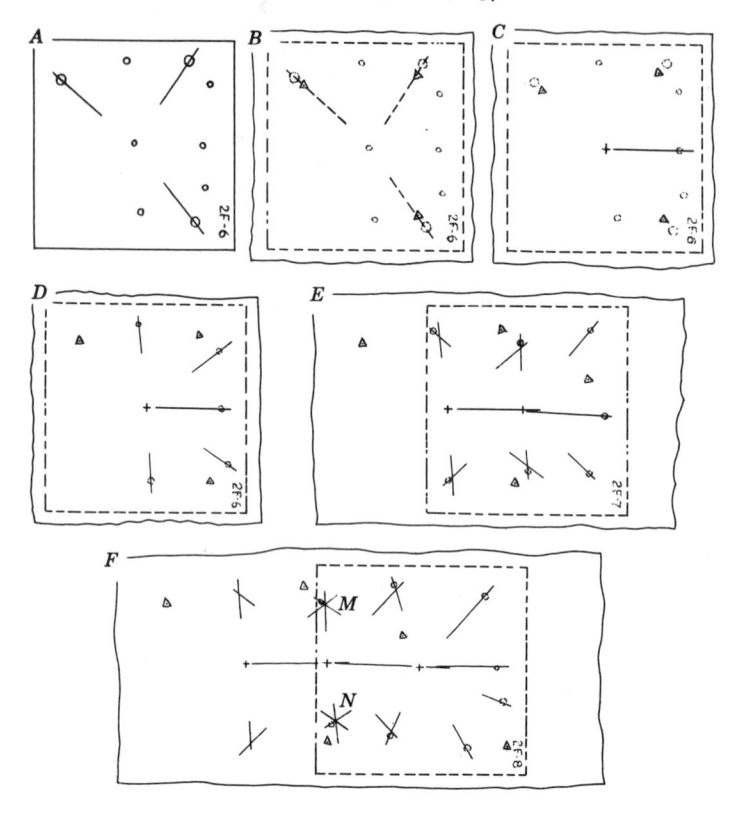

Fig. 9–7. Steps in marking radial lines and intersections on an overlay. Solid lines and symbols represent marks on the overlay, while dashed lines and symbols represent marks on prints under the overlay.

ond photograph center should lie exactly on the center line drawn for the first photograph.

5. Tape down the third photograph and orient and mark the overlay as before. The radial lines to pass points that lie in the overlap area of the first and second prints (as M and N in Fig. 9–7F) should now make three-line intersections.

6. Plot the other photographs of the flight strip by the same procedure, and then plot adjoining flight strips.

When completed, the overlay will show a series of flight lines with small crosses marking photograph centers and a large number of intersecting rays that represent pass points. Most detectable errors are caused by inexact plotting; however, if a photograph is severely tilted, its points will fall somewhat off the locations made by adjacent photographs, and small triangles will be formed at three-ray intersections. If the surrounding prints are controlled adequately, the reason for this should be apparent, and the rays from the tilted photograph can be disregarded.

9–10. Compiling Points from Uncontrolled Photographs

It is possible to use the radial line method for photographs that do not include enough control points to be oriented by the method just described. If the photographs do not combine severe tilt with high relief, the method will be reasonably precise. A single tilted photograph can throw an uncontrolled flight line out of orientation, and the wing points and centers must then be adjusted by using adjacent flight lines. These adjustments may take as much time as intersecting control points in the field; however, if the map need not be precise, they can be estimated quite readily.

To start the compilation, all photographs with three or more control points are plotted on the overlay. The remainder of the compilation consists primarily of filling in between these anchor photographs. The photographs between the closest two controlled pictures should be compiled first, and the longest series of uncontrolled photographs last.

The photographs can be compiled as follows:

1. Orient a photograph with three or more control points, using the procedure given in step 1, Section 9–9.
2. Draw a cross at the center of the print and draw center lines across *both* transferred center circles as well as across all pass points.
3. On an adjacent print, mark radial rays to the control points that fall in the overlap area of the controlled print.
4. Tape the marked print down and place the overlay on it so that the center line lies over the center of the print and the control points lie over the radial lines drawn on the print (Fig. 9–8A).
5. Draw the center line to the next center and draw radial rays over all pass points (Fig. 9–8B).
6. If the next photograph has no control points, draw radial rays through the two pass points intersected in the last step (R and S in Fig. 9–8B).
7. Tape this photograph down and shift the overlay until its center point lies under the center line on the overlay and the radial rays (to pass points R and S) lie under the intersections made on the overlay

(Fig. 9–8C). If the overlay cannot be oriented so the points and lines coincide, check the previous constructions carefully. Any remaining error is probably caused by tilt, and the points can usually be brought closer to coincidence by shifting the overlay slightly off the center points of the photograph or by twisting it slightly from the flight line orientation. Any residual errors should be adjusted after the adjacent flight lines are plotted.

8. Draw the center line to the next photograph and all radial lines to pass points (Fig. 9–8D).

9. Prepare and use the next uncontrolled photograph as in steps 6, 7, and 8, and continue until the compilation has gone halfway to the next controlled photograph; then begin the same procedure at this photograph, working back in the opposite direction.

If there is appreciable disagreement where the two plots meet, they should not be adjusted until adjacent flight lines have been plotted, for well-controlled photographs in these lines may resolve the errors in the first line. In Fig. 9–9, for example, a line of photographs started at center A can be brought into line at centers G and H by satisfying the pass-point intersections J and K made on an adjacent controlled pair

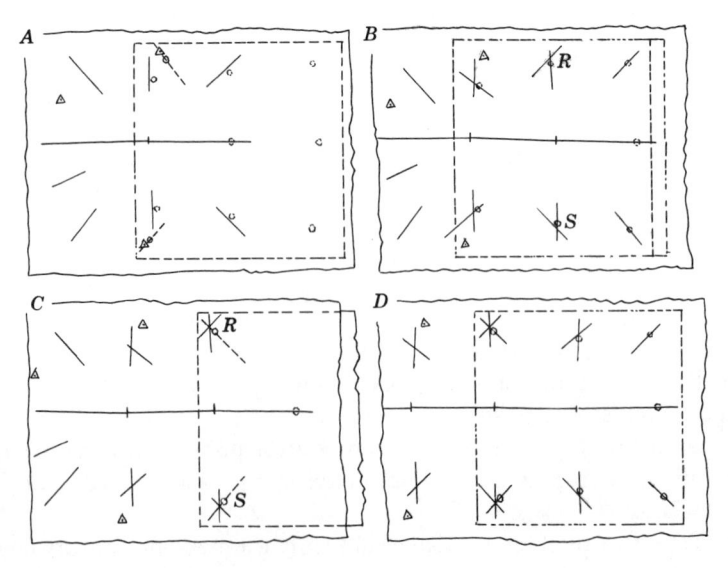

Fig. 9–8. Steps in compiling points where all photographs do not include ground control points.

Making a Geologic Map from Aerial Photographs

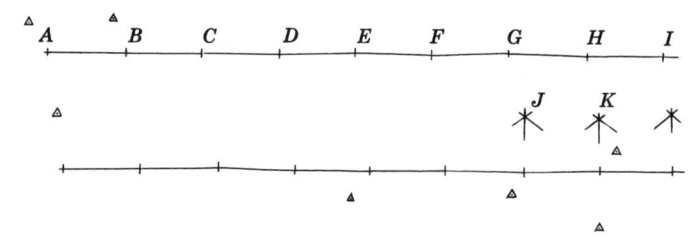

Fig. 9–9. Using controlled intersections in one flight line to improve the plot of an uncontrolled line.

of photographs. The correction of these errors can often be facilitated by transferring the centers and intersected pass points of a given flight line onto a separate strip of tracing paper. The strip is then shifted on the overlay until its pass points are brought into coincidence with those intersected on the adjacent flight strip. In some cases, the flight line on the overlay will clearly hinge out of orientation at one photograph; in other cases, the errors of the plot will be so complex that they must be corrected by trial-and-error approximations.

9–11. Compiling Photograph Data

The control points, center points, and pass points plotted on the overlay can now be used as a basis for transferring geologic features, drainage, and culture from the photographs to the overlay. The pencil rays and photograph numbers on the overlay should be erased and the points marked permanently in ink, or the positions of the points should be traced on a fresh overlay. The compilation will be facilitated if small ink circles are drawn around pass points, small triangles drawn around control points, and crosses drawn at center points. This should be done with a blue transparent ink that will not be photographed or printed when the final map is reproduced. The final overlay sheet must be strong enough to withstand considerable erasing, and pencils used on it must be soft enough (typically 1H to 3H) so they do not crease it.

Some photographs have such weak contrast or such strong vegetation patterns that it is necessary to use a stereoscope to see their drainage lines and ridge lines. These lines must be delineated on the photographs before they can be transferred, as by marking them with yellow or orange ink. It is easiest to delineate streams by placing photographs in reverse order under the stereoscope, because the peculiar appearance of the pseudoscopic image makes streams very apparent.

The transferring should be started where relief is low and where photographs have scales nearly the same as the scale of the overlay. When the overlay is oriented over one of these photographs, the center point and nearby pass points should so nearly coincide with the marks on the overlay that the features in the central part of the photograph can be traced directly onto the overlay. In cases where the scales differ moderately, the transfer can be started by placing the overlay at the photograph center and drawing the features immediately around this point. Then the overlay can be shifted so as to coincide with the nearest pass point. After the features around this point are drawn, it is shifted again to the next pass point, and so on. The features between these small areas can be completed by shifting the overlay progressively from one register to another; for example, the features halfway between two points are drawn when the overlay is set as shown in Fig. 9–10A. Features that do not lie between two points can be drawn by placing the overlay so as to satisfy the spacing of three points (Fig. 9–10B). If the three points are at about the same elevation, this can be done precisely by drawing three rays toward the feature being transferred. If their elevations differ considerably (and especially if they lie well to one side of the center of the photograph), the amount and direction of offset caused by relief must be estimated (Section 5–3).

Transferring features from photographs with little relief simply requires enlarging or reducing the photograph features to the scale of the overlay. Adjustable sketchmasters and reflecting projectors can be used to do this rapidly and accurately. Simple enlargement or reduction, however, will not be suitable for areas with considerable relief. Patient comparisons must then be made between the points and the topography, and it is necessary to use a stereoscope to check the transfers. If streams and ridge lines have been transferred in detail,

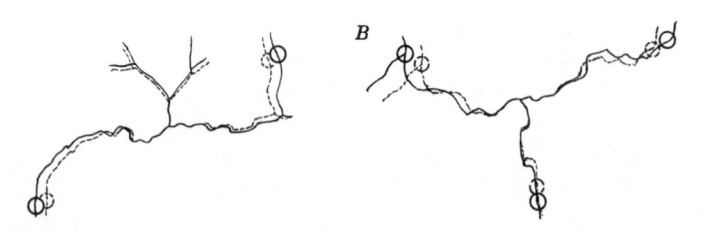

Fig. 9–10. Positions of overlay points (solid circles) relative to photograph points (dashed circles) in transferring a stream from print to overlay.

geologic features can sometimes be added accurately by examining the photographs stereoscopically and transferring by estimation.

Transferring symbols. It is usually easiest to plot structure symbols after drainage, culture, and geologic lines have been completed. The symbols can be plotted directly with a protractor if the strike direction has been lettered on the back of the prints or in the notes. If this system has not been used, the overlay must first be oriented over the photograph so that its true north direction is parallel to the north arrow on the photograph. The overlay is then shifted from one symbol to the next, *in this same orientation,* and each symbol is traced onto it.

9–12. Checking the Map and Drawing Cross Sections

The principal job in completing the map is to check the first pencil compilation carefully against the photographs; this should be done before the cross sections are compiled. It is best to do the checking with a stereoscope, working systematically through one flight line at a time. The major lines and features will already be drawn accurately, but much detail can generally be added. This is an important step since details of the drainage and culture will give other geologists their only means of locating geologic features in the field. The pencil draft and cross sections should be completed before either are inked. Suggestions for the final drafting are given in Section 11–10.

References Cited

American Society of Photogrammetry, 1952, *Manual of photogrammetry,* 2nd ed.: Washington, D.C., 876 pp.

Moffitt, F. H., 1959, *Photogrammetry:* Scranton, Pa., International Textbook Co., 455 pp.

10
Detailed Mapping and Sampling

10–1. General Nature of Detailed Studies

Detailed geologic studies usually require one or more of the following operations: topographic and geologic mapping at scales larger than 1 in. = 200 ft; cleaning or excavating of exposures; drilling; mapping of underground workings; and extensive sampling. Generally, these operations can be used to advantage wherever geologic features must be studied in a thorough, quantitative way. They are commonly used in economic studies, for example, to improve the quarrying of a gypsum body or the underground mining of a coal seam or copper deposit. In addition to their economic uses, however, they may also solve a great variety of noneconomic problems, such as the origin of a complex igneous body, the degree of alteration around an intrusion, or the effects of deformation on a sandstone. Because they are costly and time consuming, these operations are done only after preliminary studies have shown they are justified.

The regional geologic setting of local studies should be worked out thoroughly; otherwise the significance of many rock units or structures might be overlooked. The magnitude and age of most faults, for instance, cannot be determined from maps of small areas alone. Two kinds of preliminary studies should be considered: (1) regional reconnaissance of rock units and structures and (2) mapping at intermediate scales (1:24,000 to 1:62,500) around the area of specific interest. The first study defines the areas or features that should be covered by the second; together the two indicate which operations should be used in the final local study.

The local study itself should be started with mapping, for drilling, excavating, and sampling cannot be planned effectively until detailed surface and underground maps are well underway. Excavating and drilling should be overseen directly by the geologist or geologist-engineer lest critical relations are destroyed before he sees them. Besides insuring that he receives data promptly, this can increase the effectiveness, or reduce the cost, of a given program accordingly. The geologist or engineer should also supervise the sampling to make sure the samples will be as meaningful as possible.

10–2. Detailed Surface Maps and Sections

Detailed studies generally require topographic and geologic maps with scales of 1 in. = 200 ft to 1 in. = 10 or 20 ft. These maps are most useful if they are outcrop maps (Section 4–8). Excavating and drilling, for example, can be planned most effectively from outcrop maps, on which facts are clearly separated from inferred or projected relations.

Although the mapping methods described in Chapters 7 and 8 can be used for some projects, more precise methods may be needed. Control points must be numerous and accurate enough to give exact tie-ins with property lines, and control surveys of mining properties must relate underground and surface workings precisely. For properties that will be worked for some time, it is desirable to set up a coordinate system and stake out coordinate points with a transit and tape. Doing so permits accurate resetting of stakes removed during development. For both legal and geological reasons, several control points should be placed permanently enough to be reoccupied years later. A permanent marker can be made by cementing a brass bar or plate into a solid rock outcrop.

All geological features that might conceivably be useful should be plotted. Stratigraphic details that might be used for correlations are especially important. A single layer of chert nodules in a limestone, for example, may be valuable in later correlations with subsurface workings. Kinds and intensities of weathering and other alterations are pertinent to many projects, as are lithologic characteristics that affect the strength of rocks. The distribution of joints and broken or partially sheared rock is especially critical to most engineering studies. Joints are commonly plotted individually at sites for dams or other heavy construction, and special notation is given to fractures that are open or that bear altered or broken rock.

Contours that must be located within a foot or two of their true position on the ground can be mapped by walking out level lines, rather than by interpolating between scattered elevations (Section 8–13). If contours must be located more precisely in some places than others, their relative reliability can be shown by using solid lines for accurate ones and dashed lines for sketched ones. In economic studies, dumps should be contoured in enough detail so their volumes can be computed accurately. Most man-made features should be plotted to scale, and if the area is to be trenched or drilled, buried pipelines and powerlines should also be shown.

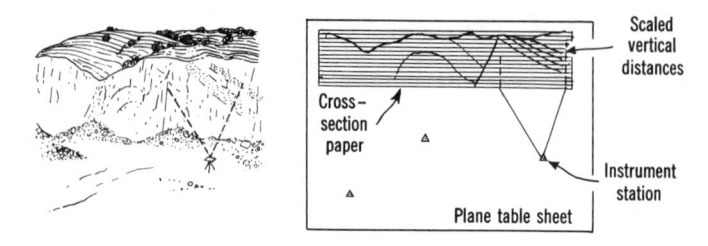

Fig. 10-1. Plotting the geologic features of a steep cut (left) on a vertical projection. The projection can be constructed directly on a plane table if a strip of cross section paper is fastened to the plane table sheet (right).

A large (24 × 31 in.) plane table board is desirable for many projects, although it is cumbersome in rough terrain. Plane table sheets should be waterproof and strong enough to take repeated erasures (Section 6-8). Stadia sights in detailed surveys are usually short; thus they can be made easily with a surveyor's leveling rod, which can be read directly to hundredths of a foot.

Detailed vertical sections. In many mining and construction projects, it may be useful to plot steep, large cuts as vertical cross sections. This can be done directly on a plane table by constructing a vertical projection (Fig. 10-1). If the slope is too steep for stadia measurements, vertical angles can be used to compute vertical distances. If this method is too slow to use at a working face, as during rapid quarrying or excavation, an enlarged photograph can be used as a base on which to plot contacts, depths of overburden, and locations of samples. Dimensions can be scaled or estimated from the print if the distances between a few key points on the photograph are measured on the ground. To be precise, however, these measurements must take into account the distances between the points and the camera (see Moffitt, 1959).

10-3. Cleaning, Excavating, and Drilling

If the outcrop map shows that exposures are not adequate to delineate critical features or to sample efficiently, excavating or drilling may be necessary. Trenching at right angles to the strike is particularly effective for uncovering steeply dipping planar structures, as a contact, a thin bed, or a vein. Pits can generally be carried to greater depths and are better suited for measuring or sampling gently dipping beds or veins, or unstratified igneous bodies. Large-scale cleaning

Detailed Mapping and Sampling

and excavating are usually done with a bulldozer or mechanized ditch-digger. Thin soil and overburden can be hosed off if water under high pressure is available. When funds do not permit these large-scale methods, trenching with a pick and shovel may be very effective. Even in the most modest projects, contacts covered by thin soil can be exposed quickly with a folding entrenching tool (Fig. 10–2).

Drilling. Drills can be used to determine the position of geologic structures, to measure subsurface units, and to sample rocks and ores. Core drills, especially diamond drills, are particularly useful for drilling long distances in hard rocks. The diamond drill consists of a tube with small diamonds mounted in the bit end. As the drill is rotated, it cuts a cylindrical core of rock, unless the rock is so fractured or friable that it crumbles. Some drills are pulled from the hole after each 5 or 10 ft have been drilled, and the core is removed and placed in a core box. Other kinds of drills permit longer intervals to be cored. As the core is examined, all rocks, ores, and geologic structures should be described in a log of the hole (Section 12–10). Where rocks are so friable or fractured that no core is recovered, the log can be augmented in part by records of the rate of drilling and water loss in the uncored interval. The finely ground sludge from the drill hole should be sieved, cleaned, and examined microscopically to identify the uncored rocks. The sludge is also collected for assays.

The diamond drill can be pointed in almost any direction and used either underground or on the surface. Occasionally, however, long holes are deflected appreciably from their initial direction, especially where they cross bedding or other structures obliquely. It is then necessary to survey the course of the hole with especially designed equipment (Collins, 1946; McKinstry, 1948).

Other drills may be used where materials are soft, where drilling distances are short, or where geological targets are simple. Hand- or

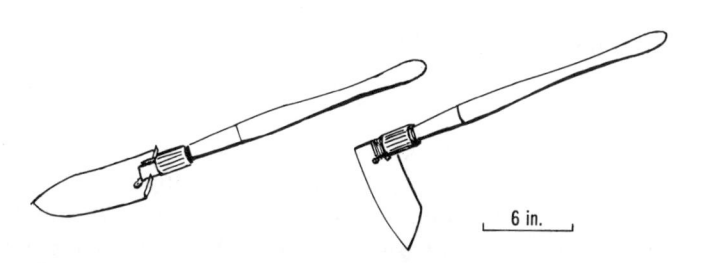

Fig. 10–2. Folding entrenching tool.

machine-driven augers will often get adequate samples of surficial or soft rocks. Miners' hammer drills and other percussion drills can be used in harder rocks (Jackson and Knaebel, 1932). Methods of logging drill holes by examining disaggregated or pulverized rock are the same as those used for well cuttings (Section 12-10).

10-4. Underground Mapping

Mines and tunnels can provide exceptionally fresh and complete exposures of rocks, ores, and structures; they should therefore be mapped regardless of whether the study is an economic one. Permission must first be obtained to enter and work on a property. If the property is not being worked, information on available maps and the safety with which workings can be entered should be obtained from an owner or lessee. When this is not possible, information can sometimes be gained from publications or state and county records. Abandoned and undescribed workings can often be entered safely if one is accompanied by someone experienced in mining.

In addition to ordinary field equipment, the following items are needed for underground mapping:

1. A hard hat and attached carbide or electric lamp.
2. A map case to hold $8\frac{1}{2} \times 11$ in. sheets, as the aluminum clip folder described in Section 1-3; this will be more serviceable if a leather pencil holder is riveted to an outside face.
3. A 100-foot cloth tape of good quality.
4. Adequately warm and waterproof boots and clothing that will not get in the way when climbing through narrow openings.

Gear should be carried so one's hands can be completely free. Because a large back-pack is likely to be cumbersome, most of the equipment should be hung from the belt or placed inside the shirt; a hunter's vest or jacket with a large inside pocket can be used for carrying samples.

Where mine maps are available, 8×10 in. work sheets that will cover the workings piece by piece should be prepared. The scale of these maps is typically 1 in. = 20 ft, 1 in. = 40 ft, or 1 in. = 50 ft. They should be traced directly in pencil or ink from up-to-date master maps. Wet workings should be anticipated by using plastic or vellum sheets, preferably ruled with a 1-inch grid. Coordinate lines and numbers should be entered on the work sheets. If there are no coordinates, a north-south and east-west grid can be numbered and carried from

Detailed Mapping and Sampling

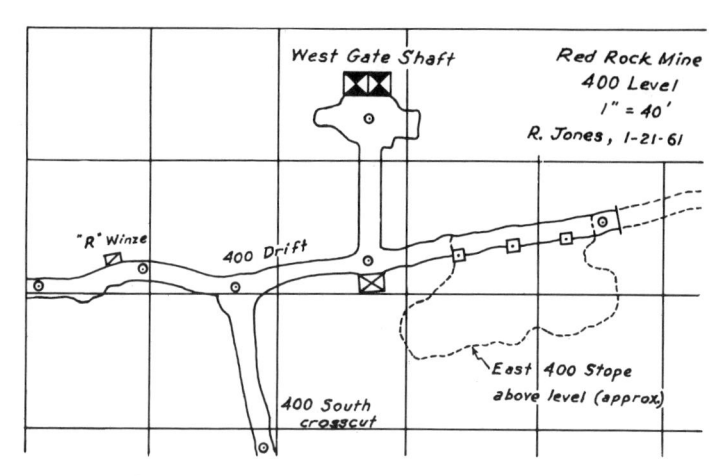

Fig. 10-3. Part of a work sheet for underground geologic mapping, with workings labeled.

one sheet to the next so that sheets can be connected readily without the office map. Transit survey points will be used for measurements underground and must therefore be traced precisely. Finally, each sheet should show the name of the workings, the scale, and the names of the geologists. A typical work sheet is shown in Fig. 10-3; the symbols used are explained in Appendix 4.

The following terms, which are often used on work sheets, should be understood before going underground:

Shaft—relatively narrow working excavated from the surface; it may be vertical or inclined.

Raise—narrow vertical or inclined working excavated upward from underground.

Winze—narrow vertical or inclined working excavated downward from underground.

Level—groups of tunnel-like workings that lie approximately the same distance below the surface in a given mine.

Adit—tunnel-like entrance to a mine.

Drift—tunnel-like working along a vein.

Crosscut—tunnel-like working that intersects a vein.

Stope—an underground working from which ore has been extracted, either above (*overhand*) or below (*underhand*) a level.

Back—ceiling of a working.

Face or *breast*—wall at the end of a working.

Before mapping starts, the workings should be reconnoitered to determine (1) the general nature of the rock units, the ore, and the principal structures, (2) how much the walls must be cleaned, and (3) the accessibility of old workings. Survey data applying to stopes and other irregular or partly filled openings should be drawn on the work sheets or on separate vertical sections. Assay maps should also be studied.

It is advantageous to start mapping where rocks are fresh and structurally uncomplicated. A hundred feet or so of walls are examined first. If they are dirty, it is necessary to scrub them with a brush and water or to chip them at closely spaced intervals with a hammer. The types of rocks, ores, and alterations are plotted on the work sheet, either with pencil patterns or colors (Fig. 10–4). Contacts, faults, and other structures are shown by the usual symbols. Mining geologists often use green for faults and red for ore. Where planar structures can be traced from one side of a working to the other, their strike can be found by standing with one's back against the plane in one wall and sighting a level compass line to the same plane on the other wall. Many strikes and dips can also be measured by method I of Section 2–7.

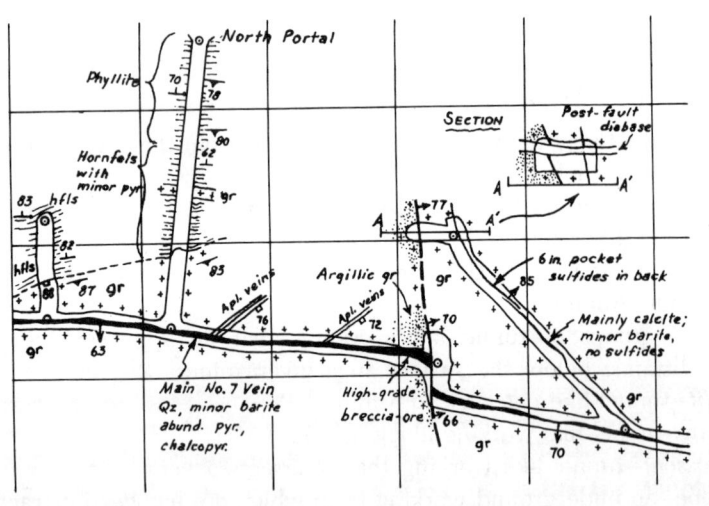

Fig. 10–4. Work sheet, showing plotted geologic features and brief notes. A flat-lying dike has been plotted in an accessory vertical section at upper right.

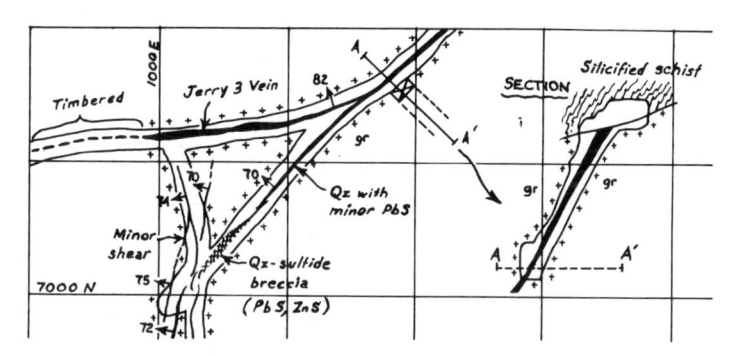

Fig. 10–5. Work sheet, showing cross section of raise.

Features are located on the map by the procedures used in all tape traverses. The measurements should be based as much as possible on established transit traverse points. These points are generally marked by numbered brass buttons (*spads*) in wooden wedges driven into the back or wall. The positions plotted on the map should refer to one datum level within a given working; this is commonly taken as 3½ ft ("waist height") above the track level. Structures that do not intersect this level (for example, flat faults exposed near the back) must be shown in cross sections or described in an accessory note that gives the elevation above the datum. These entries should be made directly on the work sheets (Fig. 10–4). Many accessory measurements may be needed to show the details of sinuous or irregular contacts or faults. Folds are shown as horizontal projections of beds in the datum plane, with plunges plotted as usual. Bifurcating faults should be checked carefully to determine whether one cuts the other or whether they are branches of one fault. Plotting should be done underground, where continuity of structures and units can be checked at once.

When walls of workings are generalized or incomplete, they should be redrawn to fit important geologic features. Features in raises, winzes, and shafts can be plotted on accessory cross sections inserted in blank areas of the work sheets (Fig. 10–5). Sections of inclined workings can be drawn readily by using the bearing and inclination of the ladder that lies along the footwall, and then measuring distances up the ladder to geologic features. Only the approximate outlines of stopes are generally shown on mine maps, though the stopes may be inclined or irregular and extend through large vertical distances. If possible, they should be plotted as a series of cross sections and horizontal sections. Data on inaccessible stopes can often be obtained from the mine engineer or miners.

If there are no mine maps, walls and geologic features must be mapped concurrently. Wherever possible the traverses should be controlled by transit-traverse points. If this cannot be done, the tape-compass traverse method described in Section 3–5 should be used, for magnetic disturbances in most mines are considerable. Horizontal angles in level or gently inclined workings may also be turned with a peep-sight alidade and traverse board. Differences in elevation can be determined with a hand level and clinometer (Sections 2–5 and 2–6).

Safety. The foreman or manager of a property should be consulted regarding the mining methods, schedule of blasting, signals for hoisting men and ore, and rules for safety. Carelessness underground can cause serious accidents. The most common dangers are those of dislodging a loose block from the back or sending a chip flying into the eye (safety glasses are recommended). Traverse stations should never be taken at the bottoms of ore chutes or in places where moving trains or skips may be a danger. Old ladders should be tried with caution, for rotted timbers may appear stronger than they are. The air in unventilated workings should be tested for carbon dioxide with a match or carbide lamp (the flame will turn orange and smoke if oxygen is scarce). For the safety of others, tracks and walkways should be cleared at once of debris knocked from the walls or back.

10–5. Sampling

The kinds and numbers of samples needed will vary greatly with the purpose of a project and its geological situation. Plans for sampling should be based as much as possible on knowledge of sampling plans used elsewhere on similar deposits. Examples of such sampling plans are given by Parks (1949), Jackson and Knaebel (1932), Pitcher and Sinha (1958), Grout (1932), and Shaw (1956). Alternate sampling plans can be evaluated statistically to determine which is likely to give the information needed with least effort and expense. Statistical methods should also be used when determining the significance of sample data. Methods of statistical evaluation are described by Dixon and Massey (1951), Hoel (1960), and Wallis and Roberts (1956).

Spot samples. Spot samples are single, relatively small specimens, for example, the hand specimens used to identify and interpret rocks during mapping. Well-chosen spot samples assist in gaining an understanding of geologic events and in laying plans for more thorough sampling. Unlike the random samples used for some purposes, spot samples should be selected at particular places to answer specific questions. General suggestions for collecting spot samples are:

1. Collect samples that are truly representative of the units being studied.
2. Collect at contacts or where structural and stratigraphic relations can best be determined.
3. Collect oriented samples where directional relations may be important.
4. Collect where new textures or small-scale structures are superimposed on older ones.
5. Collect fresh materials if possible, but do not slight important rocks because they are weathered.

Testing hypotheses with serial samples. Several samples collected in one or more series can often be used to test given hypotheses. The contact between two igneous rocks, for example, might show gradations suggesting emplacement of one magma against another; critical mineral reactions in the gradation could be tested by series of samples extending across the contact zone. Serial samples might also be used to determine changes in a soil profile, to study the differences between adjacent rich and lean segments of an ore body, or to compare compositions within graded beds. Because the samples are collected to test and augment existing evidence, they should be selected after comprehensive examination of outcrops. They will generally be connected on the basis of some known or highly probable structural relationship; nonetheless, the collector must guard lest he bias his selections. Because the numbers of samples must always be limited, the samples should be distributed to give the maximum amount of information. Where a local, thorough picture of variations is required, the samples should be concentrated on a few traverses across the sample unit. Generally, however, it is preferable to collect a few samples from each of many traverses across the unit, for the lateral as well as the vertical variations may then be estimated. Flinn (1959) has evaluated several serial sampling plans statistically.

Sampling to find variations. When heterogeneity is obscure, or when it does not show clear geographic trends, samples can be used

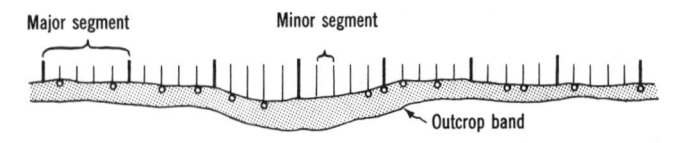

Fig. 10–6. Sampling the linear outcrop of a unit on a nested plan. Small circles indicate specimen localities.

to determine if there are significant variations within a unit. Grain sizes in a sandstone, for example, may seem to vary sporadically at scattered localities, yet it may be important for petroleum or groundwater exploration to determine if there are regional trends in these variations. Such problems require sampling on both regional and local levels, for the significance of regional variations can be understood only when the degree of local variation has been established. Some units are so homogeneous at each outcrop, and within any one-acre area, that small regional variations can be established with great confidence. In other units, local heterogeneity is so great that major regional variations can be detected only by using large numbers of samples. The sampling plan used in such studies must insure that samples will be selected randomly; otherwise statistical evaluations will not be valid. These requirements are met by the general method called *nested sampling*. This method is illustrated by the example given in Fig. 10–6, which shows the linear outcrop of a rock unit. The outcrop line has been divided into major and minor segments, which form the basis for sampling. Sample localities are marked in all (or most) major segments by randomly selecting two minor segment points. Two samples are collected at all (or most) localities. If rock cannot be exposed at a given point, the nearest possible site is used. After the samples have been analyzed for whatever data may be useful, the data are evaluated statistically. Krumbein and Tukey (1956) have described how several kinds of data can be evaluated simultaneously; they have also described examples of several kinds of nested sampling plans.

Sampling for bulk composition. Bulk composition of a body is determined by collecting enough small, random samples so that their average closely approaches the composition of the entire body. If the body is fine-grained and homogeneous, only one small sample is needed. If it is heterogeneous, the number of samples needed increases rapidly with the degree of heterogeneity. The bulk composition

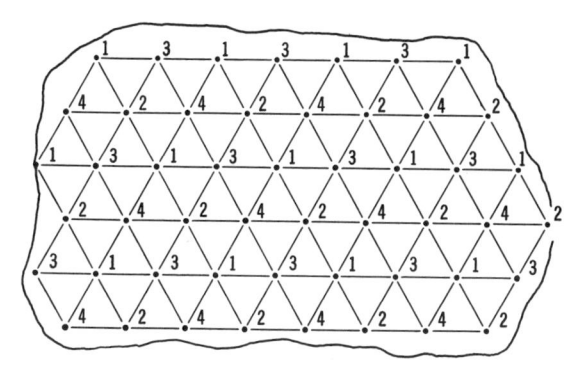

Fig. 10-7. Sampling plan for determining the bulk composition of the outlined body, using four groups of equi-spaced samples.

of many bodies can be determined only within broad limits; thus the limits allowable for a given study should be considered at the outset. The precision of the analytical methods that will be used should also be considered (see Ahrens, 1950; Fairbairn et al., 1951).

To keep to a minimum the number of samples analyzed, the body should first be evaluated by comparisons with other sampled bodies. Samples must be distributed so as to provide a statistically valid (random) measure of the entire body; they could, for example, be spaced evenly on the horizontal coordinates and levels of a mine or at intersections of a rectangular grid on the surface. Materials must be collected as closely as possible to the prescribed points. If they must be distributed unevenly, it is necessary to weight each sample according to the distances from adjacent samples (Parks, 1949, ch. 4).

The number of samples analyzed can sometimes be minimized by collecting the samples in several groups, each distributed evenly over the body (Fig. 10-7). One group of samples is analyzed first, and their average composition (mean) and some measure of the degree to which they differ (as their standard deviation) are computed. The standard deviation will show the approximate reliability of the average composition. If the average is too indefinite, a second group of samples is analyzed and a new total mean and standard deviation computed. This procedure is repeated until the overall mean can be shown to lie within limits that are small enough to satisfy the purpose of the project.

Sampling for distribution of compositions. When a large number of samples can be used, it may be possible to determine the distribu-

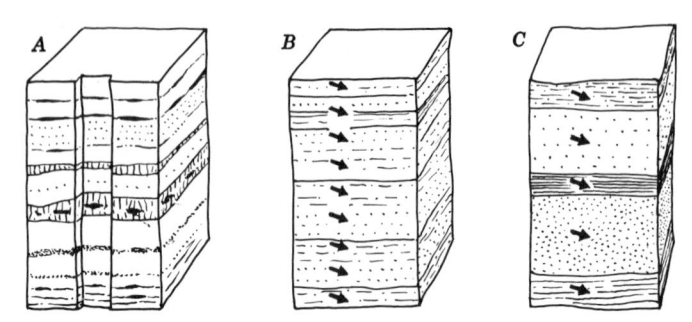

Fig. 10–8. Three methods of collecting an average sample of a gross unit. (A) Channel sample. (B) Chips collected at a specified interval. (C) Chips collected so as to represent sub-units within the total unit.

tion of compositions or values within a body. This is often done in mines to locate high-grade trends or "shoots" of ore. The method may also be used to help determine the causes of compositional variations within rock bodies. The samples may be collected on the same plans as those used for determining bulk composition; however, it is now necessary to collect them so that the bulk composition at each locality is represented reliably. If a bed or vein is homogeneous from top to bottom, only one small sample is needed from each locality, and lateral variations within the body can be determined with great exactness. In the usual case, however, the bed or vein will be variable in any single section, and sufficient material must be collected to give a reliable average at that particular place. The greater the variations, the more thoroughly each section must be sampled. Where variations are extreme, it may be necessary to cut a channel of constant width and depth across the unit (Fig. 10–8A). This is commonly done for heterogeneous ore veins. Where variations are moderate or are restricted to comparatively few components, a representative sample can be obtained by taking chips spaced at a predetermined interval (Fig. 10–8B). This will be a random sample across the unit at that place. Finally, where a few homogeneous subunits constitute the total unit, specimens may be taken from each subunit (Fig. 10–8C). Each specimen of these *stratified samples* must be weighted according to the thickness of the subunit from which it came.

Before beginning sampling, it may be advantageous to test the degree of heterogeneity within several sections of the unit. This can be done by collecting channel, chip, and stratified samples of typical sections and comparing analyses made from them. The cheapest or easiest method of obtaining adequate samples can then be chosen.

Sampling methods. Surfaces should be cleaned before cutting or chipping samples. Even fresh-appearing materials can change appreciably if exposed for some time underground or on the surface. Channels are generally made 1 in. deep and 2 to 4 in. across. If the rock is hard, the channels must be cut with a hammer weighing 2 to 3 lb and either a cold chisel or moil (a piece of drill steel pointed at one end).

Sample localities should be plotted on maps or sections and described fully in the notes, with a drawing showing the dimensions and character of the unit sampled. The source and number of the sample should be marked clearly on the outcrop so that the outcrop can be resampled if necessary. The sample should be placed in a dust-proof bag that is labeled clearly. The samples may also be registered in a sample book that is numbered in duplicate with the sample numbers. Samples collected in bulk, as channel or chip samples, are crushed and divided into smaller portions for analysis and storage. This division should, if possible, be made with a mechanical splitter.

Bias of samples. Sample data can be biased and therefore misleading if specifications for sampling are not followed closely. Channels cut across irregular surfaces must be collected or weighted so as to represent the average across the unit (Fig. 10–9). Channel samples must not include fragments from outside the prescribed limits, and chip samples must be collected at the predetermined interval, regardless of spectacular materials that do not lie at an interval point.

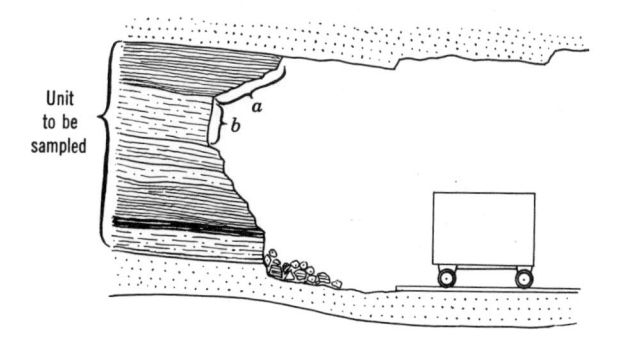

Fig. 10–9. Diagrammatic section through a working with uneven walls, showing why some parts of a channel sample have to be weighted relative to other parts (compare segments a and b).

If only fresh rocks are collected, the bulk will be biased against rocks containing unstable minerals, and if only outcrops are sampled, the bulk will be biased relative to weak rocks. A particularly large error can occur when diamond drill holes are sampled only from the core rather than from the core, fragmental materials, and sludge.

Bias can also be introduced by modifying samples after they are collected. Possibilities for fraud in mine sampling are manifold; some examples are described by Hoover (1948).

References Cited

Ahrens, L. H., 1950, What to expect from a standard spectrochemical analysis of common silicate rock types: *American Journal of Science*, v. 248, pp. 142–145.

Collins, J. J., 1946, Some problems involved in the interpretation of diamond-drill-hole sampling and surveying: *Mining Technology*, American Institute of Mining and Metallurgical Engineers, Technical Publication 1842, pp. 1–28.

Dixon, W. J., and F. J. Massey, 1951, *Introduction to statistical analysis:* New York, McGraw-Hill Book Co., 370 pp.

Fairbairn, H. W., et al., 1951, *A cooperative investigation of precision and accuracy in chemical, spectrochemical and modal analysis of silicate rocks:* U.S. Geological Survey, Bulletin 980, 71 pp.

Flinn, Derek, 1959, An application of statistical analysis to petrochemical data: *Geochimica et Cosmochimica Acta*, v. 17, pp. 161–175.

Grout, F. F., 1932, Rock sampling for chemical analysis: *American Journal of Science*, v. 24, pp. 394–404.

Hoel, P. G., 1960, *Elementary statistics:* New York, John Wiley and Sons, 261 pp.

Hoover, T. J., 1948, *The economics of mining:* Stanford, Stanford University Press, 551 pp.

Jackson, C. F., and J. B. Knaebel, 1932, *Sampling and estimation of ore deposits:* U.S. Bureau of Mines, Bulletin 356, 155 pp.

Krumbein, W. C., and J. W. Tukey, 1956, Multivariate analysis of mineralogic, lithologic, and chemical composition of rock bodies: *Journal of Sedimentary Petrology*, v. 26, pp. 322–337.

McKinstry, H. E., 1948, *Mining geology:* New York, Prentice-Hall, 680 pp.

Moffitt, F. H., 1959, *Photogrammetry:* Scranton, Pa., International Textbook Co., 455 pp.

Parks, R. D., 1949, *Examination and valuation of mineral property*, 3rd ed.: Cambridge, Mass., Addison-Wesley, 504 pp.

Pitcher, W. S., and R. C. Sinha, 1958, The petrochemistry of the Ardara aureole: *Quarterly Journal of the Geological Society of London*, v. 113, pp. 393–408.

Shaw, D. M., 1956, Geochemistry of pelitic rocks (Part III: Major elements and general geochemistry): *Bulletin of the Geological Society of America*, v. 67, pp. 919–934.

Wallis, W. A., and H. V. Roberts, 1956, *Statistics, a new approach:* Glencoe, Ill., The Free Press, 646 pp.

11

Preparing Geologic Reports

11–1. General Nature of Geologic Reports

A complete, well-written report is an extremely important part of a geologic project; without it, the best possible field study will be of little value to other persons. The general nature of the report will depend on the nature of the project. Some reports are general descriptions of relatively unknown areas; many others are more detailed accounts of certain rock associations or economic deposits; still others are recommendations regarding engineering structures, land management, or exploration for mineral commodities. To fulfil its purpose, each report will differ somewhat in organization, length, and style. Regardless of special emphasis, however, the basic purposes of a report are (1) to describe accurately what has been observed and (2) to synthesize and explain geologic relationships and events. The descriptive parts of the report should be as complete yet as concise as possible. In most cases this requires maximum use of maps, sections, tables, and graphs. These illustrations are often so important that they should be prepared first and the report written around them.

The explanation or conclusions of the report must be synthesized directly from facts; they should not merely express the writer's point of view. By presenting his data without bias of opinion, the writer acknowledges that some readers may arrive at somewhat different conclusions than his own. The intrinsic value of a report therefore depends mainly on the field study it describes. Its usefulness, however, can be increased by careful organization, clear writing, and well-thought-out illustrations.

11–2. Organizing and Starting the Report

The report should be organized and started in the field or as soon after the field season as possible. The first complete draft must be finished in an office where there are drafting and library facilities and where fossils, rocks, and minerals can be identified exactly. If possible, the map and other detailed illustrations should be prepared before the report is written. Most literature will have been read before and during the field season, but pertinent items should be reread before writing the report. Field notes should be paginated in chronological order and then reviewed systematically. Some geologists find it useful to reshuffle or cut note pages so that all items pertaining to a certain part of the report are brought together; others prefer to leave their notes intact and to make lists or card files of items pertaining to particular sections of the report.

The main divisions of the report should be set down first, and then its subdivisions and details filled in in outline form. The outline that follows presents the main parts and principal subdivisions of a typical geologic report.

I. Introduction.
    a. Purpose of project, including review or explanation of any specific problems.
    b. Dates and general procedure of the project.
    c. Acknowledgments of help received.
    d. Geographic setting.
        1. Location (a small index map will generally suffice).
        2. Accessibility, if not obvious from maps.
        3. Nature and distribution of principal geographic features.
        4. Vegetation, climate, and land use (generally stated briefly).

II. Regional geologic setting.
    a. Nature and distribution of principal rock systems, series, or formations.
    b. Major structures, in chronological order wherever possible.
    c. Summary of regional geologic history, if appropriate.

III. Rock units.
    a. Introduction to the general nature, thickness, and grouping of the stratigraphic sequence, including brief descriptions of general stratigraphic problems, if any.
    b. Systematic description of rock units, starting with oldest.
        1. Nomenclature (explanation of name used for unit, with reference to past usage).
        2. General lithology, distribution, shape, and thickness of unit.
        3. Detailed description of lithology, including lateral variations.
        4. Definition of contacts, if not included in lithology.
        5. Fossils, if any.
        6. Age and origin of unit.

IV. Structures.
    a. Brief introductory description of trends and interrelations of principal structural features.

b. Unconformities (unless described adequately under rock units).
   c. Folds, in order of importance or age, or both.
   d. Faults, as in order for folds, including pertinent relationships to folds or other structures.
   e. Structures formed in and around intrusive bodies.
V. Geomorphology.
   a. Descriptions of older erosion surfaces and other modified features.
   b. Present erosion surfaces, stream patterns, and erosive processes.
   c. Interpretation of structural or climatic changes shown by land forms.
VI. Geologic history: chronologic interpretation of processes, structural events, and paleogeography.

The introductory sections should pertain directly to the main subject. Climate, vegetation, and land use can usually be described adequately in a few sentences, unless the region is especially remote or the report is concerned chiefly with economics. Descriptions of geographic features can be kept concise and clear by including a page-size map of the region. Special field or laboratory methods should be described in an additional section of the introduction, and another section may be needed to define important terms whose usage is not widely standardized.

The section on regional geology provides an important framework for the detailed descriptions of the report. This is a section that will require a thorough winnowing of the literature. When items unsupported by adequate data are used, their source and probable reliability must be noted. A separate section presenting a chronological review of geologic work done in or near the area may be added in some cases, but a critical evaluation of important contributions is generally preferred. A page-size map showing the main geologic features of the region is always helpful, and it may be combined with the geographic map used in the introduction. When the main body of the report is completed, major structures and rock units should be reviewed to determine whether they correlate accurately with the descriptions of the regional geology.

Large numbers of rock units that cover a considerable span of geologic time can be organized on the basis of systems or combinations of systems. Accordingly, metamorphic rocks should be presented in the order of their premetamorphic age. Intrusive igneous rocks are best presented in chronologic order with all the other rocks, although they may be described in a separate section. Rock units that are formalized for the first time by the use of a geographic name must be defined according to accepted rules (see Sections 11–4 and 12–2).

For all but the simplest areas, the section on structure must be organized with special care to describe both the individual features and their interrelationships. This may require using a different order than that shown in the outline; for example, a strict chronologic order may be better in some cases. Data shown clearly on the geologic map need not be described at length in the text. If the main map is too cluttered to show the geographic distribution of the major structures clearly, a page-size structure map should be used in context with the structure section.

Most general reports conclude with a section on geologic history, but this may be followed by sections on economic deposits, metamorphism, igneous activity, glaciation, systematic paleontology, or other pertinent subjects.

11–3. Clarity of the Report

The one basic requirement for writing the report is that both data and ideas be presented clearly. This means that sentences must be constructed with care, and sentence order must be logical. Moreover, words must be chosen so as to give the precise meaning intended. One way to achieve logical sentence order is to expand an outline to sentence level. Subheadings can be used to orient the reader's thought to major topics, or each major paragraph can be introduced by a topic sentence. There are many further ways to improve the clarity and precision of a report. These are described fully in guides to general exposition and scientific writing (as Perrin, 1950).

The meaning of scientific terms must be clear to the readers for whom the report is prepared. For general reports, the less technical of two possible terms is preferred. If an important term has more than one meaning, its usage should be defined in the introduction, in context, or in a footnote. *Roget's International Thesaurus* is a valuable aid in selecting nontechnical words, and the American Geological Institute's *Glossary of Geology and Related Sciences* provides brief definitions of many technical words. *Suggestions to Authors of the Reports of the United States Geological Survey* contains many valuable instructions regarding both word usage and report writing.

Repetitious use of words and phrases may be necessary if one word, or phrase, serves best to define a feature or situation. Moreover, it is generally desirable to keep parallel the construction of sentences that describe similar things, as well as the organization of paragraphs or sections that deal with similar units of the geology. In the section dealing with rock units, for example, the various subtitled parts should be in the same order, and the detailed descriptions within each of the subtitled parts should be as nearly as possible in the same order. This enables the reader to compare the rock units mentally as he reads; it also makes it easier for him to find parts he wishes to reread.

Regardless of the grammatical correctness of sentences, scientific clarity requires that facts be demarcated clearly from interpretations. It may be advisable to keep the two separated by using only descriptive materials in the main body of the report and saving all theorizing for the concluding sections. In this way the writer can avoid the error of drawing conclusions from points of "evidence" that were originally introduced as ideas, not facts.

11-4. Use of Special Terms

Terms such as geographic names, cartographic and stratigraphic names, fossil names, and rock names must be chosen and used with special care to convey the meaning needed. Even minor modifications in their arrangement, punctuation, or spelling may be misleading. Moreover, typists cannot be expected to detect mistakes in these special terms.

Geographic names. Quadrangle maps of the U.S. Geological Survey show almost all names correctly. Errors are likely to occur only on old maps or on recent maps where it was not possible to check the names of minor outlying features. Inquiries in the area may resolve some of these errors; however, these inquiries must be thorough because local usage tends to vary with time and individual feeling. Forest Service maps and county land plats may be helpful in areas not covered by quadrangle maps.

Because most readers will not be acquainted with the geography of the area described, *all place names used in the text should appear on at least one map in the report.* Furthermore, descriptions of rock units and structures should not depend on an endless chain of place names, especially if the names may be difficult to locate on maps. Descriptions are followed more easily if locations are based on directions and distances from a few prominent towns, peaks, and streams, or simply on the gross shape of the area mapped, or on the shape of any well-known part of it.

Lithologic and time-stratigraphic names. Use of lithologic unit terms (group, formation, member, lentil, tongue, bed) and time-stratigraphic unit terms (system, series, stage, zone) must be applied consistently according to accepted rules (Section 12-2). Formerly, the first but not the second term of lithologic unit names was capitalized (as Sandholdt formation, Lucia shale), but it is now recommended that both terms be capitalized unless the name is used in an informal sense (see examples in Sections 12-2, 13-2, 14-2, and 15-2).

Spelling and usage of formalized names must be checked both in the original description and in subsequent publications where units may be redefined. Names of units in North America described prior to 1936 have been listed by Wilmarth (1938), while names introduced in the period 1936-1955 have been listed by Wilson, Sando, and Kopf (1957). The index of units prepared by Wilson, Keroher, and Hansen (1959) will also be useful. Still more recent usage should be determined by checking the bibliographies of North American geology published by the U.S. Geological Survey, by consulting recent publications, and by making inquiries to state geology offices or other agencies.

In order to avoid duplication, geographic names for new units must not be published until they have been checked against files of used and reserved names by the Geologic Names Committee (U.S. Geological Survey, Washington 25, D.C.).

Fossil names. Errors in form and spelling of fossil names may be indecipherable or may even allocate the specimen to the wrong genus or species. Certain accepted rules should be followed. Both generic and specific names are italicized (underlined in typed manuscript) where they appear in the text, but not where they are used in lists or tables. Generic names are capitalized, while specific and varietal names are not. The name of the founder of the species always follows the specific name in lists; although it should be used in the text, it is customarily omitted when the full name appears in a list or table. The founder's name is capitalized but not italicized. Generic names may be abbreviated to a capital initial (as *T. inezana* for *Turritella inezana*) if they are used in such context that the generic meaning is obvious (as in discussions of various turritellas).

Identifications of genus or species may be doubtful because of poor preservation, variations within a species, or incompleteness of the published record. In such cases the reliability of names should be indicated by appropriate designations, as shown in these examples: *Acila* cf. *A. decisa* (specimen of genus *Acila* not well preserved but looks much like *Acila decisa*); *Acila decisa?* (genus definite but species only a good guess); *?Acila decisa* (both generic and specific names only a good guess); *Acila* sp. (genus definite but species indeterminate); *Acila* n. sp. (a well preserved but unnamed species of the genus *Acila*); *Acila* aff. *A. decisa* (a well preserved *Acila* that may be a new species but is similar to *A. decisa*). Where a paleontologist has made determinations or suggestions, his statements must be quoted or transcribed with care. Identifications by paleontologists must be credited, not only as a matter of courtesy but also to make the record complete.

The classifications and descriptions of new species require an extensive library and the help of a paleontologist. Schenk and McMasters (1956) described these procedures briefly, with special reference to the rules for binomial nomenclature.

Rock names. A few simple rules should be followed in composing rock names consisting of more than one word. Hyphens are inserted only between *nouns* used in the same sense, as in *olivine-augite andesite, rhyolite tuff-breccia,* and *chlorite-albite-epidote schist* (not in *olivine basalt, lapilli tuff,* or *quartz diorite*). Similar *adjective* modifiers are separated by commas, as: *calcareous, argillaceous siltstone,* and *black, green-weathering andesite.* Where two or more mineral modifiers are used to form rock varieties, it is useful to place the name of the most abundant mineral next to the rock name. For example, a *quartz-biotite-chlorite schist* contains more chlorite than biotite and more biotite than quartz (but this rule is by no means universal).

Some terms are used loosely during field work or are used with a restricted meaning thought suitable to a given project (common examples are *migmatite, fine-grained, granite, graywacke, shale*). The writer must consider whether his usage of such terms is correct, or, if various meanings have been given them, whether he has defined his usage precisely enough.

11–5. Front Matter for the Report

The front matter consists of all materials appearing before the main body of the report, generally a title page, a table of contents, a list of illustrations, and an abstract.

Title page. In addition to the title of the report, the title page should bear the name of the writer, the date or dates of the project (or, in some cases, the report), and the name of the organization, if any, for which the work was done. The title itself should be as short as possible but should state clearly the nature or content of the report. Titles of general reports should begin with *Geology of* . . . , followed by the geographic name of the area described. It is unnecessary for geographic names to be compounded to delimit an area completely, for the area can be located readily by turning to the report.

Table of contents. The table of contents is prepared from the final copy of the manuscript, when all section headings have been decided upon. In many cases the lowest rank of headings may be omitted from the table of contents if they are repetitive (for example, the subheadings under each rock unit).

List of illustrations. The list of illustrations gives page numbers or other suitable reference to all text figures and plates, including those folded separately at the back of the report. Wherever possible, titles or captions of illustrations are shortened for the list. For example, a figure caption that reads "*Contact between Lucia Shale and The Rocks Sandstone in upper Reliz Canyon, showing thin sandstone beds and concretions in shale*" could be listed as "*Contact between Lucia Shale and The Rocks Sandstone.*"

Abstract. The abstract is a very brief version of the report. It is all some readers will ever see (or read) and should therefore be as informative as possible. Abstracts are generally 150 to 600 words long, although much shorter ones may be adequate for some papers and 1,000 to 4,000 words may be needed to abstract exceptionally long or thorough reports. The length of the abstract also depends on the nature of the paper and on the specifications set by the publisher. Three suggestions for writing abstracts are:

1. Each sentence must be informative. Such statements as "The sedimentary rock units are described, their fossils named, and their ages delimited" are superfluous because the reader assumes the report will cover these things. Generally, only a few more words could give useful data, for example, "The sequence is: 600 ft of unnamed Cretaceous shales, 1,200 ft of Laron Sandstone (Eocene), 800–1,400 ft of Cutler Formation (Upper Eocene and Oligocene?), and 3,000 ft of lavas that intertongue to the south with the Justin Formation (Miocene)."

2. Data should be presented in the same order as in the report or as nearly so as possible. The reader who is interested in only a part

of the report can then go quickly and easily from abstract to table of contents to text.

3. Each major section of the report should be summarized in at least a sentence, and if the report is long or thorough, each major section should be summarized in a paragraph. Favorite topics should not be emphasized so much that other topics that may be of interest to many readers are omitted. The main descriptive materials should receive the most thorough treatment. New lithologic units, unusual associations of rocks or minerals, and new fossils or minerals should at least be mentioned in the abstract, regardless of the main purposes of the report. Where new or unusual methods have been used, they should be described briefly.

11–6. Form of the Report

All manuscripts should be typed in acceptable double-spaced format. Quotations should be checked with care and the citation of their source must include a page number. Data and ideas condensed from other reports are not placed in quotation marks but must be accompanied by a full citation of their source. Footnotes other than bibliographic references are troublesome to insert in a typed manuscript and may disrupt the continuity of the reader's thought; therefore, they should be used only where necessary. Short statements or definitions that are incompatible with the text can commonly be placed in parentheses in context.

Lists and tables. In many cases data can be presented more clearly and concisely in lists or tables than in the text. Lists are used for data that can be organized in one or two columns (as fossil collections, stratigraphic sequences, or well logs), whereas tables are used when both horizontal and vertical divisions must be shown (as several stratigraphic sequences, serial mineral assemblages in metamorphic zones, or modal compositions of various rocks). Well logs and sedimentary sequences should be run one column to the page, but other lists may be run in two or three columns, as space permits. Tables are divided by horizontal and vertical lines, and the typed insertions should be single spaced. Titles for tables must indicate both the kind of data presented and its geographic (or other) source. Labels for the various columns may be abbreviated; if they must be incomplete, appropriate explanatory notes should be added beneath the table.

Bibliography and list of references cited. A *bibliography* is a list of references that includes all works pertaining to the subject of the report, whether they are actually referred to in the report or not. A *list of references cited* includes only the works referred to in the text of the report. The advantage of the bibliography is that it gives a complete picture of background materials for the report.

The list of references cited or bibliography is placed at the end of the report, and the items are arranged in alphabetical order by author. Two or more works by one author are listed in chronological order. Items by co-authors appear in order of the first co-author's name and after his single entries. These rules are illustrated by the examples that follow, and the order and punctuation shown is that which has been standardized for most geological publications in America.

>Alling, H. L., 1926, The potash-soda feldspars: Jour. Geology, v. 34, p. 591–611.
>——, 1936, Interpretative petrology of the igneous rocks: N.Y., McGraw-Hill Book Co., 353 p.
>Calkins, F. C., 1930, The granitic rocks of the Yosemite region: p. 120–129 *in* Matthes, F. E., Geologic history of the Yosemite Valley: U.S. Geol. Survey Prof. Paper 160, 137 p.
>Calkins, F. C., and Emmons, W. H., 1915, Philipsburg, Mont.: U.S. Geol. Survey Geol. Folio 196, 26 p.
>Gilbert, G. K., 1890, Lake Bonneville: U.S. Geol. Survey Mon. 1, 375 p.
>Mayo, E. B., 1959a, Banerjee's study of the Oracle granite (Abstract): Geol. Soc. America Bull., v. 70, p. 1735.
>——, 1959b, Recent concepts of ore localization in southern Arizona (Abstract): Geol. Soc. America Bull., v. 70, p. 1735.
>Muller, S. W., 1945, Permafrost or permanently frozen ground and related engineering problems: U.S. Army Engineers, Strategic Eng. Study, Special Rept. no. 62, 231 p. (Reprinted by J. W. Edwards, Inc., Ann Arbor, Mich., 1947).

Explanatory items are inserted in parentheses (as in the last two examples) or put in italics (as in the item by Calkins).

Citations based on a list of references or a bibliography give the author, date, and the specific pages referred to, as shown in this example:

>The formation was described in a general way by several early workers (Smith, 1902, p. 14–16; Johnston, 1908, p. 4–5; Caster, 1909, p. 7), but Williams (1928, p. 9–14) was the first to define it precisely.

If only a few references are used, they may be entered as footnotes rather than in a list at the end of the paper. This has the advantage of presenting the reference directly where it applies. Footnote references are usually arranged in this order: author, name of article or book, name of journal or (if a book) place of publication, name of publisher, volume number, date, and pages.

11–7. Planning Illustrations for the Report

The map and other detailed illustrations should be prepared first so the report can be written around them. To select additional illustrations, the various field sketches and photographs should be reviewed and set in order. As writing proceeds, ideas for illustrations can be sketched on a scratch pad. These ideas are culled down to a final choice after considering the costs and methods of reproduction. If the report is to be published, the writer must determine at an early date the publisher's specifications as to size, kind, and number of illustrations. Besides meeting these specifications, the illustrations must present data or relations that could not be presented more clearly in the text; they should never be mere decorations. The prime basis of selection is that of usefulness. The writer need not be a skilled artist to prepare a useful illustration.

If illustrations are to be prepared by draftsmen or artists in one step, sketches or other copy submitted to them should be fully labeled and detailed because it may be impossible to change final inked copies. Instructions to draftsmen should be placed on the sketch to which they refer. Time allowing, it is better for the draftsman to prepare a pencil copy from the writer's sketch so that the writer can make changes on this before final copy is prepared. The draftsman can thereby improve the illustration without altering its accuracy.

Once the writer and draftsman have decided what the illustration is to accomplish, they can prepare clear copy by considering each of the following points:

1. Layout of the illustration should be uncrowded and well balanced.
2. Design of details and patterning should be simple.
3. Dimensions (scale) and directions should be indicated clearly.
4. Titles, legends, and captions should be as complete as space allows.
5. If the illustration will be reduced, line weights and letter heights must be suitable.

When the final choice of illustrations is made—particularly when the final copy is in hand—the text should be re-examined to make sure that it coordinates correctly with the illustrations and does not repeat their content needlessly.

Duplication processes. Since many illustrations must be limited because of the processes that will be used to duplicate them, these processes must be considered when the illustrations are planned. Some of the factors to consider are: size of copy, costs, legibility of final copy, accuracy of duplication, durability of copy (both to light and handling), and nature and color of print surfaces (can they be colored and inked?). The possibilities vary so with local facilities that it is necessary to contact blueprinters or engravers and discuss items specifically. They will be able to suggest processes that will give the best results, and they may also be able to suggest how a map can be enlarged or reduced efficiently during the drafting stages. The following notes apply to the processes available most widely.

Ozalid (diazo-ammonia) process. Typically blue or black lines on white base, but sepia tones are available locally; blue lines may fade on some paper stocks; made from a transparent drawing; various papers, including card stock, available locally; widths typically up to 42 in. or 54 in. and any length; one of least expensive processes for less than 50 prints.

BW or Bruning (diazo) process. About the same as for Ozalid prints; white base may age yellowish or brownish in some paper stocks.

Vandyke, blueprints, brown-line prints. Various wet processes, typically made in vacuum frame; require negative of original to give positive copy; more expensive than those listed above, but high-quality papers available; prints typically permanent; scale likely to be somewhat different from original.

Photostat and other photographic prints. Various paper sizes and surfaces, depending on facilities; some surfaces not suitable for coloring or ink work.

Lithographic, photo offset printing. Regular press printing, produces excellent copy on various papers; size limitations vary greatly with local press equipment; usually economic only where more than 30 to 50 copies needed.

Xerography, duplimat. Excellent black-line prints from any good line copy; cheaper than photo offset printing, but image sizes limited to approximately 9½ × 12½ in.

11–8. Kinds of Illustrations

Detailed geologic maps, cross sections, and columnar sections are generally the most important illustrations in a report, and they will be considered in Sections 11–10 and 11–11. Other kinds of illustrations are described briefly in this section. Ridgway (1920) has given many additional suggestions.

Small-scale maps. Most reports require small-scale maps that show area locations, geographic features, or generalized regional geologic features. These maps are most useful if they are page size or smaller, and are bound in the part of the report to which they refer. They should be simplified by removing or subordinating nonessential geographic features. Place names should be shown clearly. It is easiest to make these drawings for about 50 percent reduction.

Photographs. No illustration is as convincing as a good photograph, but only the exceptional outcrop or hillside can be photographed effectively. Critical features generally lack contrast or are obscured by patterns of shadows or vegetation. These difficulties can sometimes be overcome by photographing the outcrop at a certain time of day or by using flash attachments or colored filters. Some prints can be clarified greatly by inking contacts and labels directly on them. Photographs should not be altered in such a way as to modify geologic features, but complicating shadows or other misleading effects may be retouched. When only a few copies of a report are needed, color photographs can be used to illustrate features that would not be clear in black and white prints.

Drawings of outcrops and specimens. Drawings can eliminate disturbing and superfluous material and can clarify relations by showing cut-aways and enlargements (as Fig. 15–11). Drawings from field sketches or photographs are made most easily by tracing directly from the original. Three-dimensional objects are generally shaded as though light were coming from the upper left corner of the drawing. Ink dots (stipples) are used to shade rough surfaces (as Fig. 13–4), while parallel lines are used for smooth or striated surfaces (as Fig. 13–7). Drawings can also be shaded with lamp black, carbon pencils, or watercolor wash, but they can be reproduced only by photographic processes.

Isometric diagrams. Isometric diagrams show structures in such a way that the reader can measure distances parallel to the three mutually perpendicular coordinate-axes on which the drawing is based (Fig. 11–1). The scales on the three axes are the same, and the two horizontal axes are typically placed at angles of 60° to the vertical axis (lined papers of this type are available from suppliers of drafting materials). Other angular arrangements can be constructed. Detailed instructions regarding both isometric (axonometric) and oblique dimensioned projections are given by Hoelscher and Springer (1956, chs. 16, 17).

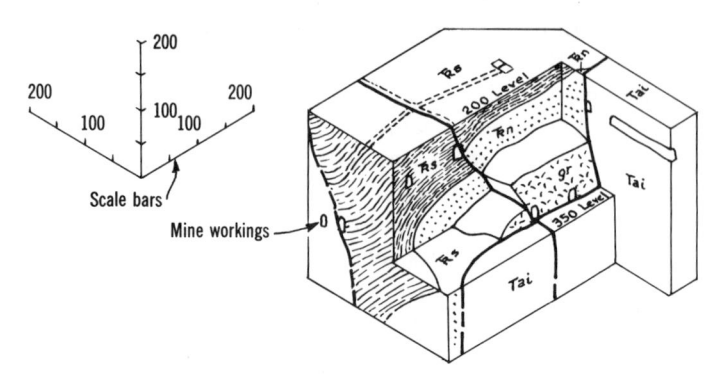

Fig. 11–1. Isometric diagram of part of a mine, showing the scaled axes on which it is based.

Perspective drawings. Block diagrams and other three-dimensional objects should be drawn as perspective projections in cases where their natural appearance is more important than is scaling dimensions from them. The ideal perspective projection is that seen by the human eye or the camera lens, and it results from the simple fact that distant objects appear smaller than those close by. In making a sketch of an outcrop or landscape, the observer draws in approximate perspective without thinking about it, but a correctly dimensioned drawing requires making measurements and projecting them into the drawing according to certain rules. Constructions according to the so-called one-point and two-point perspective systems have been described by Lobeck (1924, 1958). Hoelscher and Springer (1956, ch. 18) have given instructions for all types of perspective drawings, including those made by the three-point system. Three-point perspective should be used to draw objects whose vertical dimension is as long or longer than their other dimensions.

Graphs and curves. Various simple graphs and curves can be used to present data that are related to two variables, and more complicated diagrams can be used to show a third variable. The points from which curves are drawn should appear on the final illustrations, preferably with numbers that refer to lists or other sources from which the data were taken. Cross-ruled paper for graphs and curves should be checked for accuracy, and plotting should be done with a sharp, hard pencil. Graphs that make comparisons by showing areas or volumes may be misleading if not thought out and explained with care.

Spherical projections. Stereographic and equal-area spherical projections can present large amounts of data relating to structural orientations. The references cited in Section 15–11 describe some of these possibilities fully. The final illustration must be labeled sufficiently so that the reader can orient it with the geologic map and sections or with other relevant structural data.

11–9. Drawing Methods

In addition to the materials noted below, a large pad or roll of inexpensive tracing paper will be useful for sketching layouts. If the illustrations will be reduced, a reducing glass can be used to check line weights and letter heights.

Layout. Illustrations should be composed and arranged before detailed work is done on them. After their various parts are sketched on scratch paper, they can be assembled on a drawing board, and the overall composition can be made by placing a piece of tracing paper over the assemblage and composing the illustration in pencil. The distribution of labels and other lettering should be determined at this stage.

Finished line work. The finished drawing may be made directly on a piece of top-quality tracing paper or linen placed over the layout sheet, or it may be made on a piece of opaque drawing paper after transferring the layout with carbon paper (or by blacking the back of the layout sheet with a soft pencil). Engineering drafting methods are used where illustrations require straight or smoothly curving lines, as described in books on engineering drawing (as Hoelscher and Springer, 1956, ch. 3). Most natural objects should be inked free-hand over the pencil guide lines; these lines can be drawn neatly if made a little at a time. Freely curving lines that must have a precisely even thickness (as contours and contacts) can be drawn with a contour pen or a pen from a lettering set. In order to reproduce well, line work must be black, not gray.

Shading. Stippling can be done evenly if the dots are made in small spiraling groups (Fig. 11–2). *Stipple board,* a heavy white paper with a bumpy surface, gives a roughly stippled effect when rubbed with a black crayon (as Fig. 15–1). Line-shading of flat surfaces should be

Fig. 11–2. Steps in spiral-technique for making even stipple patterns.

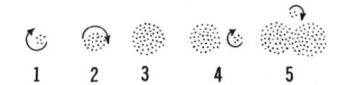

done with parallel, even lines, while the lines on curving or irregular surfaces should be made to curve back away from the observer. If a fine, supple pen point is used (as Gillott's 290), the lines can be pressed thick or thin to increase the effect of depth.

Delicately shaded drawings should be made on a hard white paper with a satin-smooth surface, as *Bristol board*. Carbon lamp-black can be rubbed on this paper to shade broad areas evenly, while a Wolff carbon pencil can be used for more intricate effects. Drawings can also be shaded with water washes and a brush, but this technique requires considerable practice. A machine eraser, razor, or china white poster paint can be used to touch out errors or add highlights.

Printed patterns. Machine-printed patterns (as Zip-a-tone and Craftint) are available in numerous patterns of lines and dots, as well as in geological patterns. They should be used wherever a smooth, flat effect is needed, as on geologic maps. Full directions for applying them can be obtained from suppliers. When used on maps, the patterns must be chosen so as to contrast suitably with one another, but must not be so bold or dark as to obscure other features.

Coloring illustrations. Colors can clarify complex illustrations greatly, and can be used with little added expense if only a few copies of a report are needed. Line-work, as for faults or other structures, should be done with colored inks or pencils. Formational areas on maps can be colored evenly by rubbing the long side of a colored pencil point in small circles to build the color value up to that desired. Large areas can be colored evenly by rubbing the pencil on a sheet of fine sandpaper and applying the colored powder to the map with a bit of folded cloth. Water colors cannot be used because they wrinkle all but the heaviest papers, but printer's inks, artist's oil colors, and some wax-base pencils can be applied lightly by dissolving them in benzine or any other suitably volatile solvent. These dilute stains should be applied with a brush or soft cloth and the excess fluid wiped off the surface at once. The resulting appearance is typically excellent, but the colors cannot be erased and some mixtures cannot be inked or penciled. It is therefore necessary to experiment with materials before using them on expensive copy.

Manufactured color sheets (as Zip-a-tone, Craftint) can be used to give a flat effect where only one copy of an illustration is needed. Many of these colors, however, are too bright or opaque for detailed geologic maps.

Lettering. Lettering guides and machines provide a means to do professional-appearing lettering, and full instructions come with these instruments. Long titles, legends, and similar items should be lettered first in pencil in order that they will be spaced correctly.

If an electric typewriter with suitable type-style is available, lettering that looks like printing can be prepared quickly. The typed copy is cut and attached to the illustration, and if special opaque tape or transparent sheets (as Copy-zip) are used, no glue or paste are needed. Prepared letters and words can also be obtained on a Zip-a-tone base. An advantage of this method is that spacing can be done perfectly and quickly.

Hand lettering must be used for fine characters, for words that must curve with streams or other features, or for special letter styles. Correct styles for lettering maps can be seen on any quadrangle map of the U.S. Geological Survey, and standard lettering procedures are given in all books on drafting and commercial drawing.

Enlarging and reducing. Enlargements and reductions that must be accurate and detailed should be done photographically. Illustrations that do not require precise details can be enlarged or reduced by first applying a square grid over the surface of the original copy and then transferring the drawing square by square to a larger or smaller-scale grid. Proportional dividers can be used to transfer details within each square. Camera lucidas, adjustable projectors, and pantographs can also be used to enlarge or reduce illustrations accurately.

11-10. Detailed Geologic Maps and Cross Sections

The first step in preparing a detailed geologic map is to determine the limitations imposed by methods of duplication or publishing. These limitations will apply mainly to the size (or scale) and the use of colors. Because over-sized maps are difficult to use in the field as well as at a table, maps and sections should be as small as clarity allows. If colors can be used, all or almost all of the data on field sheets can be shown clearly at one-half to one-third the field scale.

After the methods of duplication are decided upon, the field sheets should be studied to determine what will be included on the final map. A general rule is to show as much as possible, within the limits of legibility. Contacts may have to be simplified here and there, and closely spaced structure symbols averaged, but data must not be removed or averaged just to make a map look even. Readers know that observations can be more complete in some places than others, and

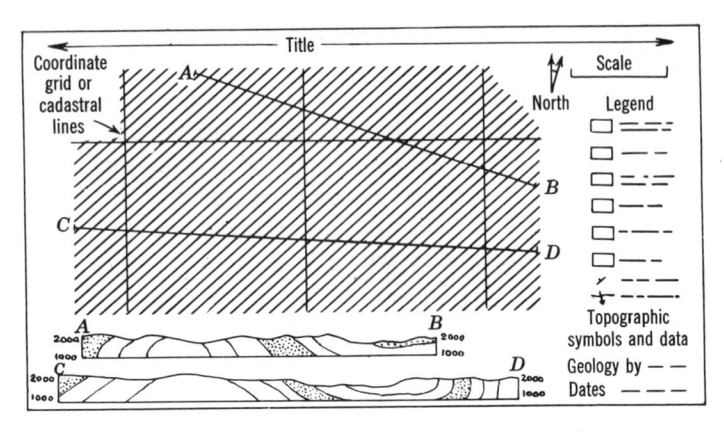

Fig. 11–3. Possible layout for a detailed geologic map and cross sections.

they will appreciate a full record where it can be given. When exceptional amounts of structural data have been plotted, as with several foliations and fracture systems, it may be necessary to prepare an accessory *structure map;* however, the possibility of showing these data in colored inks on the main map should first be investigated.

Composing the legend and layout. When the field sheets are examined during the planning stage, a list should be made of all rock units, structure symbols, and special culture symbols that will be shown in the legend of the final map. These items should be penciled out in order and to scale on scratch paper. The title, scale bars, north arrow, and other accessory items should also be penciled to scale. The field sheets should then be laid out so as to join, and the various penciled items arranged with them until an optimum balance and use of space are achieved. The final cross sections should be selected at this time so that they can be drawn on the same plate as the map (Fig. 11–3). The map should be oriented so that north is toward the top, unless it is costly or ungainly to do so.

Tracing paper can now be placed over the assembled parts and a pencil sketch made of the layout. This sketch serves to guide the remaining steps of the drafting.

Preparing drafting sheets. If more than one color will be used in printing the map, separate sheets must generally be prepared for each color separation. They must match exactly at all stages of drawing and printing. The separates should be cut from the same roll or stock of paper, linen, or acetate. They must be oriented so that their grain is parallel and must be left to season until measurements show they have reached an even regimen of scale-changes. They are then keyed to one another by small crosses or circles placed near their corners, and they are kept in the same room and in the same manner (flat, rolled, or hanging) until sent to the printer.

Plotting base grid. When map sheets have been seasoned suitably, a map grid exactly like that used on the base maps, plane table sheets, or photograph compilation should be plotted and drawn on each sheet with a sharp, hard pencil and a steel straightedge. In cases where the geology was mapped on a published base, a longitude-latitude grid may be traced directly from a fresh quadrangle sheet. In other cases a control grid must be constructed according to the procedures given in Section 7–11, and control points may be plotted as described in Section 8–5. The grid should not be inked on any of the sheets at this stage.

Drawing the map. The field sheets can now be taped to the drafting board one at a time, a map sheet oriented over them by reference to the grid or control points, and the various data traced in black ink. If opaque paper is used, the map may be traced over a light table with a low-temperature light source. The transfer can also be made with a reflecting projector (as the Saltzman), but this requires penciling the map first and inking it afterwards (and checking carefully against the original as the inking is done). A great advantage of the projector method is that the map can be reduced at this stage.

Where copy is being prepared for color printing, one separation must

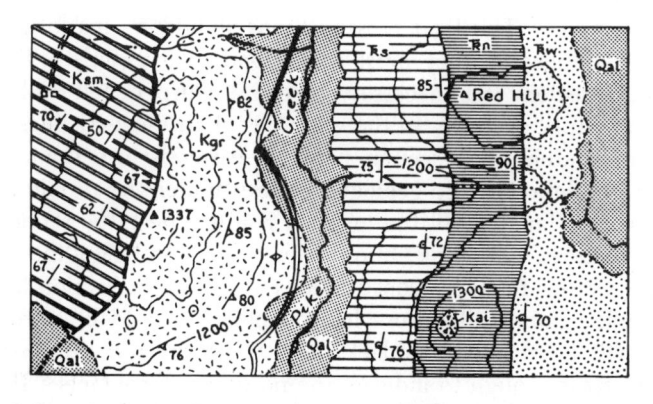

Fig. 11–4. Use of line weight and patterns to show various kinds of features and data in black ink.

be made for each color that will be used, and only the items that will appear in that color are drafted on that particular sheet. For example, a separate sheet might be made for drainage (to be printed blue), another for contours (to be printed brown), and so forth.

When all features and patterns will be printed in one color, widths and styles of lines must be selected so that the final copy will be clear. The patterns applied to rock units will indicate which lines are contacts and which are contours (Fig. 11-4). The order of inking is important if erasures are to be kept to a minimum because many features may cross or overlap one another. The following order may be used: (1) structure symbols without numbers, (2) faults, (3) contacts, (4) cross-section lines, (5) numbers for structure symbols, (6) culture, (7) drainage, and (8) contours. Geographic names should be penciled in at an early stage lest they be forgotten. They can be shifted as necessary so as not to cover important features, and they should be inked after the contours. The grid should be inked next, but if the map is crowded, it should be supplanted by short lines at the margins of the map. Letter symbols for rock units should be inked last. The order of inking implies that later items are broken to accommodate earlier ones, as shown in Fig. 11-4.

Most maps are completed by inking the title, legend, scale bars (scale bars should be used whether or not a fractional scale is given), north arrow, and other accessory items. Prepared patterns (Zip-a-tone, Craftint) should be applied after the map surface has been cleaned with a soft eraser and brushed thoroughly. Great care must be used not to thin the ink work by erasing, particularly where lines will be covered by applied patterns.

Cross sections. The final cross sections should be made after the map is inked because they must correspond to it exactly. To make an accurate section it is necessary to use a sharp pencil and to key the section paper precisely to the end ticks of the section lines on the map. Transparent cross-section paper is easier to use than opaque paper. Completion of the geologic features under the profile line requires a careful analysis of the mapped features near the section line, for many features will project into the line of section below the surface. Some suggestions for these constructions are given in Section 3-8.

Lithologic symbols like those in Appendix 5 should be used when sections will be printed in black only. Patterns for adjacent units should contrast enough to indicate the positions of contacts. If colors are used on the map, the cross sections may also be colored, but the inked lithology should then be simplified. Unit symbols, fault arrows,

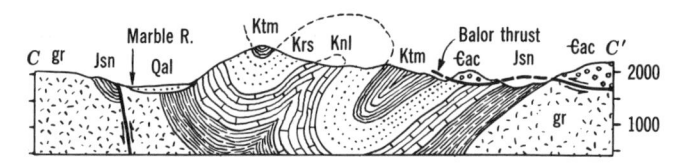

Fig. 11-5. Cross section, showing accessory symbols.

and other details, as shown in Fig. 11-5, should be added to complete the sections. The projected lines above the section should be used only where needed to clarify a well-controlled structure.

Elevations should be shown at the ends of the section. If the sections are assembled on a separate plate from the map, they must bear a scale bar, complete title, and a legend or a clear reference to the legend of the map.

11-11. Stratigraphic Illustrations

Stratigraphic data are generally presented in detailed columnar sections that may represent either single measured sections or an average sequence for a given area or region. The basic procedure for constructing columnar sections is given in Section 3-8. The drawing is made at the smallest scale that will show the units or beds considered pertinent. Scales will typically range from 1 in. = 100 ft to 1 in. = 10 ft.

The formal units (formations, and so on) or the principal unnamed divisions should be blocked out first after computing true thicknesses from the survey notes or maps (Section 12-8), then the lithologic details are added by starting at the top and working down. If the section was measured from the base up, this must be done with care in order not to transpose data. In cases where the field measurements were made in far greater detail than can be used on the column, the field notes should first be transcribed into a sequence of somewhat generalized units that can be drawn to scale. Lithologic descriptions should be added to most detailed sections, but lithologic symbols may be adequate where several small-scale sections are used on one plate.

Serial sections and panel diagrams. If a number of stratigraphic sections have been measured, they can be plotted together to illustrate lateral variations in thickness and lithology. If the sections are located along a simple linear feature, as the outcrop of a homocline, they can be plotted in a geographic series, spaced in proportion to the distances

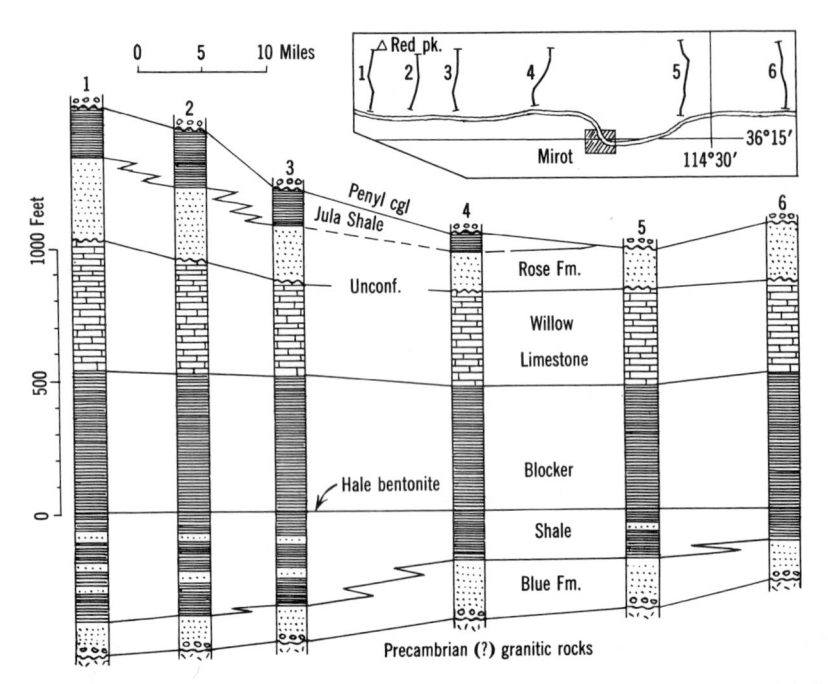

Fig. 11-6. Serial sections, with traverse lines shown on location map. Simplified greatly compared to most full-scale illustrations.

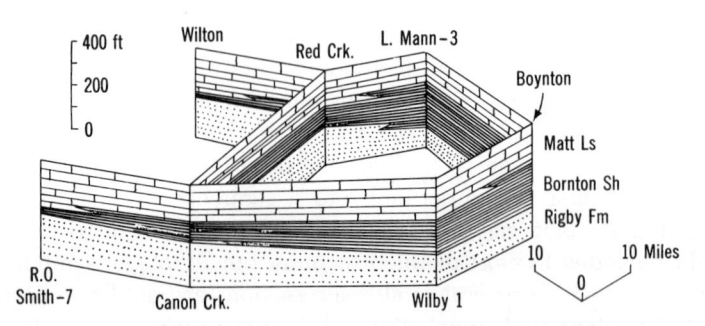

Fig. 11-7. Panel diagram based on drill holes. Simplified greatly compared to most full-scale illustrations.

between traverses (Fig. 11–6). They should be assembled on the drawing on the basis of one horizon (a stratigraphic time-plane without vertical dimensions), and contact lines should be projected between them to show the continuity of units.

If measured sections are scattered geographically, they are commonly plotted as columns on an isometric projection of the area. The continuity of the units can be shown by drawing continuous planar sections between adjacent columns, to make what is called a *panel* (or *fence*) *diagram* (Fig. 11–7). All measurements parallel to the vertical axis must be to scale, as required by the isometric projection.

Where subsurface data are abundant, it may be possible to construct structure maps and isopachous maps of certain units or contacts, and if a survey has covered a large area, facies maps may also be constructed from petrologic and fossil data (see, for example, Bishop, 1960).

References Cited

American Geological Institute, 1957, *Glossary of geology and related sciences, a cooperative project:* Washington, D.C., American Geological Institute, 325 pp. (with supplement, 1960).

Bishop, M. S., 1960, *Subsurface mapping:* New York, John Wiley and Sons, 198 pp.

Hoelscher, R. P., and C. H. Springer, 1956, *Engineering drawing and geometry:* New York, John Wiley and Sons, 520 pp.

Lobeck, A. K., 1958, *Block diagrams and other graphic methods used in geology and geography*, 2nd ed.: Amherst, Mass., Emerson-Trussell Book Co., 212 pp.

Perrin, P. G., 1950, *Writer's guide and index to English:* Chicago, Scott, Foresman, 833 pp.

Ridgway, J. L., 1920, *The preparation of illustrations for reports of the United States Geological Survey:* Washington, Government Printing Office, 101 pp.

Roget, P. M., 1946, 1958, *Roget's international thesaurus:* New York, Thomas Y. Crowell Co., 1194 pp. (earlier editions published under title *Thesaurus of English words and phrases*).

Schenk, E. T., and J. H. McMasters, 1956, *Procedure in taxonomy*, 3rd ed. (revised by A. M. Keen and S. W. Muller): Stanford, Stanford University Press, 119 pp.

U.S. Geological Survey, 1958, *Suggestions to authors of the reports of the United States Geological Survey*, 5th ed.: Washington, D.C., 255 pp.

Wilmarth, M. G., 1938, *Lexicon of geologic names of the United States (including Alaska):* U.S. Geological Survey, Bulletin 896, pts. 1 and 2, 2396 pp.

Wilson, Druid, G. C. Keroher, and B. E. Hansen, 1959, *Index to the geologic names of North America:* U.S. Geological Survey, Bulletin 1056-B, 622 pp.

Wilson, Druid, W. J. Sando, and R. W. Kopf, R. W., 1957, *Geologic names of North America introduced in 1936–1955:* U.S. Geological Survey, Bulletin 1056-A, 405 pp.

12

Field Work with Sedimentary Rocks

12–1. Interpreting Sedimentary Rocks

Although many problems relating to sedimentary rocks can be solved directly by mapping formations, some require additional studies. Systematic observations on textures, compositions, and small-scale structures are generally essential to interpreting sedimentary processes and events. The compositions and shapes of the grains of a sandstone, for example, can indicate not only the kinds of rocks from which it was derived but also how much they were weathered and how rapidly they were eroded. Cross-bedding and linear structures in the sandstone can record the direction as well as the nature of the currents that deposited the grains, while the thicknesses and textures of individual beds can cast light on the rate of sedimentation and burial. Fossils in the sandstone may give a measure of its age and the conditions under which it was deposited. Finally, the distribution of cements and the alterations of minerals along beds and fractures can provide evidence of important post-depositional processes.

Studies of these various features must be as systematic and quantitative as possible. In general, the following requirements must be satisfied:

1. The stratigraphic frame of reference for measurements must be well understood.
2. Rock names and descriptions should be quantitative and based as much as possible on characteristics that have genetic meaning.
3. Small structures must be observed to determine conditions of sedimentation as well as stratigraphic sequence.
4. Methods of sampling and measuring must be precise enough to permit accurate correlations of data.

If deformation is extreme, especially where rocks are strongly lithified, the principles and methods described in Sections 15–5 to 15–11 may also prove useful.

A single comprehensive study is generally preferred to several partial ones, for some data can be interpreted only in the light of others. Many projects, however, must be limited in scope, and must therefore be organized carefully in order to meet their special purposes effectively. General geologic mapping will provide the best means for selecting and planning detailed or specialized studies. If such mapping cannot be undertaken, the area of interest should be reconnoitered thoroughly, and, if possible, trial studies should be made in several subareas.

Sedimentary petrology is a rapidly expanding field, and the reader is urged to consult not only books such as those by Krumbein and Sloss (1951), Pettijohn (1957), and Dunbar and Rodgers (1957), but also recent issues of geological journals.

12–2. Lithologic and Time-Stratigraphic Units

Field work with sedimentary rocks requires knowledge of two kinds of stratigraphic units: lithologic units, which are defined on the basis of physical characteristics, and time-stratigraphic units, which are generally defined on the basis of the fossils they contain. Suggestions for classifying each kind of unit are presented briefly in the paragraphs that follow.

Lithologic units. The general nature and use of lithologic units (also called *rock-stratigraphic units* or simply rock units) are described in Section 4–6. These units are considered *formal* when they have been named and defined according to the rules of the Stratigraphic Code and have been allocated to one of three ranks of units. The *formation* is the rank of lithologic unit used most widely. When a region is first mapped, or when obsolete mapping is revised, lithologic units are set up as formations wherever possible. Formations must be cartographic and therefore thick enough to plot accurately at map scales of about 1:24,000. A *group* is a lithologic unit consisting of two or more superjacent formations. Upon detailed mapping, a heterogeneous formation may be made into a group by dividing it into new formations.

Irregular or heterogeneous formations can also be divided into lithologic units of lesser rank. These units are called *members* if they have wide lateral extent, *lentils* if they are lens shaped, and *tongues* if they are wedge-shaped projections from the main body of the formation. These smaller units need not be cartographic. It is occasionally useful to map a distinctive *bed* or *lens* within a formation or member (see *key beds*, Section 4–6); however, these units are not commonly formalized by geographic names. Single beds may also be used for mapping the contacts of some formations.

Rules for naming and describing units must be considered early in field work. A summary of the principal rules is given below; however, the full code of the American Commission on Stratigraphic Nomenclature (1961) should be consulted before a new unit is described in the literature.

1. The name and definition of the unit must appear in a scientific publication that is widely known and available to geologists.
2. A statement that the unit is new must accompany the definition.
3. The definition must include a full description of the unit (Section 11-2).
4. The definition must cite a *type locality* where the rocks are well developed and well exposed.
5. The name must consist of a geographic name and a term for the dominant lithology (as Dakota Sandstone); or, if the lithology cannot be expressed in a single word, the term for the rank of the unit is used (as Tobin Formation).
6. The geographic name should be that of a permanent feature at or near the type locality and must not duplicate a name used for another unit.

A detailed *type section* is often described for the type locality, but this should be omitted if the unit has such lateral variations that a single detailed section would be misleading. It is often more useful to include a geologic map showing the continuity of the unit over a large area.

Established units should not be changed in rank or redefined without thorough field studies because each change in nomenclature makes the published record more difficult to use. Sometimes, however, it may be necessary to subdivide old, heterogeneous units into new formations or to redefine the contacts or critical features of formations that were defined loosely. When these changes are considerable, it may be best to abandon old names and use new ones.

Formal names should not be used for rocks that are poorly exposed or are found as fragments of complete formations. The contacts of these rocks may, for example, be faults or may be covered by surficial deposits. Such units may be useful in mapping, but if named formally, they will probably require subsequent redefinition. Their usefulness is in no way impaired by informal names, as *shale of Bradford Creek, Cretaceous rocks of Mill Valley, Cooper Point beds.*

Time-statigraphic units. A time-stratigraphic unit is a sequence of strata that accumulated during a particular interval of geologic time. The interval is defined and generally recognized by the fossils the strata contain. Table 12-1 shows the equivalence between the ranks

Table 12-1. Equivalence Between Time and Time-Stratigraphic Terms

Time Interval	Time-Stratigraphic Unit
1. Period (as *Triassic Period*)	System (as *Triassic System*)
2. Epoch (as *Late Triassic Epoch*)	Series (as *Upper Triassic Series*)
3. Age (as *Norian Age*)	Stage (as *Norian Stage*)

of geologic time terms and the ranks of time-stratigraphic units. The *Stage*, the smallest unit used widely for regional and interregional correlations, is established by a careful faunal study of a stratigraphically continuous, well-exposed sequence of beds. The stage is named after the locality where the initial study was made. Many stages used in America are named after local rock-stratigraphic sequences, but European stages are being used more and more for American as well as for world-wide correlations. Exceptionally thick and fossiliferous rocks can often be subdivided locally by detailed stage classifications; in the Tertiary of the Pacific Coast region, for example, a sequence of local stages has been based on microfossils (Kleinpell, 1938; Mallory, 1959).

Faunizones are generally the smallest units used in a time-stratigraphic sense. They are considered by some to be purely *biostratigraphic* units because they do not have a usable time value. They are, however, the basis on which most stages are delineated; thus they comprise a very important part of time stratigraphy. The kind of zone used most widely is based on the range of an assemblage of fossils. The zone is named after a distinctive fossil of this assemblage, but not necessarily one that is restricted to the zone (as the *Bulimina corrugata Zone*). Conditions of deposition limit the use of a given faunizone; zones based on pelagic foraminifera, for example, cannot be used to classify nonmarine rocks.

The methods of establishing time-stratigraphic units have been described briefly by Schenck and Muller (1941, p. 1424):

> First, the stratigrapher carefully studies a continuous section of similar facies; collections of fossils are carefully allocated in this section. Second, the species are identified, and their stratigraphic ranges in this section are accurately determined. Third, these ranges are analyzed so as to show a certain grouping of the strata; these are the arbitrarily delimited, provisional, time-stratigraphic units. Each such unit contains species or genera restricted to it; some fossils have their lowest strati-

graphic occurrences in it; and some may occur above and below as well as in the unit. Other species which occur first in the immediately overlying beds range no higher than the superjacent unit. Fourth, these time-stratigraphic units are tested by determining a similar distribution of fossils at some other more or less distant locality. Repeated testing proves the validity of the units for the entire geologic province. The sequence of time-stratigraphic units established on such a basis would serve as a yardstick, and its application to more distant areas may be justified if supported by valid paleontologic evidence.

With regard to field procedures, the following points should be noted especially:

1. *The sequence of strata should be continuous.* Unconformities and faults must be detected by mapping around each provisional section.

2. *Fossils must be located accurately.* The precision of measurements must be well within the limits of the faunal changes used.

3. *The rocks must be of one facies.* Lithologic studies must determine whether physical conditions were uniform; reworked materials should be noted with care so that fossils of two ages will not be mixed.

Relations between lithologic and time-stratigraphic units. The time interval represented by a vertical section through a formation may vary considerably from place to place. Even where the interval is constant over a large area, the contacts of the formation are not likely to coincide with the boundaries of standard stages or series. Lithologic and time-stratigraphic units cannot, therefore, be used interchangeably. Where fossils are so abundant that ages of rocks can be determined at many outcrops, it may be possible to plot lines showing boundaries between time-stratigraphic units. These lines must, however, be distinguished clearly from contacts between lithologic units.

In some regions, extensive "formations" or "groups" may have been defined on the basis of age, or "series" or "stages" may have been set up and traced on the basis of unconformities or lithologic changes that are not necessarily time horizons. It is then necessary to sort out and understand the existing nomenclature before beginning a new field study.

12–3. Naming and Describing Sedimentary Rocks

Because of the history of sedimentary petrology, there is no single method for naming rocks with descriptive terms that also have genetic meaning. Some names, as *limestone,* have only a compositional meaning; others, as *sandstone,* have only a textural meaning. Most sedimentary rocks, however, can be given names that are consistent and meaningful within any one rock group. Several such systems are presented briefly in this section. Not all the terms used are universally standard, and therefore field notes should include *actual quantities* that express textural and compositional characteristics. The basic characteristics are: grain size, degree of sorting of detrital grains, shapes of grains (especially degree of rounding by abrasion), fabric of shaped grains, porosity, color, composition of the detrital fraction, nature and amounts of authigenic materials (as cements), and degree of recrystallization or replacement.

Grain size. The primary basis for naming detrital rocks is grain size. The size scale proposed by Wentworth, used widely in America, is the basis for the classification shown in Table 12–2. The numbers in the table are the limiting sizes of the various size classes. The pebble and sand sizes are divided into equal classes to facilitate determining the degree of sorting, as explained below. Pebbles can be measured with a scale, while sand sizes are best determined by comparing a given rock to a sand-size card. These cards can be made by

Table 12–2. Classification of Detrital Sediments by Grain Diameter

Exact Size Limits	Approximate Inch Equivalents			Name of Loose Aggregate
>256 mm		>10	in.	Boulder gravel
64–256 mm	2.5	–10	in.	Cobble gravel
32– 64 mm	1.2	– 2.5	in.	Very coarse pebble gravel
16– 32 mm	0.6	– 1.2	in.	Coarse pebble gravel
8– 16 mm	0.3	– 0.6	in.	Medium pebble gravel
4– 8 mm	0.15	– 0.3	in.	Fine pebble gravel
2– 4 mm	0.08	– 0.15	in.	Granule (or very fine pebble) gravel
1– 2 mm	0.04	– 0.08	in.	Very coarse sand
½– 1 mm	0.02	– 0.04	in.	Coarse sand
¼– ½ mm	0.01	– 0.02	in.	Medium sand
⅛– ¼ mm	0.005	– 0.01	in.	Fine sand
1/16– ⅛ mm	0.002	– 0.005	in.	Very fine sand
1/256– 1/16 mm	0.00015–	0.002	in.	Silt
<1/256 mm		<0.00015	in.	Clay (clay-size materials)

sieving sand into the standard Wentworth divisions and gluing a pinch of each size fraction to a card graduated as Table 12–2.

Degree of sorting. Degree of sorting measures the thoroughness with which a sediment has been winnowed or reworked by transporting agents. It is a valuable indicator of rate and environment of deposition, and it has been used to classify sandstones. Size sorting expresses the degree to which the grains approach being the same size. This is the measure of sorting used most widely; however, it should be remembered that small grains of heavy minerals are sorted out with larger grains of light minerals.

To determine the degree of sorting in the field, the rock should be examined on freshly broken surfaces and on weathered (but clean) ones. It should be examined both parallel to and at right angles to bedding. As this is done, an estimate is made of the range of grain sizes that includes the great bulk (here 80 percent) of the detrital materials. For sandstones, a sand-size card (see *Grain size,* above), can be used to estimate the lower size limit of the coarsest 10 percent of the rock and the upper size limit of the finest 10 percent of the rock. These limits may then be compared with the size scale (Table 12–2) to determine the number of size classes between them. This number gives a measure of the degree of sorting, and it may be recorded in the notes or converted to sorting terms such as those suggested in Fig. 12–1. This figure should be used for rocks consisting mainly of sand or small pebbles; finer-grained materials cannot be classified precisely in the field, and coarse conglomerates require special considerations.

By far the greatest errors in estimating degree of sorting occur when dark sand grains are mis-identified as very fine-grained matrix mate-

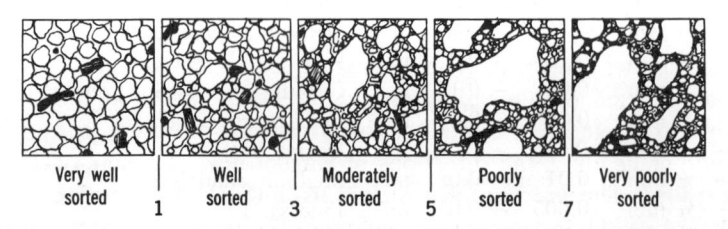

Very well sorted	Well sorted	Moderately sorted	Poorly sorted	Very poorly sorted
1	3	5	7	

Fig. 12–1. Terms for degrees of sorting. The numbers indicate the number of size-classes included by the great bulk (80 percent) of the material. The drawings represent sandstones as seen with a hand lens. Silt and clay-size materials are shown diagrammatically by the fine stipple.

Field Work with Sedimentary Rocks

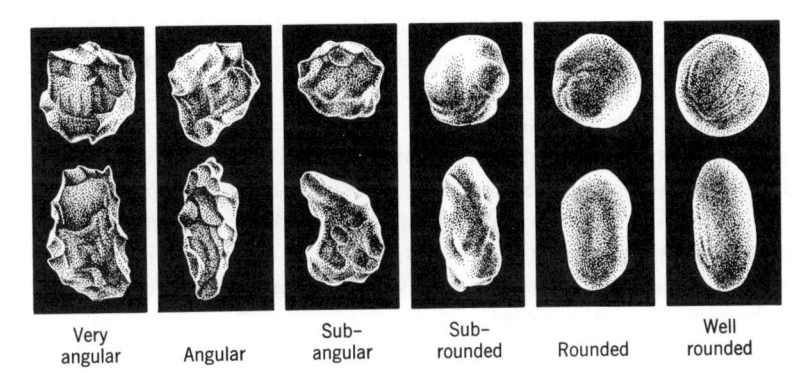

| Very angular | Angular | Sub-angular | Sub-rounded | Rounded | Well rounded |

Fig. 12–2. Terms for degree of rounding of grains as seen with a hand lens. After Powers, M. C., 1953, *Journal of Sedimentary Petrology*, v. 23, p. 118. Courtesy of the Society of Economic Paleontologists and Mineralogists.

rials, especially where grains have recrystallized. An examination of associated rocks can often resolve this difficulty (coarse-grained laminae may show the grains more clearly); however, petrographic work may be needed.

Rounding. Abrasion or rounding of grains is used to distinguish *conglomerates* from *breccias*, in which gravel-size grains are angular. Unlithified aggregates of angular gravel-size clasts are commonly called *rubble*. Generally, the degree of rounding in finer-grained rocks is not used to name them; however, it should be estimated and included in descriptions of sandstones. Rocks that contain both angular and well-rounded grains are of particular interest because they may have been derived in part from former sedimentary rocks. The grain models shown in Fig. 12–2 can be used to estimate degrees of grain rounding.

Fabric. Platy and linear grains may be oriented so as to give a directional fabric. The fabric is *planar* where platy grains are more or less parallel and *linear* where elongate grains are parallel. Fabrics can affect such directional properties as permeability to fluids and response to deforming forces. Fabrics can also be used to estimate attitude of bedding and directions of current flow (Section 12–4).

Porosity. The *porosity* of a rock is its amount of pore space, generally expressed in percent by volume. Permeability to the transmission of fluids depends on the porosity and on the size and interconnected nature of pores; consequently, these properties should be described as completely as possible in field notes. The origin of the larger pores can often be determined by hand lens study. In calcareous rocks, for example, some pores can be seen to be irregular channels formed by fracturing, leaching, or alterations (as dolomitization); other pores may be partly filled openings in fossils. Porosity in detrital rocks may depend on sorting and the degree to which grains have been packed and cemented. Porosities are commonly increased or decreased greatly by weathering. Fresh outcrop samples can be used to estimate porosity at depth; however, the freshest rocks at a given exposure may be misleading since they are generally those with the lowest original porosity.

Color. Color can be used effectively in mapping and correlations if its causes are understood. Rock color depends not only on the color and grain size of the original sediments but also on their cementing materials, the degree to which they have recrystallized, and, especially, the degree to which they are weathered. The secondary origin of colors can often be detected because the pigmenting minerals occur along fractures, cut across bedding, or are related to the surface of the ground. Even moderate weathering may change colors so much that fresh colors can be seen only in deep cuts, tunnels, or well cuttings.

Color terms should be standardized by comparing rocks to color charts, such as the rock-color charts that can be purchased from The Geological Society of America, or the soil-color charts of the Munsell Color Company (see *Soils,* Section 12–11). Colors for both dry and wet samples should be recorded.

Naming mixtures of materials. Detrital rocks made up of grains of various sizes derive their name from the size class that is most abundant (except for certain conglomerates and breccias). The names in Fig. 12–3 may be used for various mixtures of detrital and nondetrital materials. Mixtures containing substances not shown in the diagram can be named in a similar way, as *ferruginous sandstone, gypsiferous claystone,* and so on.

Sandstones. Rocks consisting mainly of sand-sized grains are especially useful in field studies because their textures and compositions can be determined readily. Furthermore, an average hand specimen is ample for petrographic and permeability studies. Names for sandstones should be based both on degree of sorting and on composition, as in the system proposed by Gilbert (Williams, Turner, and Gilbert, 1954). In this system, the name *arenite* is used for relatively well-sorted sandstones, and the name *wacke* for more poorly sorted sandstones (Fig. 12–3). These rocks may be classified further on the basis

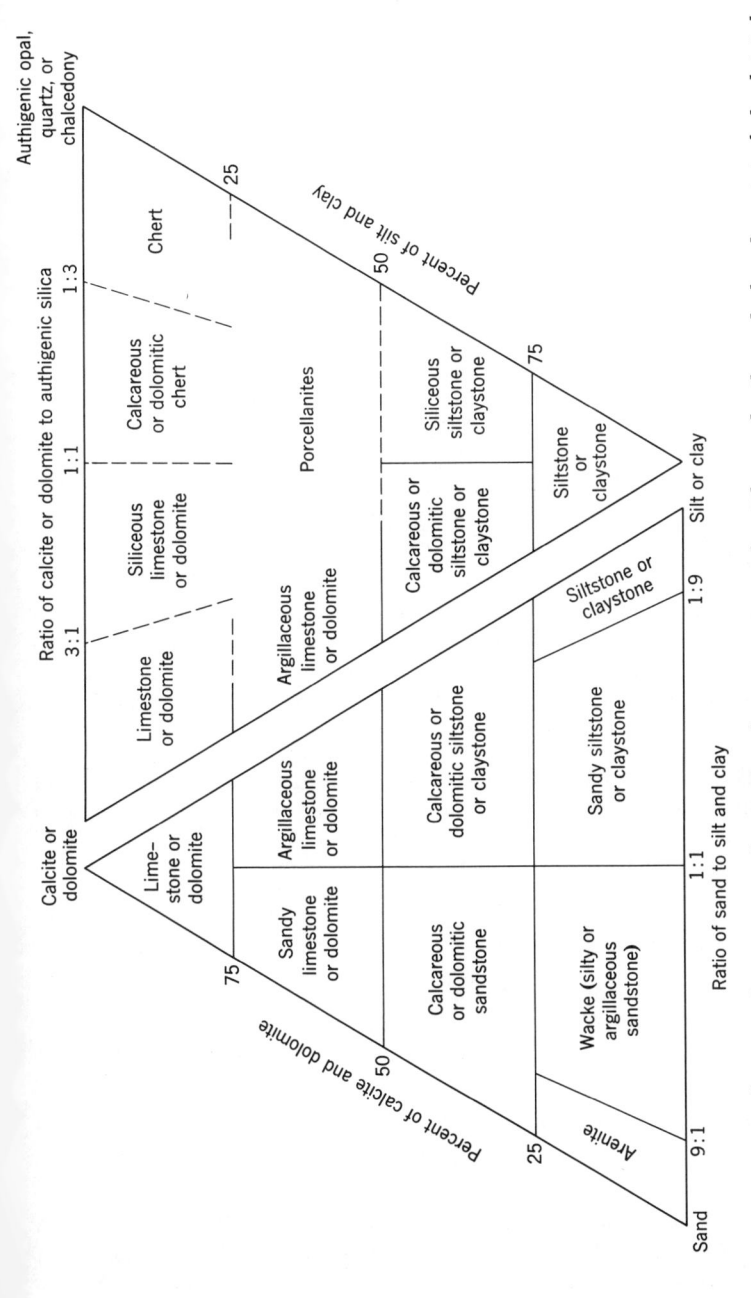

Fig. 12-3. Rock names for mixtures of various sedimentary materials. The sand, silt, and clay do not include detrital grains of calcite or dolomite. Mainly after Gilbert (in Williams, Howel, F. J. Turner, and C. M. Gilbert, 1954, *Petrography*: San Francisco, W. H. Freeman and Co., p. 270).

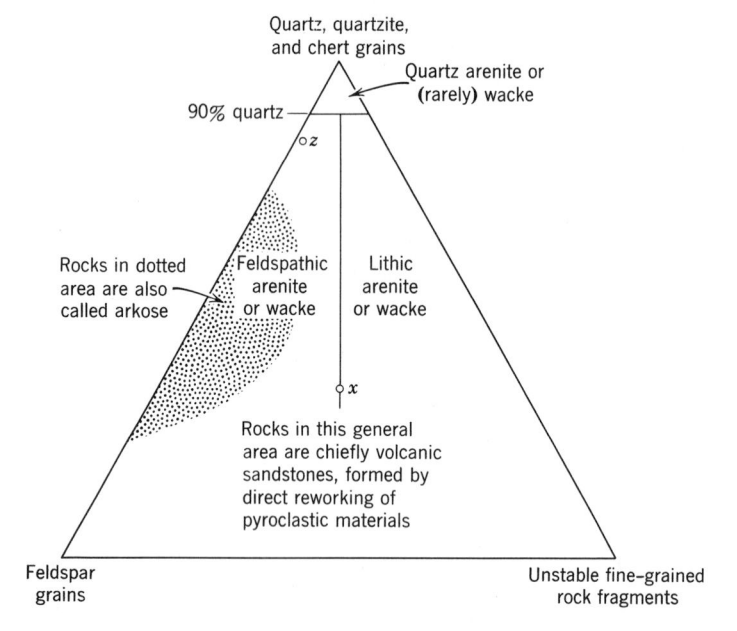

Fig. 12–4. Names for various sandstones, based on composition. The name for a rock can be found by (1) determining the amounts of the various mineral and rock grains in it, (2) summing these amounts into the three groups shown at the corners of the triangle, and (3) using the proportions between the three groups to estimate a point in the triangle. A poorly sorted sandstone with equal amounts of quartz, feldspar, and slate grains, for example, would fall at point x (a lithic, feldspathic wacke). A well-sorted sandstone whose grains are all quartz except for 15 percent of feldspar and a few percent of lithic grains would plot at point z (a feldspathic arenite). If some other mineral forms more than 10 percent of the sand, adjectives such as *biotitic* should be used. From a diagram supplied by C. M. Gilbert (see also Williams, Turner, and Gilbert, 1954, pp. 292–293).

of the composition of the detrital grains (Fig. 12–4). Sandstones in which calcitic sand grains are abundant will be considered in the section on limestones. The term *graywacke* is most suitable for strongly indurated dark-colored wackes.

Fine-grained detrital rocks. Although all fine-grained detrital rocks are often called *shale* in the field, it is possible to classify them more usefully. The basic size terms, *siltstone* and *claystone,* can be applied with confidence to most well-sorted rocks. Well-sorted siltstones can be identified with a hand lens or by their gritty feel (especially when rubbed against the teeth). Claystones, on the other hand, are aphanitic to waxlike. They can be cut or scraped smoothly with a knife and have a smooth, soapy feel when wet. Nearly equal mixtures of silt and clay may be difficult to identify without a microscope.

The term *shale* (as, *silty shale, clay shale*) is best reserved for rocks with prominent bedding cleavage (*fissility*). Many shales also show thin laminations. *Mudstone* is a useful field name for mixtures of silt and clay that have a blocky or spheroidal fracture and are typically unlaminated. Finally, *argillites* are highly indurated (generally recrystallized) claystones and siltstones that break into hard, angular fragments. They can be distinguished from calcareous or dolomitic mudstones because they do not react and crumble in acids.

Conglomerates and breccias. Field studies of conglomerates and breccias must be especially thorough, for these rocks are generally too coarse to sample for laboratory analysis. Mixtures of gravel size and finer materials may be named as shown in Fig. 12–5. Where the fine-grained materials of the matrix make up about 25 percent or less of the rock, the larger fragments, if fairly well sized, will touch one another. The concept of a well-sorted conglomerate is therefore somewhat different from that of a well-sorted sandstone, where the matrix

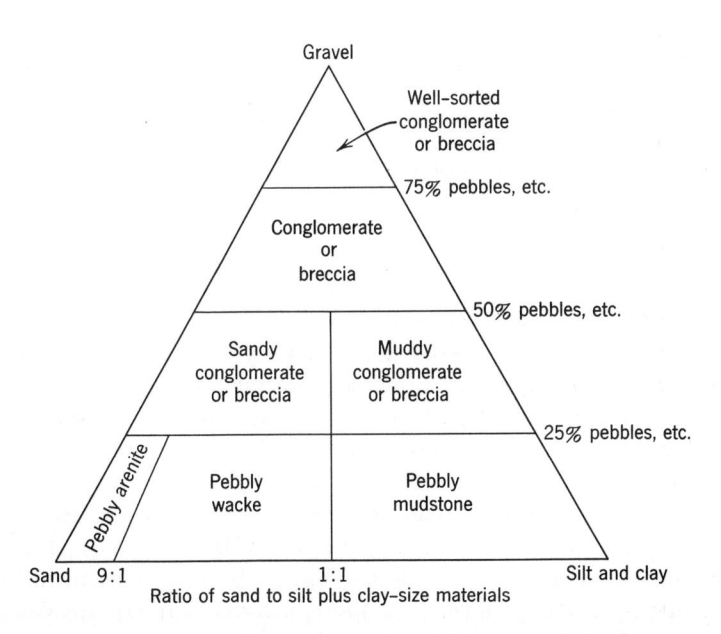

Fig. 12–5. Names for sedimentary rocks containing gravel-size fragments.

may be far less abundant. Both the matrix and the gravel-size fragments should be described. Most pebbles can be identified as rocks, and therefore conglomerates are particularly valuable in determining sources of sediments. The large fragments should be tabulated by systematic counts, as by marking off an area on an outcrop and identifying all the pebbles and cobbles that fall within its perimeter. Fragments that are physically weak (as claystones) or chemically unstable (as limestones) can provide evidence as to the nature of the transporting agent or the amount of transport. Surface effects such as polish, scratches, sandblast fluting, chemical corrosion, or abraded relics of old weathering-rinds can be used to interpret conditions of transport and deposition. Mixtures of well-rounded and angular fragments, or presence of broken rounded pebbles, may indicate reworking of older conglomerates. Conglomerates and breccias can form in many ways, and accounts such as those by Pettijohn (1957) or Dunbar and Rodgers (1957) should be read before studying these rocks.

Limestones and related rocks. Although they are commonly recrystallized or replaced to varying degrees, limestones should be named as much as possible on the basis of primary textures. This can be done by examining many samples with a hand lens, both on freshly broken and on slightly etched, weathered surfaces. The textural constituents of the rock should first be classified into the following groups:

1. Abraded biogenic grains, limestone fragments, and accretional grains (oöids, pisoliths).
2. Unabraded fossils and fecal pellets.
3. Fine-grained carbonates (originally calcitic or aragonitic mud).
4. Noncalcareous detrital grains.
5. Cements (especially phanerocrystalline calcite).

Materials that have been worked and deposited by currents will fall dominantly in groups 1, 3, and 4, and rocks consisting of these materials can be named on the basis of their size sorting (Table 12–3). In addition to their textural characters, these rocks may show laminations or cross-bedding. They are commonly interbedded or intermixed with size-sorted, noncalcareous rocks, and if the limestones have recrystallized greatly, this relation may provide the only means of determining their origin. Limestones with gravel-size clasts can be called limestone conglomerates or calcirudites, but if the large clasts are fossils, such names as *shelly calcarenite* and *spirifer calcarenite* are generally more useful.

Table 12–3. Names for Detrital Limestones, Based on Degree of Sorting

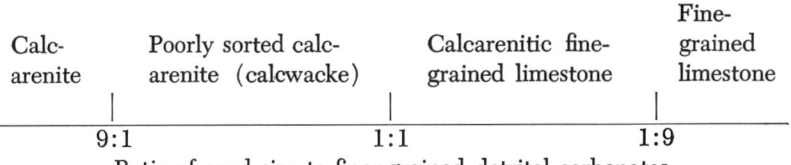

Calc-arenite	Poorly sorted calc-arenite (calcwacke)	Calcarenitic fine-grained limestone	Fine-grained limestone
|	|	|	|
9:1		1:1	1:9

Ratio of sand size to finer-grained detrital carbonates

Limestones consisting essentially of groups 2 and 3 have probably accumulated after little transport, and their degree of sorting is of little consequence compared to the nature of the fossils they contain. Where shelly fossils dominate, the general term *coquinite* may be used, and it can be modified by fossil names, as *spirifer coquinite*. The name *pelletite* is useful for rocks consisting mainly of pellets, but these structures are generally too small to be seen without a microscope.

Calcite is by far the most common cement, and it typically forms rather coarsely crystalline aggregates that are so transparent they appear dark on smooth surfaces. This cement is especially abundant in well-sorted detrital limestones, accumulations of fossils with hollow parts, and in openings formed by brecciation or leaching. It is an important indicator of the original porosity and therefore of the potential porosity of a rock. Other cements and authigenic minerals may be designated by adjectives such as *siliceous* and *gypsiferous* (Fig. 12–3).

The degree of recrystallization and dolomitization should be recorded in the notes and plotted on maps. Alteration can thereby be related to original sedimentary structures or to fracture systems and igneous bodies. Altered rocks should be classified primarily on their original characteristics (as *dolomitized calcarenite*), but if primary features are eradicated over large areas, they may be classified on the basis of color, sizes, and shapes of secondary grains, or relative amounts of dolomite and calcite (Table 12–4).

Table 12–4. Classification of Limestones and Dolomites, by Composition

Limestone	Dolomitic limestone	Calcitic dolomite	Dolomite
|	|	|	|
9:1		1:1	1:9

Ratio between the minerals calcite and dolomite

12–4. Beds and Related Structures

The layered or bedded nature of sedimentary rocks is produced by differences in texture, composition, or color of the original sediments. These differences are caused by changes in conditions at the site of deposition or at the source from which the sediments were derived. Bedding features are therefore of great value in interpreting the origin of sedimentary sequences. They may also be the most useful single basis for tracing or correlating formations from one place to another.

Bedding provides a valuable structural datum because it is parallel to the surface on which the constituent grains accumulated. This surface, however, need not necessarily have been horizontal or planar. Furthermore, although bedded rocks commonly cleave parallel to their bedding, the platy fragments may be either thicker or thinner than the beds themselves. Prominent secondary cleavage can thus be mistaken for bedding. Secondary features such as rhythmic stains or cemented layers can also be more apparent than primary bedding.

Before bedded rocks can be described at the outcrop it is necessary to determine what constitutes a single bed. Genetically, a bed represents a depositional episode during which conditions were relatively uniform. Practically, a bed is a layer that is sufficiently distinct from adjoining layers so that it can be measured and described. Where each bed is nearly homogeneous or grades evenly from one extreme at the base to another at the top, the choice of beds presents no problem. Where thicker beds contain thinner beds or laminations, it is

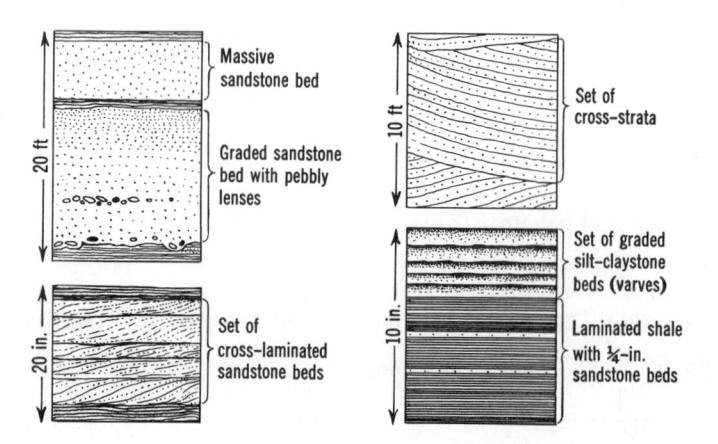

Fig. 12–6. Various beds and sets of beds.

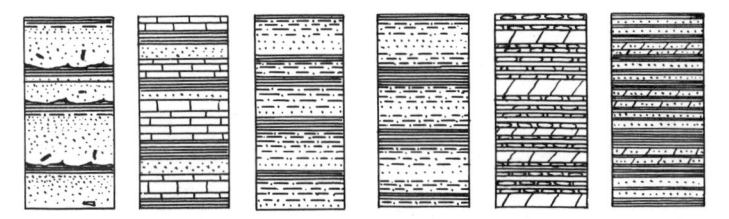

Fig. 12–7. Three cyclic repetitions (left) and three rhythmic repetitions in bedded rocks. The lithologic symbols are explained in Appendix 5.

necessary to determine which units have the greatest genetic and practical value. Commonly, the thinner beds are subordinate features, caused by moderate fluctuations in the conditions that produced the thicker, more distinctive layers. Less commonly, the thinner layers are as clear and genetically significant as the thicker composite layers. McKee and Weir (1953) suggested the useful term *set* for these groups of closely matched thinner beds. Figure 12–6 shows examples of simple and complex beds and sets of beds.

Repeated sequences of beds. Repeated occurrences of certain kinds of beds or sequences of beds give some sedimentary units their most striking character. These repetitions imply cyclic or rhythmic changes in sedimentation processes or conditions and should therefore be studied and described. Cyclic repetitions have a sequential order such as A, B, C, A, B, C, etc., while rhythmic repetitions have a sequential order such as A, B, C, B, A, B, C, B, and so forth (Fig. 12–7). Although complex cycles and rhythms are rarely perfect, they are uniform enough in some units to have genetic significance.

Shapes of individual beds. Where exposures are adequate, beds can be classified descriptively as *tabular, lenticular, linear, wedge shaped,* or *irregular.* If quantitative comparisons are needed, the lenticularity or linearity of a bed may be expressed by the ratio of its maximum thickness to its lateral dimensions. Typically, the lateral dimensions of all but thin or very lenticular beds cannot be determined with confidence. Still, even limited data may prove useful; for example, notes may read, "beds 1 in. thick can be traced for 200 ft" or "beds appear tabular for as much as 32 ft."

Cross-bedding. Cross-bedded units may be classified and compared by the following geometric criteria: (1) whether or not the base of a set of cross-beds is an erosional surface, (2) gross shape of the set, (3) pattern of beds within the set, and (4) thicknesses of sets

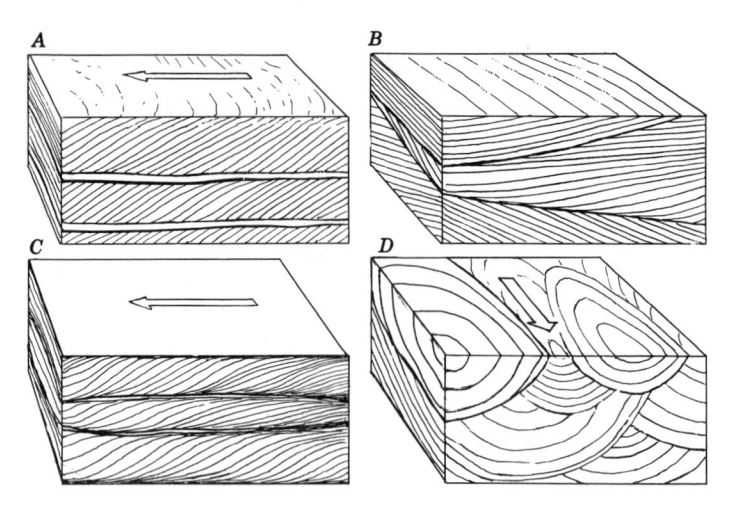

Fig. 12–8. Cross-bedded rocks. (A) Tabular sets with diagonal patterns. (B) Wedge sets, showing considerable erosion between each set. (C) Tabular to lenticular sets with tangential patterns; typically, these are laminated marine beds. (D) Symmetrical trough sets with distinctly linear axes; typically, these are large-scale fluvial features. The arrows indicate current directions.

and individual beds (or laminations). A few selected examples and terms are shown in Fig. 12–8, adapted in part from classifications by McKee and Weir (1953) and Lowell (1955). Cross-bedding is particularly useful in determining current directions, as discussed below under *Internal features.*

Thickness. Thickness of beds is a particularly useful aspect of their morphology. Actual thicknesses should be recorded in all cases because the limiting dimensions of terms such as *thickly bedded* and *thinly bedded* have not been standardized widely enough (see Ingram, 1954). It is here recommended that the term *laminae* (or *laminations*) be used for thin subordinate layers within a bed (Fig. 12–6). Thicknesses of laminae typically vary with grain size. Beds of coarse-grained sandstone are likely to have laminae ¼ to 1 in. thick, whereas a bed of clay shale ¼ in. thick might have anywhere from a few to many laminae.

The maximum, minimum, and average thickness of beds at a given outcrop should be estimated, and where more than one kind of rock or bed is present, thicknesses should be recorded for each. Thicknesses of bed sets should be recorded, as well as typical thicknesses of laminae or other subordinate bedding features. In all cases, however, the precision of the estimates or measurements should be gauged suitably to the regularity of the beds and the purpose of the study, for measuring beds can become involved and time consuming.

Internal features. A bed may be internally homogeneous (*structureless* or *massive*); it may be graded, or it may be laminated in a variety of ways (Fig. 12–6). Internal features can be used to interpret conditions of deposition as well as stratigraphic tops of beds. The dips of cross-strata or cross-laminae give a valuable measure of the current direction. This direction should be plotted on the map as an arrow rather than a strike and dip symbol. Wherever possible, the average direction of current flow and the degree of variance from this average are determined by taking readings on several beds at each outcrop. If the readings vary no more than about 10° from the average, a single arrow may be plotted. If the variance is greater, the individual readings may be plotted as a radiating set, or may be recorded for more thorough analysis in the office. Plotted dip arrows should represent cross-bedding of one type, especially where rocks are not of marine origin. Trough-type sets, for example, show a great range of dip directions; where exposures are adequate the trough itself is the most useful linear element. Lowell (1955) has described a field study that utilized various types of cross-bedding.

Where fine sand or silt accumulates on a current-rippled surface, it may build up a strikingly laminated bed by accumulating on the downstream slopes of the ripples (Fig. 12–9A). This type of bedding gives an excellent measure of current flow and depositional conditions, but unlike true cross-bedding, the layers of superimposed cross-laminations dip upstream.

Folding and other internal flow structures in beds can sometimes be used to ascertain the direction of slope of the original deposits (Fig. 12–9B). The nontectonic origin of these structures can generally be proven because the beds are intercalated in sequence with undeformed beds. Commonly, the upper surfaces of deformed beds have been eroded moderately before the overlying beds were deposited. Undeformed animal borings can also be used to determine the early deformation of beds (Fig. 12–9C). Slope and current directions need not necessarily be parallel, and therefore a distinctive arrow should be used to plot each on the map.

Abundant filled animal borings ("worm tubes") are characteristic of some sequences. The slumping of fine laminations into large borings may give an indication of tops of beds (Fig. 12–9D).

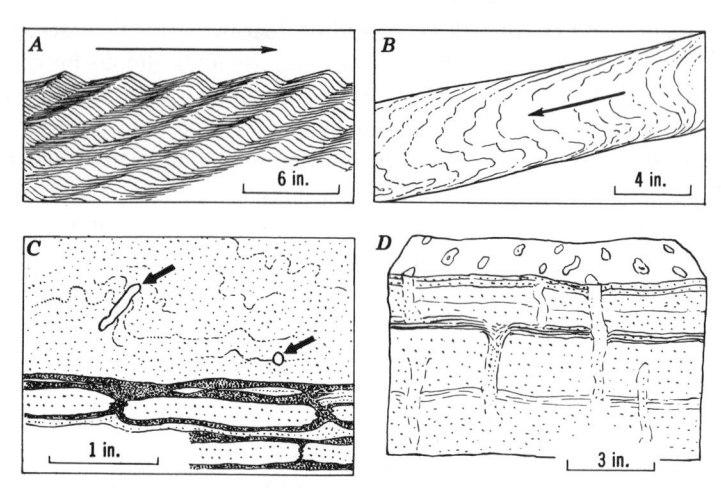

Fig. 12–9. (A) Current-rippled laminations that give an overall impression of cross-bedding. Current direction shown by arrow. Terrace of Truckee River, east of Reno, Nev. (B) Flow-folded cross laminations in Cretaceous sandstone bed near Pt. Conception, Calif. Both the original current and the subsequent flowage were from right to left. See also Prentice (1960). (C) Filled animal borings (arrows) cutting folded laminations. Note pull-apart structures in thinner sandstone beds. Drill core of Upper Pennsylvanian rocks from Runnels Co., Texas; courtesy of M. K. Blaustein. (D) Filled animal borings in Cambrian Pipe-rock (quartzite), at Moine thrust east of Loch Stack, Scotland.

Internal fabric of beds. Planar fabric in a detrital bed is parallel to the surface on which the grains were deposited, and linear fabric indicates the current direction across this surface. Pebbles, fossils, chips of mud or shale, and fragments of carbonized wood are particularly useful elements of fabric because they are easily seen in outcrops (Fig. 12–10A and B). Fossils shaped like saucers tend to be turned convex side up by a current, and they may also show imbrication (Fig. 12–10C). Where they are oriented with their convex sides down, they may have settled into position and been covered directly, perhaps during the evolution of a turbidity current (Fig. 12–10D). Elongate fossils can also be oriented strongly by currents; acutely pointed shells like those of turritellas, for example, tend to lie with their apices pointing upstream.

Detrital micas, chlorite, feldspar plates, and small leaf fragments commonly lie parallel to bedding, and elongate plant fragments and mineral grains may form a linear fabric. These fabrics usually impart a planar fissility to beds that are otherwise structureless; thus they help determine bedding directions in very thick, homogeneous beds. Grain lineations can often be measured on surfaces that parallel the bedding.

Reorienting linear data. Current arrows and slope arrows measured in folded or faulted sequences should be reoriented as nearly as possible to their original positions before being used to analyze ancient slopes and currents. In most unmetamorphosed sequences, the correction can be made by rotating the tilted beds back to their original (typically horizontal) position. This rotation is made about an axis that lies parallel to the axis of the fold on which the beds occur. Where outcrops are rugged and great precision is not required, this can be done in the field by rotating a piece of lineate rock into its approximate original position and then taking a compass reading on it. The correction can also be made by reading the orientation of the linear structure as it lies in the outcrop and then using a stereographic projection to make the rotation (Phillips, 1954; Badgley, 1959). Beds that have been folded strongly more than once, or that have been caused to

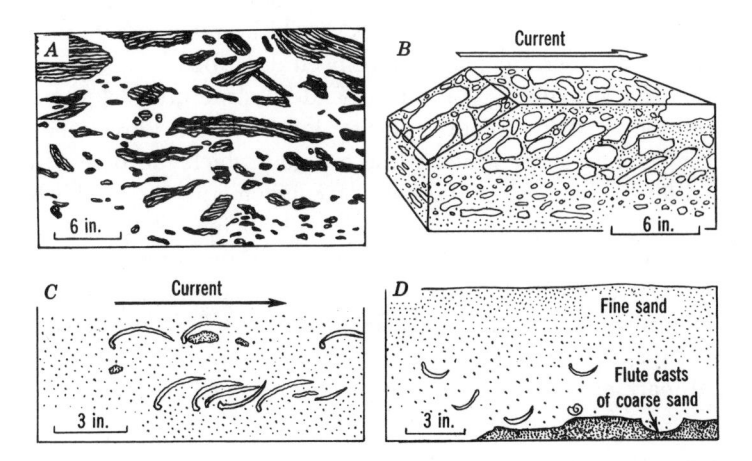

Fig. 12–10. Arrangements of large fragments in cross sections of beds. (A) Bedding-plane orientation of mudstone fragments in Upper Cretaceous sandstone of Santa Ana Mts., Calif. Note reverse size-grading of fragments. From a photograph by C. A. Hopson. (B) Imbricate structure. Planar cobbles tend to dip upstream and linear cobbles tend to plunge upstream. Generalized from dissected terrace gravels of Pacheco Pass area, Calif. (C) Stable position of large clam shells relative to current. (D) Orientation of clam shells in a graded Paleocene sandstone bed in the Arroyo Seco, Monterey Co., Calif.

flow or shear during folding, can be reoriented only after a detailed study and analysis of the deformation. In any case, reoriented structures should be plotted or recorded in such a way that they will not be confused with structural data that have not been reoriented.

12-5. Surfaces between Beds

Surfaces between beds may be smooth and essentially planar or n show features caused by rippling, current cut-and-fill, compactiv loading, downslope movements, movements of animals, desiccation, or combinations of these processes. These features are classified below on a genetic basis. Because the origin of some of the features is conjectural, each case should be examined with care.

Ripple marks. Ripple marks may be classified as *asymmetric* or *current* ripples and *symmetric* or *oscillation* ripples. Current ripples generally lie at right angles to the local current direction, their steeper side facing downstream (Fig. 12–11A). Oscillation ripples lie at right angles to the direction of wave propagation (Fig. 12–11B), although they frequently show complex shapes caused by interference or superposition of waves. The pointed crests of well-formed oscillation ripples can be used to determine the tops of beds. A number of observations and directional readings should be made where possible, for ripples with aberrant shapes have been described.

Cut-and-fill structures. Linear cut-and-fill structures are especially useful because their trends indicate local current directions, and their shapes may show which is the top and which the base of beds. Cut-and-filled channels at the base of conglomerate or pebbly sandstone beds are among the largest and most obvious of these features, and they may form in either fluviatile or marine sediments. Smaller markings are typically formed where currents laden with silt or sand scour depressions in argillaceous sediments, or where pebbles, shells, wood, or displaced seaweed drag or bounce against the bottom. The original markings can be seen in cross sections of beds but are rarely

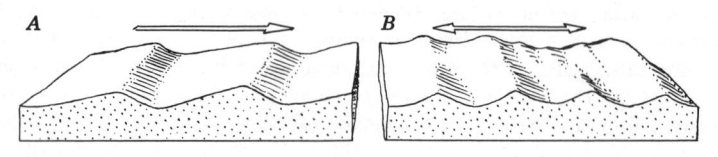

Fig. 12–11. Current and oscillation ripple marks in sand. Arrows indicate direction of current and of wave propagation.

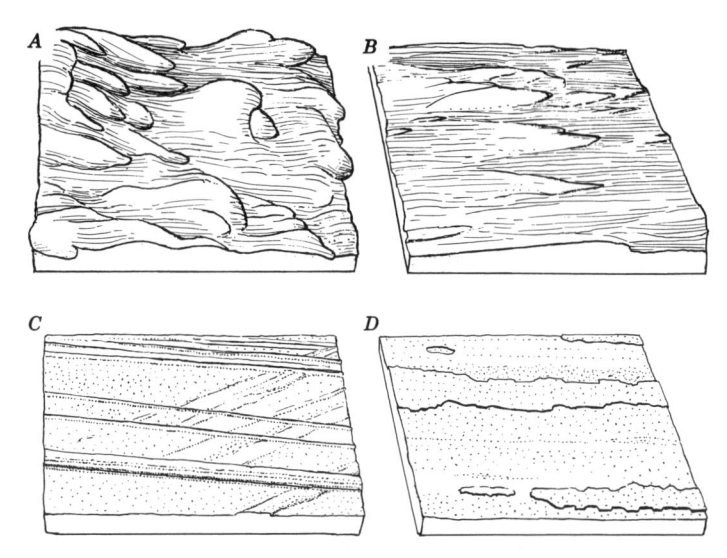

Fig. 12–12. Linear structures formed by currents. (*A*) Flute casts with prominently beaked ends on bottom of Eocene sandstone beds west of Stanford, Calif. Courtesy of E. A. Schmidt. (*B*) Simple flute casts on bottom of Upper Triassic sandstone bed from Santa Rosa Range, Nev. (*C*) Current groove casts, one set superimposed on another; Upper Triassic near Izee, Oregon. From a photograph by W. R. Dickinson. (*D*) Subdued current lineations on *top* of sandstone bed. Generalized from various sandstone flags. See also Stokes (1953). All blocks are 1 ft across.

preserved in three dimensions because the underlying shale or mudstone is almost always eroded more readily than the overlying sandstone. The structures occur commonly, however, as reversed three-dimensional *casts* on the base of the overlying sandstone beds. The terms used most often relate almost entirely to these reversed forms. *Flute casts* are elongated in the direction of the current and have a roughly triangular, spoonlike, or beaklike shape, with the more prominent or pointed end directed upstream (Fig. 12–12*A* and *B*). *Groove casts* are simpler elongate features that give only the orientation of the current (Fig. 12–12*C*). Crowell (1955), Kuenen (1957), and Wood and Smith (1959) have described these features in detail, particularly emphasizing their relation to turbidity currents.

Another type of current lineation that forms in laminated or thin-bedded sandstones consists of subdued lines or slabby, linear plates and low ridges on bedding cleavage surfaces (Fig. 12–12*D*). This

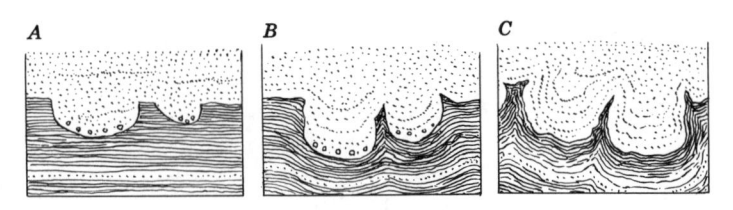

Fig. 12–13. Idealized sections of cut-and-fill and loading structures. (*A*) Simple cut-and-fill in laminated shale. Note concentrations of coarser materials. (*B*) Cut-and-fill with moderate loading, as shown by bending of laminae. (*C*) Loading structures with little or no cut-and-fill.

linear structure is probably formed by accumulation and streamlining of grains into trains and windrow-like ridges by currents. Although it may not be a cut-and-fill structure, it is included here for sake of comparison with flute and groove casts.

Load casts. Where water rich muds are overlain by a layer of coarser detritus, there is a tendency for the coarser materials to pocket or load downward into the mud where the surface between them is somewhat depressed. The original depressions can be formed by current cutting, by rippling, by movements of animals, or by unequal compaction, and therefore the downward-loaded bodies of sand may have any shape from distinctly linear to equidimensional or irregular pockets. They are called *load casts* because they are typically preserved on the base of sandstone beds. Crowell (1955, p. 1360) suggested that linear load casts may result directly from cut-and-fill where a turbidity current is heavily charged with detritus. If laminations or large sand grains can be found, relations such as those shown in Fig. 12–13 can indicate the relative degree of cutting and loading. In any case, load casts can be used to determine stratigraphic sequence.

Structures formed by creep, drag, and sliding. Sediments can be folded or disrupted by creeping or sliding downslope, and materials deposited from dense turbidity currents may be contorted as the currents drag against them. These movements commonly overturn projections of mud between sand pockets (Fig. 12–14A). With greater movement, the pockets may be complexly folded or rolled (Fig. 12–14C). Internal laminations are an important key in determining the sense of these movements (Figs. 12–9B and C). Gravity sliding may fold, disrupt, and interject sediments on such a large scale that the effects can be seen only at large outcrops (Fig. 12–14B).

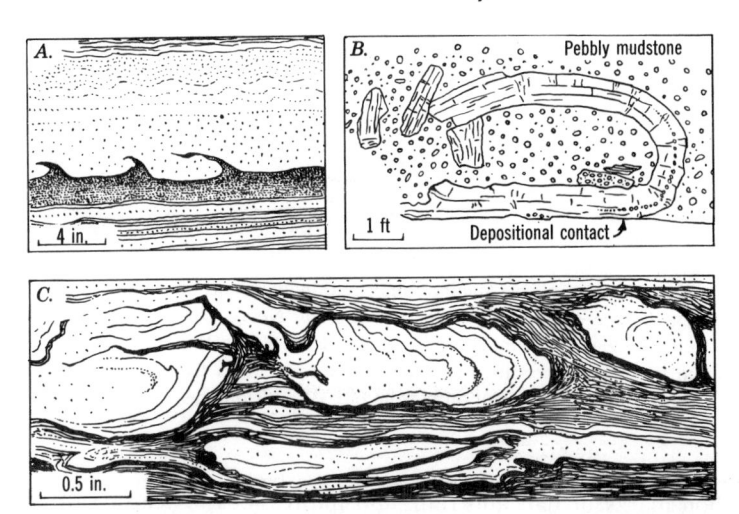

Fig. 12–14. Sections showing creep and flowage features, with sense of movement toward left. (A) Asymmetric sand pockets at base of graded, laminated, and rippled sandstone bed. Upper Cretaceous of Santa Lucia Range, Calif. (B) Slump overfold of mudstone bed; Grindstone Crk., Glenn Co., Calif. After Crowell (1957, p. 1000). (C) Flowage folds, rolls, and pull-apart structures in sandstone beds in argillaceous materials; there is a moderate lineation at right angles to the page. Large core of Upper Pennsylvanian rocks from Runnels Co., Texas; courtesy of M. K. Blaustein.

Animal markings. There are a great variety of animal markings, but by far the commonest in marine beds are roughly cylindrical bodies that are 0.02 to 0.5 in. in diameter and appear to be the fillings of tubes, burrows, or probe holes in loose sand and mud. They are useful indicators of tops of beds because they are typically preserved on the bottoms of sandstone or siltstone beds that lie on shale or mudstone beds (Fig. 12–15A and B). Less commonly, marks are made by animals that crawl on the bottom or by the percussion of swimming animals that vault against it. Shallow water or subaerial sediments may bear the tracks of larger animals (Fig. 12–15D). The positive and negative faces of these various marks can be recognized in most cases and can be used to determine the tops of beds. They also provide a means of determining the relative age of deformation of soft sediments (Fig. 12–9C).

Marks in desiccated sediments. Alternately desiccated and wetted sediments may show mud cracks, tracks of terrestrial animals, per-

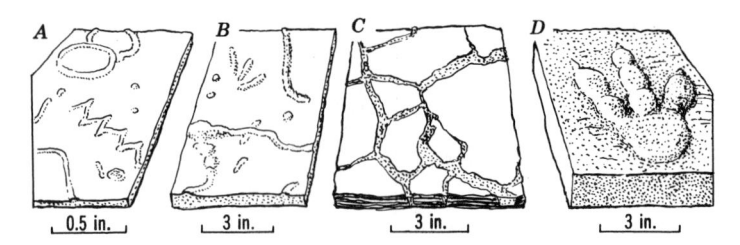

Fig. 12–15. (A) Distinctive small animal markings on bottom of fine-grained sandstone flag; Upper Triassic of Santa Rosa Range, Nev. (B) Tubular prominences on bottom of Paleocene sandstone bed from Santa Lucia Range, Calif. (C) Mudcracks in silty claystone, filled by sandy claystone; from Colorado Plateau region. (D) Cast of dinosaur footprint on bottom of red sandstone bed from Portland, Conn.

cussion marks of hail and rain, and impressions of ice or salt crystals (Fig. 12–15C). As with other surface markings, these forms are most commonly preserved as casts on the base of silty layers that overlie the layers of mud on which the marks were made originally.

12–6. Unconformities

Unconformities are surfaces of erosion or nondeposition that are eventually covered by younger strata. These surfaces have two important stratigraphic values: the amount of strata removed by erosion, and the time interval during which no sediments were deposited. These values may be consistent over large areas or may vary appreciably in short distances. They can be determined by detailed mapping and by faunal studies of the rocks just above and below the unconformity. A third important value associated with many unconformities is the structural discordance of the overlying and underlying rocks; this discordance gives a measure of the deformation associated with the unconformity.

Most major unconformities show some angular discordance; maps of lithologic units provide the best means of detecting them. Angular relations at single outcrops, however, may be misleading; they may have been caused, for example, by sliding of sediments or by faulting at a low angle to bedding. Faults can often be distinguished by a zone of sheared or brecciated rock, as well as by other means (Section 4–11); however, some angular unconformities have been modified by local shearing so as to look like major faults. Angular unconformities that have been obscured by a later period of folding can

Field Work with Sedimentary Rocks

sometimes be detected by comparing structures of the younger sequence with those of the older (refolded) sequence (Sections 15–7 and 15–11).

Unconformities without appreciable angular discordance can be detected by the following means:

1. *Fossil record.* Unconformity is proven by superimposed faunas of distinctly different ages.

2. *Contrasts in sediments.* Units formed under contrasting conditions indicate unconformities. Examples are: nonmarine beds overlain by marine beds, cross-bedded sequences overlain by sequences of graded beds, calcarenites overlain by fine-grained limestones with unabraded nektonic fossils.

3. *Fossil soils.* Subaerial weathering profiles may be preserved as clay-rich, reddened, or ocherous horizons, and locally even as humic layers. These materials may also be incorporated in the basal layer of a marine sequence. More rarely, old subaerial surfaces may be indicated by lag gravels, sand-blasted pebbles, attached fossil plants, surficial travertine deposits, or fossil caliche layers (Section 12–11).

4. *Local angularity.* Cross-cutting relations may indicate disconformity if cut-and-fill in loose sediments can be disproven (Section 12–5). A solid substratum may be indicated by encrusting or rock-boring animals. Fragments of the underlying bed are significant if they are not argillaceous or calcareous. Pocketed surfaces between limestone beds are significant if they bear residual clay or chert nodules, but irregular contacts in themselves are not (consider the irregular forms of biogenic reefs).

5. *Nondeposition.* Submarine surfaces on which accumulation is very slow may grade laterally to disconformities. These surfaces are indicated by thin layers rich in glauconite, bones, phosphatized shells or nodules, fecal debris, silicified shells, or manganiferous nodules. Where weathered, the surfaces may be indicated by discolored to distinctly ocherous zones.

12–7. Tops and Bottoms of Beds

Known sequences of stratigraphic units provide the surest basis for determining tops and bottoms of beds. Until this sequence is established, however, tops and bottoms should be determined by criteria such as those described in the three foregoing sections. If there is any question as to the dependability of one criterion, others should be sought. Features that do not have clear genetic significance can some-

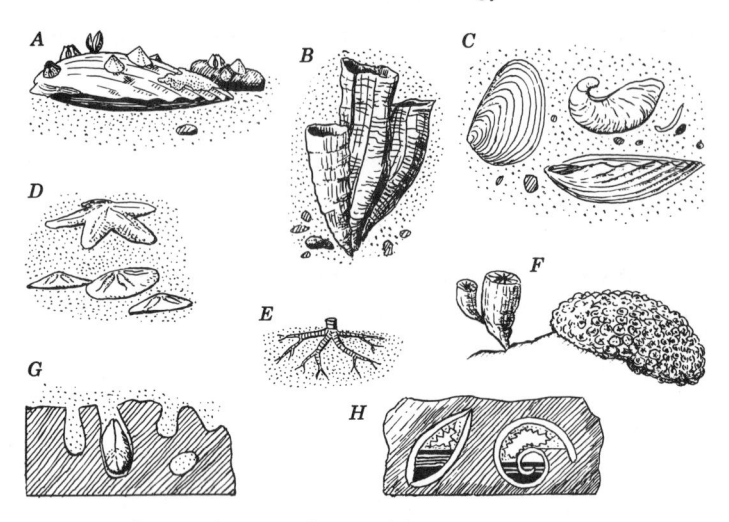

Fig. 12–16. Fossils as indicators of tops of beds. (A) Bryozoa, barnacles, and other encrusting animals on exposed tops of shells and cobbles. (B) Rudists or rudist-like pelecypods in growth position. (C) Shells of *Schizothaerus* (Tertiary) or *Pholadomya* (Mesozoic) may be found standing in living positions (left), whereas inequivalved pelecypods will lie on their more convex valve if buried alive. (D) Starfish and echinoids such as sea urchins tend to lie with their flat ventral (oral) side down. (E) Root-like structures of crinoids branch downward. (F) Solitary corals attach at base and branch outward and upward; colonial corals and calcareous algae tend to grow outward and upward to give a convex form facing upward (but not on steep or overhanging surfaces of reefs). (G) Animal holes drilled in rock substratum tend to open upward. (H) Partial fillings in cavities of buried shells indicate original dip of deposits. Condensed and modified from *Superposition of strata* by S. W. Muller (1958, Stanford University), reproduced in *Geotimes*, 1959, v. 3, nos. 5 and 7.

times be used locally after being tested empirically. A well-exposed section might show, for example, that abundant animal borings occur everywhere on the bottoms but nowhere on the tops of sandstone beds in a given formation. This criterion could then be used to determine tops of beds at isolated outcrops of this formation, though it might be unreliable for beds of another formation.

Graded bedding and cross-bedding are used so widely to determine sequence that a few precautions are necessary. Although grading from a coarser base to a finer top is by far the most common, the opposite relation has been observed. Thin silt-and-clay beds (varves) and argillaceous sandstone beds that grade upward to more argillaceous materials are by far the most dependable. The grading of well-sorted sandstones and conglomerates is always suspect. Cross-bedding is generally more difficult to interpret in finer-grained marine beds than in coarser rocks, mainly because distinct cross-cutting relations may be scarce in the finer-grained rocks. The base of a set of fine-grained laminae may curve into tangency with the underlying beds; however, the tops of cross-laminated beds may show the same relation (Fig. 12–8C).

Tops of beds as well as ecologic conditions can be determined where a large number of unbroken fossils are found in living positions (Fig. 12–16). Partial sediment fillings in hollow fossil forms indicate both the tops of beds and their approximate initial dip (Fig. 12–16H). Shrock (1948) has described additional criteria for determining tops of beds.

Where beds are so structurally disturbed that top-and-bottom criteria must be sought at many outcrops, structure symbols should indicate where these criteria have been observed. A small dot or other mark may be added to strike lines where beds are known to be upright (Appendix 4). The symbol for overturned beds would then be used only where beds are known to be overturned, while the ordinary bedding symbol would be used where beds could be either upright or overturned.

12–8. Measuring Stratigraphic Sections

Measuring and describing rock sequences is a basic stratigraphic procedure. Stratigraphic sections with accurate fossil locations are used to interpret geologic history, to work out complicated structures in nearby areas, to establish lithologic and microfossil control for drill-holes, to serve as type sections for rock units, and to help solve regional problems.

Localities for measuring detailed sections should meet these requirements: (1) the structure must be simple or must be so well known that complications can be resolved, (2) the sequence must be well exposed, and (3) the general nature of the rock units, especially the fossiliferous ones, must be known. These requirements can usually be met when an area has been mapped at least in part. In cases where the structure is known to be simple and the stratigraphic sequence is complex, however, it may be necessary to measure some sections before mapping can be started. These preliminary measurements must be planned by reconnaissance and photogeologic studies.

The method used to measure a section should be selected on the

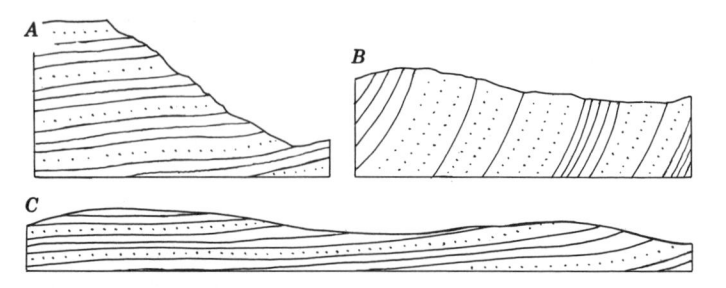

Fig. 12–17. Ideal geometric relations for measuring strata (top) compared to a region where low dip, low relief, and variable thicknesses make measuring difficult.

basis of the degree of detail required, the physical nature of the terrain and outcrops, and the time, funds, equipment, and personnel available. Measurements and descriptions of thick unfossiliferous formations, for example, need not be nearly as precise as those used to correlate fossiliferous units with subsurface data. The precision of the measurements is determined partly by the surveying methods and partly by the geometric relations between the ground surface and the dip of the rocks. Where the stratigraphic sequence is crossed at high angles by the ground surface, precise measurements can be made far more quickly and easily than when the two are nearly parallel (Fig. 12–17). The amount of detail to be recorded and the spacing of fossil and rock samples must be considered carefully when the project is planned; in general, it is wiser to collect too much than too little.

Measuring sections from maps or photographs. Sections may be measured and described by plotting a sequence of contacts and points on a topographic map or aerial photograph. Because the scales of most maps and photographs are too small for detailed work, this method is used mainly to reconnoiter more complete studies, to measure gross cartographic units, and to check total thicknesses accumulated by more detailed measurements.

Compass-pace surveys. This method is described in detail in Chapter 3. It is generally adequate only where the ground surface is level or slopes gently, as along a road, a smooth ridge, or by a waterway. Its great advantage is that it can be executed by one man. Details can be accumulated more quickly and accurately if a tape or graduated pole is used to measure beds in cuts.

Measurements with the Jacob's staff. A *Jacob's staff* is a pole about 5 ft long graduated to feet and tenths of feet. It is used to measure

Field Work with Sedimentary Rocks

thicknesses of beds directly at the outcrop, and with it one man can readily accumulate a detailed stratigraphic section where units are well exposed and where they meet the ground surface at high angles. The measurements are started by placing the staff on the lowest exposure and orienting it at right angles to bedding (Fig. 12–18A). The observer sights parallel to the dip of the beds across an even mark on the staff, and notes the point on the outcrop that falls in his line of sight. The stratigraphic thickness to this point is equal to the distance subtended on the staff. The observer then moves uphill and places the staff either on the point sighted *or on any other point on the same bedding surface,* and he continues to measure the beds by repeating the steps just described. If the staff can be oriented by estimation so that it is within 8 or 10° of normal to the beds, the error in a 5-foot measure will be 0.1 ft or less. For greater precision, it is necessary to orient the staff and line of sight by using a sighting bar and clinometer (Fig. 12–18B), or by holding or attaching a Brunton compass to the staff. Dips of beds must be checked frequently, a procedure that is facilitated by having one man do the measuring while a second checks attitudes and describes the lithology.

Exposures are rarely continuous enough to be measured in a straight line directly upslope, and even where they are, offsets must be made to fossil localities. The Jacob's staff method enables these offsets to be made easily. It is only necessary to walk along the trace of the last bedding surface sighted and then to resume measurements at the most useful outcrops (Fig. 12–18C). Because of the zigzag nature of the course, important specimen localities must be described in detail.

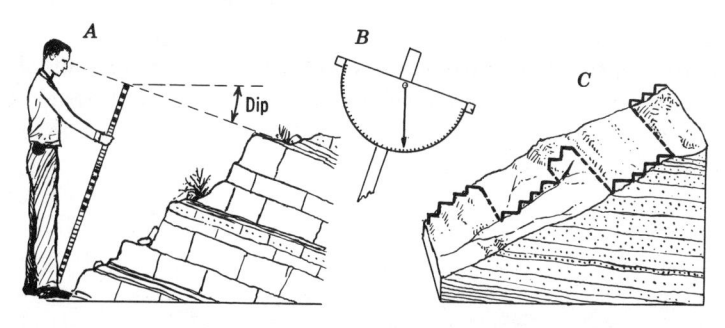

Fig. 12–18. Measuring beds with a Jacobs staff. (A) Holding staff at an outcrop. (B) Enlarged view of sighting bar and simple clinometer made from protractor and pendulum pointer. (C) Stepwise course of measurements, showing lateral offsets.

The localities can be collected easily again if marked with brightly colored tape or painted numbers.

Plane table methods. The plane table and alidade are ideal for mapping the exact locations of all samples and data collected. Sometimes this map will be no more than a skeleton of traverse lines, stadia points, elevations, and offsets to important outcrops. Other times it will be a complete strip map showing all geologic features as well as the topography, drainage, and culture. The type area and section of a formation could be an example of the latter case.

Various degrees of precision and speed are possible with the alidade and plane table. An uncontrolled stadia traverse is often adequate if instrument stations are no more than 800 ft apart, and if the section lying between the stations is measured by taking stadia sights to lithologic breaks or to places where the dips of beds change more than 5 or 10° (Section 8–12). Between these points, the details of the beds can be located by measuring with a cloth tape or by using the stadia rod as a scale. Vertical angles should be read between instrument stations or prominent breaks in slope.

Where plane table measurements must extend for long distances (as where strata dip gently across areas of low relief), it may be necessary to expand a control base for the survey by intersecting a chain of triangles or quadrilaterals (Fig. 12–19). Stadia traverses between the intersected points can then be used to measure the section and to locate sample points. If the dip of the beds is less than 15°, as in Fig. 12–19, vertical measurements must be made with particular care because errors in vertical position become translated directly into errors of stratigraphic thickness. Compass measurements of dips in such areas are generally so unreliable that average dips must be determined by the three-point method (Section 2–7, method VI).

Measuring sections with a transit and tape. A transit and tape should be used for making measurements where an accurate and permanently

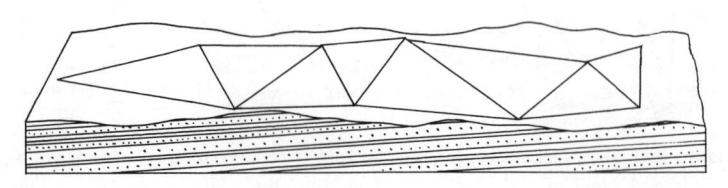

Fig. 12–19. Plane table control for measurements across gently dipping beds, allowing for many offsets without reduction of accuracy.

Field Work with Sedimentary Rocks

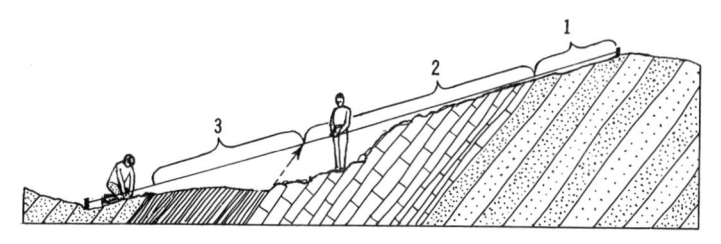

Fig. 12–20. Measuring strata on slope by reference to stretched tape. Brackets indicate three units accumulated during this measure. Note projection of limestone-shale contact to tape.

staked traverse course must be set up and where slopes are too steep and rugged for an efficient plane table traverse. A small (mountain) transit is ideal for this work, and the methods used are basically the same as those described in Section 7–13. The stratigraphic details between instrument stations may be surveyed by stadia methods or by reading vertical angles and taping distances directly on the slope, as described below.

Brunton and tape methods. Brunton-tape surveys are generally precise enough for measuring stratigraphic sections where beds dip more than 15° and intersect the ground surface at a fairly high angle. Measurements are made most easily in a downhill direction. The procedure that follows refers to a two-man party; however, one man can make the measurements if the tape is anchored at the 0 end.

1. One man holds the 0 end of the tape at the starting point while the second man carries the other end of the tape to the first break in slope.
2. The slope angle is read and checked by both men.
3. The tape is secured on the slope (as by tying it to a stake), and, starting at its 0 end, the rocks are examined and described; their positions are recorded according to measurements (slope distances) read directly from the tape (Fig. 12–20).
4. Strikes and dips are measured as necessary, and the angle between the strike and the section line is measured and recorded.
5. The 0 end of the tape is moved downhill to the first break in slope, and steps 2 to 4 are repeated.

Where outcrops are rugged, or where beds meet the surface at close to 90°, the tape can be used as a Jacob's staff to measure true thicknesses directly. These thicknesses must be recorded as true thick-

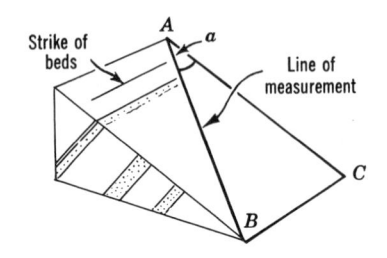

Fig. 12–21. Correction of slope distance that was measured oblique to dip of beds.

nesses. In any case, terms such as *overlying* and *above* must not be used unless they clearly refer to stratigraphic rather than to topographic position.

Corrections for slope angles and oblique sections. Unless true thicknesses are measured directly, as with the Jacob's staff, they must be calculated from the slope distances. If the measured slope distance is oblique to the dip of the beds, it must first be corrected by the following formula (based on the notation shown in Fig. 12–21).

$$AC = AB \cos a$$

The true thickness is then calculated by one of the formulas shown in Fig. 12–22, the choice depending on the relation between the ground slope and the dip.

These calculations apply in the usual case, where dips are essentially parallel through each calculated interval. If it is necessary to calculate thicknesses across the limb of a fold, where dips vary progressively between two limiting values, the method described by Hewett (1920) can be used.

The calculations are made most easily with a slide rule, using trigonometric functions (Appendix 7). They can also be determined graphically or, in most cases, by nomographs (see, for example, Mertie, 1922). The drawing or nomograph must be large, however, to give a degree of precision that is adequate for detailed work, and such determinations may take as much time as slide rule calculations.

12–9. Sampling for Microfossils

Microfossils often provide the best means for correlating stratigraphic sequences with standard stages or series. Rocks should be examined for microfossils when an area is first mapped, and samples should be taken wherever fossils can be seen. Knowledge gained from reading or from discussions with micropaleontologists is particularly valuable

a. Slope and dip are opposed and angle of slope (y) plus angle of dip (x) is < 90°

$BC = AB \sin(x+y)$

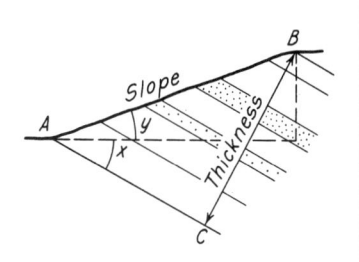

b. Slope and dip are opposed and angle of slope plus angle of dip is > 90°

$BC = AB \cos(x+y-90°)$
or $BC = AB \sin[180°-(x+y)]$

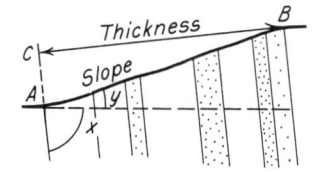

c. Slope and dip are in the same direction, with dip > slope

$AC = AB \sin(x-y)$

d. Slope and dip are in the same direction, with dip < slope

$BC = AB \sin(y-x)$

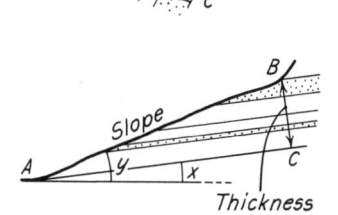

Fig. 12–22. Formulas used for the various possible combinations of direction and amount of ground slope and dip of beds.

at this stage. If it is known, for example, that only the smallest calcareous foraminifera will be useful in a certain part of the section, the finest claystones are collected in preference to silty shales that might show large arenaceous foraminifera. Ideally, the sample localities at this stage are scattered widely, both stratigraphically and parallel to the strike of the beds. If there are thick, fine-grained units in which no fossils have been seen, several samples of the most promising (generally, the finest grained) materials should still be collected.

The faunal determinations from the first group of samples may solve many stratigraphic problems, but they usually also demonstrate a need for additional sampling. More samples may be needed along certain contacts or faults, or where relatively thin parts of a section appear to contain many faunal zones or several stages. Unusually well-exposed sequences of fossiliferous rocks might be sampled in great detail, both stratigraphically and laterally, to test the validity of tentative faunal zones or local stages (Section 12–2).

The stratigraphic measurements described in Section 12–8 are generally used as a basis for systematic collecting. The following suggestions apply to the samples themselves:

1. If calcareous fossils will be used, the samples should be fresh (although moderately weathered rocks are sometimes useful because they are friable).

2. Samples must be large enough; unless fossils are obviously abundant, 10 lb of material may be needed.

3. Locations must be plotted precisely on a map or described completely in the notes.

4. The stratigraphic position must be known. It will be determined automatically if the section is being measured; otherwise the stratigraphic distance to the nearest contact or key bed should be measured directly.

5. The source of the sample should be marked at the outcrop so that more material can be collected later.

6. Sample bags must be silt-tight and marked on the outside so they will not have to be reopened until they are in the laboratory.

12–10. Logging Wells and Drill Holes

Well logs are descriptions or graphic representations of the strata cut by a well or drill hole. Where beds are horizontal, these logs are essentially columnar sections. Logs should be made when the well is being drilled because the procedures of collecting, labeling, storing and reopening samples tend to increase the possibilities of contamination and mis-identification. The geologist should oversee these procedures if possible, and he should also make sure that as much material is saved as possible. Before sampling a drilling well, he should determine the exact drilling methods, the typical rates of drilling in various units, and what is the expected performance of sampling equipment available. These aspects of well logging have been described in LeRoy's *Subsurface Geologic Methods* (1950), as well as in subsequent journals.

Logging cores. Large amounts of core can generally be logged by the following steps:

1. Make a graphic log, either with lithologic symbols or colors, that blocks out the main rock units and contacts and also indicates occurrences of fossils, key beds, and shows of petroleum or zones of high porosity.

2. Fill in the uncored gaps in this record by reference to the driller's log and other accessory logs.

3. Compare the graphic log with the expected stratigraphic or structural sequence.

4. Re-examine the cores as necessary to make more detailed descriptions of rock units, to inspect critical intervals, contacts, and structurally anomalous relations, and to collect samples.

Samples are generally collected at certain prescribed intervals. They should also be collected systematically at contacts, key beds, faults, or other features important to stratigraphic interpretations. Lithologic samples are most useful if split axially to the core. The footage should be lettered in waterproof ink or paint on the side of the core, reading so that the sample is right side up. If bedding is not obvious in the sample, the bedding direction should be painted as a band around it. If possible, the well number and name should also be lettered on the sample.

Microfossil samples should be collected both at prescribed intervals and at critical horizons. Hand lens examinations are necessary to select the most useful materials since many sandy or silty rocks are dark colored in cores. Mudstone and shale are particularly susceptible to fragmentation and contamination during coring and storage, and therefore broken materials must be selected with care. Any question as to their reliability must be marked prominently on the label, as well as in log descriptions. Samples should be placed in strong, silt-proof bags and labeled as to well name and number, core number, footage or interval, rock name, formation (or other unit name), and probable stratigraphic position, as determined from the core.

The data in lithologic descriptions of cores should be recorded in order of importance. The amount of detail needed will vary with the project; somewhat more should be collected than appears necessary. Colors of cored rocks are much more diagnostic than those of surface outcrops, and these colors can be recorded precisely by using detailed rock color charts (p. 216). Horizons of secondary coloration should be noted because they may indicate faults or unconformities. Porosities should be described completely.

Logs from cuttings. Lithologic logs of most wells are made by examining the cuttings brought up in the circulating drilling muds or fluids. These materials are evolved continuously while drilling is in progress, and samples are generally taken at certain time intervals or at every 5 or 10 ft of depth drilled. Cuttings must be washed thoroughly; this is done by repeatedly decanting rinse washes until they are free of clay-size materials. The samples are then dried, placed in dust-proof bags, and labeled with the well description, driller's name, and footage or time interval. If the footage has been corrected by taking into account the circulation time from bit to surface, this should also be noted on the label.

The size and appearance of cuttings vary greatly with drilling methods and local lithology. Cuttings from the bit are generally mixed to some degree with fragments that have been caved or abraded from higher parts of the hole. Picking contacts or key beds is not difficult if rock units are thick and distinctive. The upper contact of a formation is drawn in the log at the first appearance of a cutting or of disaggregated grains from that formation. Where rocks grade into one another, or where they are intercalated through considerable thicknesses, the local section must be known in detail.

Moderately friable clastic rocks are usually disaggregated by the bit; thus a binocular microscope should be used to examine cuttings. Samples stained with petroleum or asphaltic residues should first be cleaned with a volatile solvent. Accessories needed for most examinations include dilute HCl, chemicals required for staining tests, flat evaporating dishes for examining materials, dissecting needles and forceps for separating fragments and testing hardness, and a grid ocular for measuring grains and estimating their amounts. Graphic logs and brief descriptions are customarily entered on strips of paper

Field Work with Sedimentary Rocks

3 in. wide and 40 or 60 in. long. More detailed notes and comments may be recorded in an accessory notebook that is organized in the same order as the plotted log.

12–11. Surficial Deposits and Related Landforms

Surficial deposits and related landforms are studied to determine recent geological history, to develop resources of groundwater and other economic materials, and to plan for various engineering projects. In areas where surficial deposits are widespread, studies of bedrock may require a considerable knowledge of the overlying deposits. Geologic maps of such areas are preferably outcrop maps that show both the bedrock exposures and the various units of surficial materials (Section 4–8).

Some cartographic units of surficial materials have been defined on a lithologic basis; others are lithologically composite and have been defined on the basis of apparent age. Lithologic units may be characterized by their structures, textures, compositions, gross shapes, associations with other units, or by the nature of the surfaces on which they were deposited. Common and important lithologic units are: alluvium, colluvium (accumulations formed by downslope movements of rock and soil), till (or boulder clay), glacial drift (till and other closely related materials, undifferentiated), glacial outwash deposits, fluvioglacial terrace deposits, loess, dune sands, beach deposits, volcanic ash and lapilli, peat, soils, lag gravels, caliche, and clays residual to solution of limestones.

Sequences of similar cartographic units, such as several alluvial deposits or tills, are subdivided on the basis of unconformities (Section 12–6). Where sequences are not complete or are only partly exposed, the relative ages of similar units can be distinguished by the degree of weathering of rock fragments or mineral grains. Tills and drift of various ages, for example, can commonly be distinguished by the amount of leaching of limestone fragments or fossils, by the amount of decomposition of such rocks as basalt and diabase, or by the degree of disintegration of granitic rocks.

Relation to landforms and agents. To trace and correlate surficial deposits, it is generally necessary to study landforms associated with them. Old alluvial or fluvioglacial deposits, for example, can often be traced out or matched by mapping the terraces related to them. Conversely, structural studies based on series of terraces may require correlations of the sediments that lie under their surfaces. In some cases, recent deposits and landforms can also be evaluated in light of the agents that formed them. Some marine deposits, for example, can be related to the terraces on which they lie, to the ancient sea cliffs from which they were eroded, and to the present level and regimen of the sea. Stream terraces and their deposits often correspond to existing streams or to slightly antecedent stream systems.

The following steps can be used to correlate landforms and surficial deposits: (1) study of aerial photographs and topographic maps to delineate the more apparent landforms, (2) reconnaissance of the area to determine which surfaces are underlain by surficial deposits, (3) study of the deposits to determine their origin and relative age, and (4) mapping or walking out landforms by matching their topographic positions and their surficial deposits. Correlation of closely spaced, well-preserved shorelines and terraces may require carrying accurate lines of levels for considerable distances, as with a transit or spirit level. Stanley (1936) has given an example of such a study. Other kinds of terraces may have so much relief that precise matching of levels is either misleading or impossible. Their relative ages may, however, be determined by such criteria as: (1) relative topographic position and continuity, (2) type of sedimentary cover and degree to which it is weathered, (3) degree of dissection and removal of loose detritus, and (4) degree of warping. Detailed profile studies based on maps and photographs can be used to make tentative correlations of these irregular surfaces, but the surfaces should then be walked out and mapped. Field methods must also be used to determine if imperfections in a given surface were caused by initial dip, later deposits of sediments, or by erosion of original sediments. Frye and Leonard (1954) and Howard (1959) have given critiques of the terminology and interpretation of terraces associated with stream deposits. An example of the correlation of alluvial terraces on a local and regional scale has been described by Leopold and Miller (1954); Bradley (1957, 1958) has described field and laboratory studies of marine terraces and their deposits.

Soils. Soil profiles give a means of correlating landforms and their associated sediments, and contacts between surficial units are commonly drawn on the basis of ancient soils. Moreover, bedrock units must often be mapped on the basis of soils (Section 4–7). Systematic descriptions of soils as well as soil mapping and sampling methods are given in the U.S. Department of Agriculture's *Soil Survey Manual*

(1951), and this book should be consulted for terms and information relating to soils and soil profiles.

The total soil profile can be a complete gradation from surface soil to unaltered rock or sediment, or it can consist of various layers that may be defined sharply or may grade to one another. Field descriptions should give the nature, thickness, and regularity of each of these layers and gradations. Where soil profiles are being compared or correlated from one place to another, it is useful to classify the principal layers, customarily called the *A*, *B*, and *C horizons* of the profile. The upper or *A* horizon is characterized by one or more layers in which organic materials have been concentrated and from which soluble materials and colloids have been leached; it is typically gray to brown, friable, or very rich in plant materials. The *B* horizon consists of a layer or layers in which materials (especially colloids) transported from the *A* horizon are concentrated. Additions of clay and hydrated iron or aluminum oxides make it more compact and more red, yellow, or brown than the *A* horizon; however, additions of organic materials from above make some *B* horizons gray or gray-brown. The *C* horizon lies under the soil proper and consists of disintegrated rock or loose sediments that are more or less weathered and are clearly related to the overlying soil. In many areas, the upper part of the *C* horizon is characterized by concentrations of calcium carbonate and other soluble salts, which may be so abundant as to form a hard caliche layer.

The relative ages of soils and the land surfaces on which they lie can sometimes be judged on the basis of thickness, degree of leaching of carbonates, degree of disintegration of certain rocks, or by the degree of argillization and the compactness of the *B* horizon. These criteria must be modified and weighted, however, where downslope creep and sheet wash have thickened soils on slopes and in valleys and have thinned them on ridges, or where variations in underlying rocks affect soil characteristics. In buried ancient soils, only the *B* horizon or some part of it is generally preserved. Soils at different outcrops must then be identified or correlated on the basis of texture, clay content, color, and compactness (indicated in part by a tendency to develop columnar or platy fractures in outcrop). Color is such an important measure of oxidation and concentration in the *B* horizon that precise color terms should be used. Standardized soil color charts are available from the Munsell Color Company, 2441 North Calvert St., Baltimore, Md. These charts give a wide range of color names and symbols that can be used for rocks as well as for soils.

References Cited

American Commission on Stratigraphic Nomenclature, 1961, Code of stratigraphic nomenclature: *Bulletin of the American Association of Petroleum Geologists,* v. 45, pp. 645–665.

Badgley, P. C., 1959, *Structural methods for the exploration geologist:* New York, Harper and Brothers, 280 pp.

Bradley, W. C., 1957, Origin of marine-terrace deposits in the Santa Cruz area, California: *Bulletin of the Geological Society of America,* v. 68, pp. 421–444.

———, 1958, Submarine abrasion and wave-cut platforms: *Bulletin of the Geological Society of America,* v. 69, pp. 967–974.

Crowell, J. C., 1955, Directional-current structures from the Prealpine Flysch, Switzerland: *Bulletin of the Geological Society of America,* v. 66, pp. 1351–1384.

———, 1957, Origin of pebbly mudstones: *Bulletin of the Geological Society of America,* v. 68, pp. 993–1010.

Dunbar, C. O., and John Rodgers, 1957, *Principles of stratigraphy:* New York, John Wiley and Sons, 356 pp.

Frye, J. C., and A. R. Leonard, 1954, Some problems of alluvial terrace mapping: *American Journal of Science,* v. 252, pp. 242–251.

Hewett, D. F., 1920, Measurements of folded beds: *Economic Geology,* v. 15, pp. 367–385.

Howard, A. D., 1959, Numerical systems of terrace nomenclature: a critique: *Journal of Geology,* v. 67, pp. 239–243.

Ingram, R. L., 1954, Terminology for the thickness of stratification and parting units in sedimentary rocks: *Bulletin of the Geological Society of America,* v. 65, pp. 937–938.

Kleinpell, R. M., 1938, *Miocene stratigraphy of California:* Tulsa, American Association of Petroleum Geologists, 450 pp.

Krumbein, W. C., and L. L. Sloss, 1951, *Stratigraphy and sedimentation:* San Francisco, W. H. Freeman and Co., 497 pp.

Kuenen, P. H., 1957, Sole markings of graded graywacke beds: *Journal of Geology,* v. 65, pp. 231–258.

Leopold, L. B., and J. P. Miller, 1954, *A postglacial chronology for some alluvial valleys in Wyoming:* U.S. Geological Survey, Water-Supply Paper 1261, 90 pp.

LeRoy, L. W. (editor), 1950, *Subsurface geologic methods, a symposium,* 2nd ed.: Golden, Colorado School of Mines, Dept. of Publications, 1156 pp.

Lowell, J. D., 1955, Applications of cross-stratification studies to problems of uranium exploration, Chuska Mountains, Arizona: *Economic Geology,* v. 50, pp. 177–185.

McKee, E. D., and G. W. Weir, 1953, Terminology for stratification and cross-stratification in sedimentary rocks: *Bulletin of the Geological Society of America,* v. 64, pp. 381–390.

Mallory, V. S., 1959, *Lower Tertiary biostratigraphy of the California coast ranges:* Tulsa, American Association of Petroleum Geologists, 416 pp.

Mertie, J. B., 1922, *Graphic and mechanical computation of thickness of strata and distance to a stratum:* U.S. Geological Survey Professional Paper 129-C, pp. 39–52.

Pettijohn, F. J., 1957, *Sedimentary rocks,* 2nd ed.: New York, Harper and Brothers, 718 pp.
Phillips, F. C., 1954, *The use of stereographic projection in structural geology:* London, Edward Arnold Ltd., 86 pp.
Prentice, J. E., 1960, Flow structures in sedimentary rocks: *Journal of Geology,* v. 68, pp. 217–225.
Schenck, H. G., and S. W. Muller, 1941, Stratigraphic terminology: *Bulletin of the Geological Society of America,* v. 52, pp. 1419–1426.
Shrock, R. R., 1948, *Sequence in layered rocks:* New York, McGraw-Hill Book Co., 507 pp.
Soil Survey Staff, 1951, *Soil survey manual:* U.S. Dept. of Agriculture Handbook No. 18, 503 pp.
Stanley, G. M., 1936, Lower Algonquin beaches of Penetanguishene Peninsula: *Bulletin of the Geological Society of America,* v. 47, pp. 1933–1959.
Stokes, W. L., 1953, *Primary sedimentary trend indicators as applied to ore finding in the Carrizo Mountains, Arizona and New Mexico:* U.S. Atomic Energy Commission, RME-3043 (pt. 1), 48 pp.
Williams, Howel, F. J. Turner, and C. M. Gilbert, 1954, *Petrography:* San Francisco, W. H. Freeman and Co., 406 pp.
Wood, Alan, and A. J. Smith, 1959, The sedimentation and sedimentary history of the Aberystwyth Grits (Upper Llandoverian): *Quarterly Journal of the Geological Society of London,* v. 114, pp. 163–195.

13
Field Work with Volcanic Rocks

13–1. Volcanic Sequences and Unconformities

Volcanic rocks are mapped by the same general procedures as those used for sedimentary rocks, and tabular volcanic units can often be correlated and measured by stratigraphic methods. Fossils are rare in lavas, but fossiliferous tuffs can be used to place volcanic sequences in the standard geologic column. Moreover, radioactive age-dating makes it possible to assemble well-dated volcanic sequences.

Many volcanic units, however, are not as simple and predictable as most sedimentary units. Most volcanic rocks form subaerially and are therefore controlled by the irregularities of the topography. In Fig. 13–1, for example, two recent flows followed irregularly linear valleys, and although the flows are contemporary, they lie at such different levels that they might later be assigned different stratigraphic ages. Misleading relations can also be caused by the high viscosity of some lavas. Many flows of rhyolite and dacite, for example, have such steep fronts and sides that younger flows deposited against them give the illusion of a faulted or intrusive relation (Fig. 13–2A). If erupted under water, even fluid basaltic lavas may produce giant mounds of pillows that seem to lie transverse to bedding in adjacent sediments and tuffs.

The initial dips of volcanic rocks are also significant. If well exposed, the beds of an eroded volcano may be seen to dip away from the central

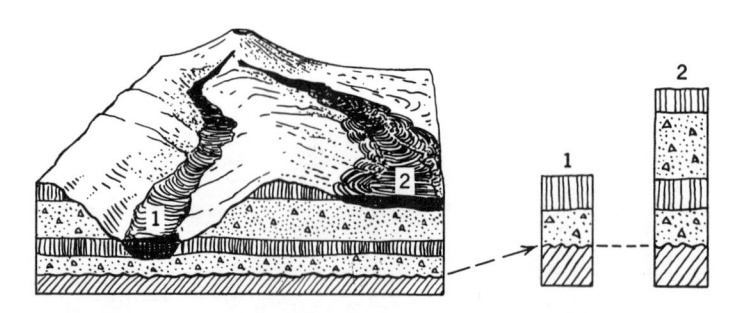

Fig. 13–1. Misleading stratigraphic relations between contemporary flows that come to lie at different levels.

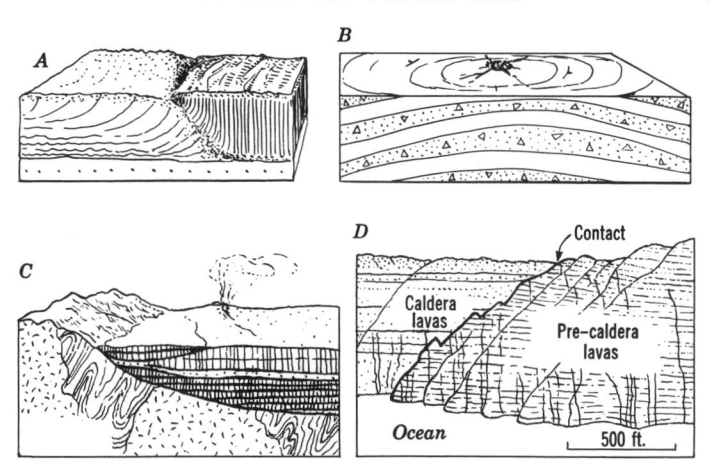

Fig. 13–2. (A) Basalt flow resting against steep edge of somewhat older rhyolite flow. (B) Anticlinal section through flank of eroded volcanic cone. (C) Onlap of lavas against steep mountain front. (D) Cliff-section on the Island of Kahoolawe, Hawaii, showing how caldera-filling lava flows are almost parallel to older lavas of the main volcanic shield. Their surface of contact (heavy line) is the eroded fault that was once the caldera wall. The irregular vertical lines are dikes. After H. T. Stearns (1940, p. 133).

conduit in all directions; however, a single section of outcrops could be mistaken for a tectonic anticline (Fig. 13–2B). Moreover, abrupt onlap of an accumulation of lavas against a steep mountain slope may suggest large-scale faulting (Fig. 13–2C), while filling of volcanic depressions such as calderas may form puzzling or misleading contacts (Fig. 13–2D). Although these stratigraphic relations can be solved readily if exposures are adequate, sketching of volcanic units and structures across large areas can cause incorrect correlations. Even where tabular-appearing marine beds are intercalated with volcanic rocks,

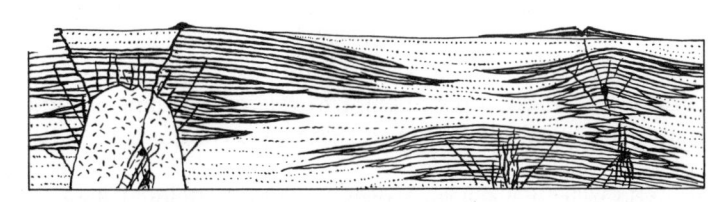

Fig. 13–3. Intertonguing relations in thick accumulations of lavas and sediments.

it is uncertain that either will project evenly for a long distance. Volcanic piles accumulate rapidly; hence associated sedimentary units tend to be lensoid or to interfinger with the volcanic rocks through considerable vertical thicknesses (Fig. 13-3).

Unconformities. Each flow and pyroclastic deposit in a "continuous" accumulation can be considered as bounded by unconformities whose time values may be anything from a few days to tens of thousands of years. The tops of some flows and ash layers are scarcely altered or modified before being covered by another deposit, while many others are weathered and eroded appreciably and some severely. Although these minor unconformities are useful in interpreting the periodicity of ancient volcanoes, they can make it difficult to detect and trace major unconformities unless large areas are mapped. Major unconformities can be recognized by: (1) persistent angular discordance, (2) truncation of volcanic necks or other volcanic intrusions, (3) layers of sedimentary rocks that contain materials foreign to the underlying volcanic sequence, (4) zones of weathering that persist across all types of underlying rocks, and (5) abrupt change in the general magma-type of the rocks, as from calc-alkaline to alkaline. The most obvious feature at any one outcrop is the zone of weathered materials, which is typically colored in reds, yellows, or greens by ferruginous clays or iron-oxide pigments. Channels filled with sand or gravel may occur here and there, as may fine-grained sediments containing vegetative materials or fossils. Because subaerial surfaces can develop as a series of pediments rather than as simple peneplains or wave-planed surfaces, some pass into irregular upland surfaces. The surfaces may also disappear laterally into lake or marine deposits.

13–2. Cartographic Units of Volcanic Rocks

Tabular sequences of plateau basalts and ignimbrites can be divided into cartographic units comparable to simple sedimentary formations. Procedures and rules for setting up and describing these units are given in Sections 4–6 and 12–2. Many flows, pyroclastic deposits, or composite units, however, may be irregular or stratigraphically unpredictable (Section 13–1); these units cannot be defined by measuring and describing isolated stratigraphic sections. Mapping must be extensive enough to relate each unit accurately to adjacent units, especially to named sedimentary formations with which the volcanic rocks interfinger.

The basic unit of a volcanic terrain is the individual flow or pyroclastic bed. The more of these units that can be mapped, the more accurate will be the interpretation of the volcanic sequence. Distinctive, relatively extensive flows or tuffs can constitute separate members within formations, as the *Wenas flow* of the *Ellensburg Formation* (Waters, 1955). A formal geographic name is usually omitted for a member or bed whose lithology is unique within a formation, as *dacite flow* of *Umptak Formation*. Informal names should be used in cases where the area mapped is small, the units discontinuous, or relations to formalized units undetermined.

Formations consisting of a number of flows or pyroclastic beds are selected on the basis of their physical individuality and the close genetic relation between their constituent rocks. The ideal volcanic formation is formed by closely succeeding eruptions of one kind of magma; examples are sequences of basalt flows or dacite tuff beds. Two units of similar lithology which are in contact can often be divided on the basis of an unconformity (but see *Unconformities*, Section 13-1). Formations or Groups that include a variety of rocks are generally delineated at major unconformities. Some can be recognized in a general way on the basis of magma-parentage (magmatic consanguinity); however, considerable petrographic study may be needed to resolve such units and cases are known where nonconsanguineous rocks are associated closely.

Intrusive bodies in volcanic areas should be mapped individually. If they cannot be plotted to scale, appropriate symbols can be used, such as colored lines for swarms of parallel dikes. Most intrusive units should be named in an informal way, as *dacite neck*, but if their genetic relation to effusive units is clear, they may be considered as members of formations, as *dikes of Koolau Volcanic Series*. The term *complex* is sometimes used for composite units of intruded, collapsed, and faulted rocks that commonly form in the central parts of large volcanoes, as *caldera complex of Wailuku Volcanic Series* (Stearns and Macdonald, 1942).

13-3. Naming Volcanic Rocks

Field names for volcanic rocks should be as complete as possible, even though a petrographic microscope or chemical analysis may be needed to classify the rocks precisely. Names should be based not only on minerals and textures but also on small-scale structures and on associated rocks. A given specimen, for example, might be so aphanitic that it could only be given the name *felsite*, but other parts of the same body might show enough recognizable grains or structures to indicate more specifically that the rock is a rhyolite lava. Associated intrusive bodies are of great value in naming aphanitic rocks because they often show gradations from glassy or aphanitic borders to phaneritic interiors which can be named confidently. Weathered surfaces of volcanic rocks should be examined because textures and structures tend to be accentuated on them (especially in partly glassy rocks), and some minerals are far easier to see when altered.

Where a large number of mineral grains (typically phenocrysts) are visible, the following notes can be used in naming rocks.

a. **Rocks with quartz phenocrysts.**

Rhyolite. Phenocrysts of potassium feldspars (chiefly sanidine) at least as abundant as those of plagioclase; biotite typically only mafic mineral; groundmass glassy and dark, or stony and light colored; deuteric silica minerals commonly present.

Dacite. Plagioclase predominant or only feldspar of phenocrysts; quartz may be scarce; typically a variety of mafic minerals; granular clots of plagioclase and pyroxenes quite common; groundmass as for rhyolite.

Quartz keratophyre. Typically pale green to light gray with prominent phenocrysts of quartz and gray or whitened plagioclase (albite); chlorite is mafic mineral.

b. **Rocks without quartz or feldspathoid phenocrysts.**

Trachyte. All or most feldspar phenocrysts are sanidine; biotite only abundant mafic mineral in potash-trachytes, but many others may be present; groundmass as in rhyolite.

Latite. Most of feldspar phenocrysts are plagioclase; a variety of mafic minerals likely to be present; usually indistinguishable from dacite that has no quartz phenocrysts.

Andesite. Plagioclase and pyroxene generally only phenocrysts, though hornblendic varieties locally common and grains of olivine or biotite possible; clots of phenocrysts typical; groundmass aphanitic and gray to black; may be indistinguishable from dacite that has no quartz or sanidine phenocrysts.

Keratophyre. Gray to gray-green stony rock; may have whitened or greenish phenocrysts of plagioclase (albite) and small grains of chlorite.

Basalt. Heavy rocks with phenocrysts of plagioclase, pyroxene, or olivine (these may be anywhere from very small to very large); groundmass of olivine-poor varieties dark-gray to black and aphanitic; groundmass of olivine-rich varieties commonly light gray and phaneritic (dark rocks that carry abundant olivine phenocrysts and are aphanitic throughout thick flows are likely to be subsilicic, alkaline rocks, as *trachyandesite*, etc.).

Spilite. Phenocrysts of augite (commonly altered) and milky or pale green albite in dark gray to gray-green groundmass; amygdales of calcite and chlorite common; non-porphyritic varieties may be variolitic (bearing globular or branching spherulites that are 0.1 to 1 in. in diameter); commonly pillowed.

c. **Rocks with phenocrysts of feldspathoids, analcite, or melilite.** (Where feldspathoid phenocrysts are scarce, these rocks may be mistaken for trachytes, andesites, or basalts. Occasional grains of the distinctively colored sodalite family minerals may indicate subsilicic compositions. Cores of flows and associated dikes should be searched for coarser-grained facies.)

Phonolite. Sanidine most abundant of phenocrysts; mafic grains scarce but a large variety possible; groundmass light to medium gray or greenish gray, except where dark and glassy.

Tephrite. Phenocrysts mainly augite, commonly with some plagioclase but no olivine; rock heavy (as basalt); groundmass dark and locally phaneritic.

Basanite. As tephrite, but with olivine.

Leucitite, nephelinite, etc. No feldspar; augite generally abundant and feldspathoid grains commonly obvious.

Pyroclastic rocks. Pyroclastic rocks are classified first on the basis of texture and second on the basis of composition. The classification of Wentworth and Williams (1932) is based primarily on textural characteristics, and a condensed version is presented below. The one major modification has been to place the limit between bombs or blocks and lapilli at 64 mm rather than 32 mm because the terms *bomb* and *block* seem to connote rather large objects.

a. Clasts larger than 64 mm (2½ in.).
  1. Angular fragments are *blocks;* aggregate is *volcanic rubble,* and lithified aggregate is *volcanic breccia.* Lithified aggregates of poorly sorted materials (with much ash) are *tuff-breccia.*
  2. Fragments that were plastic during eruption and partly shaped in flight are *bombs* (Fig. 13–4); both loose and lithified aggregates are *agglomerates* (or *agglutinate* where bombs were so soft they cohered).
  3. Fragments that were very fluid when they fell are *driblets* or *spatter;* aggregate is *driblet agglutinate.*

b. Clasts from 64 mm to 4 mm (2½ to ⅙ in.).
Fragments are *lapilli;* aggregate also *lapilli*, and lithified aggregate *lapilli tuff*. Scoriaceous and *pumiceous* lapilli are commonly rounded or drawn out during eruption; *lithic* lapilli are typically angular, while *accretionary* lapilli are pellets formed by accretion of fine ash, usually around raindrops, during eruption.

c. Clasts from 4 mm to ¼ mm (⅙ to ¹⁄₁₀₀ in.).
Loose aggregate is *coarse ash*, lithified aggregate *coarse-grained tuff*. The larger ash particles may have discernable textures, structures, and shapes, as for lapilli. If most fragments are glass, rock is *vitric tuff;* if most are crystals, rock is *crystal tuff*, and if most are rock fragments, rock is *lithic tuff;* nearly equal mixtures are *crystal-vitric tuff*, etc.

d. Fragments smaller than ¼ mm (¹⁄₁₀₀ in.).
Loose aggregate is *fine ash* or *volcanic dust;* lithified aggregate is *fine-grained tuff;* compositional adjectives used as in *c* above.

When the nature of the various clasts provides evidence of their source, the following adjectives can be used to describe them.

Essential or *juvenile.* Materials formed directly by eruption of magma.

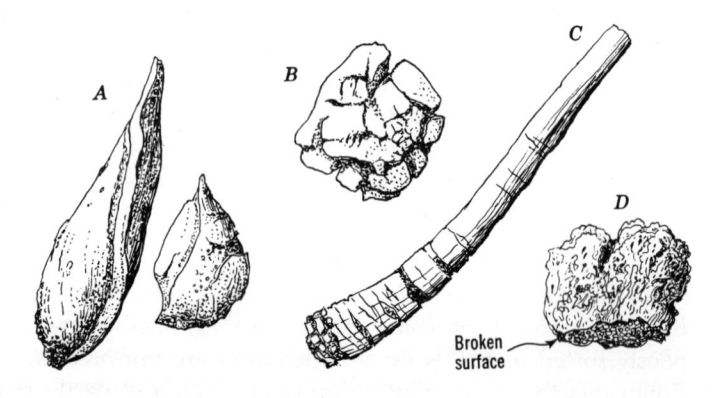

Fig. 13–4. Volcanic bombs. (*A*) Almond-shaped bombs from Silver Peak, Nev. (*B*) Breadcrust bomb (skin cracked prominently by expansion of vesicular core); from Mono Craters, Calif. (*C*) Ribbon bomb with tension cracks; Silver Peak, Nev. (*D*) Moderately shaped scoria bomb (probably the commonest type of bomb); Fujiyama, Japan; courtesy of W. M. Quackenbush. All bombs are 3 to 4 in. across.

Fig. 13–5. Welded rhyolite tuff from Bare Mt., Nev., showing black glass lenses formed by collapse of pumice lapilli, compactive effects around larger angular rock fragments, and abundant fragments of crystals. Unlike most lavas, these rocks do not show lineations because the foliation formed by flattening after the avalanches ceased to flow.

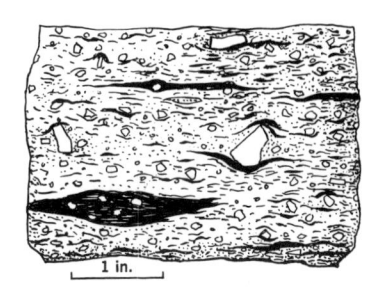

Accessory. Fragments of rocks closely related (cognate) to the eruption, as a vent-filling plug, an antecedent flow, or a pyroclastic bed of the same volcanic cone; *plutonic cognate ejecta* are coarse-grained crystal aggregates formed in the subjacent magma chamber or feeder.

Accidental. Fragments of rocks unrelated to the eruption; may be non-volcanic rocks from beneath the volcano, or fragments of extinct cones or feeders.

The term *ignimbrite* may be applied to all rocks laid down by glowing avalanches or pyroclastic flows (Section 13–6). The adjective *welded* (as *welded dacite tuff*) applies to pyroclastic materials that have been heated sufficiently to cohere. The most thoroughly welded tuffs are massive and look like vitrophyric lavas; others are light colored but with streaky or lenticular bodies of darker glass (Fig. 13–5), and still others are so devitrified or altered as to be red or light colored, stony, and without visible glass. Although most welded rocks are ignimbrites, intrusions and lavas can also weld ash or lapilli deposits.

13–4. Structures of Basic Lavas

A number of structures formed during movement and solidification of basic lavas can be used to determine the stratigraphic tops of folded sequences, as well as to interpret the overall shapes and directions of movement of individual flows. Single occurrences of some structures are significant; others, as small-scale flow structures, must be studied and plotted at a number of outcrops. Certain associations of structures may be so characteristic of a given flow or formation that they can be used in tracing units and mapping contacts. Viscosity and gas content of a lava may change as it spreads, however, so that lateral changes in some structures can be expected.

Very fluid basic lavas, as olivine basalts, generally have vague in-

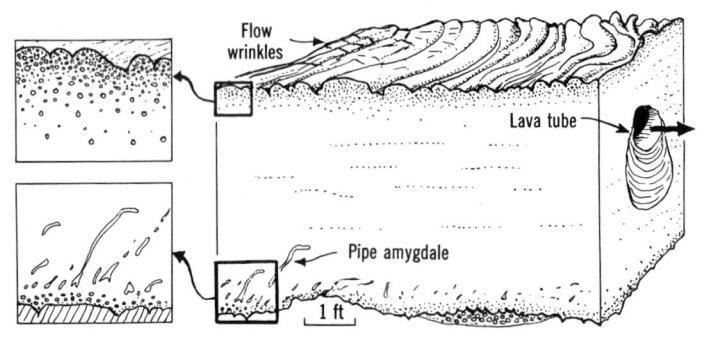

Fig. 13–6. Idealized section through flow of fluid basalt, showing partly filled lava tube, wrinkled and vesicular pahoehoe top, and pipe amygdales near base. Although the flow wrinkles can often be used to determine the stratigraphic top of a folded sequence, wrinkles can also form at the base, as shown in the drawing.

ternal flow structures consisting mainly of aligned plagioclase tablets. Small vesicles may form continuous sheets that give the rocks a banded appearance and a platy fracture, but these structures are not necessarily parallel to the primary flow direction. Larger vesicles are more abundant in the upper parts of flows, where they may be stretched parallel to the last movements in the lava. *Pipe amygdales,* however, fill tubular cavities formed by upward streaming of gases from the lower parts of the flow; thus their position or upward confluence indicates the stratigraphic base of beds (Fig. 13–6). Some of these tubes are filled with a pegmatitic residuum of the lava, others bear silica minerals, zeolites, or calcite.

Lavas called *pahoehoe* have billowy, hummocky, or ropy upper surfaces and vesicular glassy skins that give them a rough but glistening appearance. The crusts of these lavas are continuous over large areas but are commonly deformed by festooning, bulging, cracking, or collapse caused by movements of the underlying fluid lava (Fig. 13–7). Minor eruptive cones can be formed in this way; consequently they do not necessarily indicate an underlying vent. Lava tubes are common in pahoehoe flows; frequently their fillings give evidence of the tops of the flows (Fig. 13–6). At the edge of a flow, pahoehoe lava may advance as a series of thin tongues that override each other for distances of a few feet or a few tens of feet and then solidify. The base of each tongue tends to be cuspate or irregular, while the top is rounded, and therefore these accumulations can be used to determine

Field Work with Volcanic Rocks

Fig. 13–7. Tumuli (left), pressure ridge with squeeze-out of lava (center), and hornitos (right) on a pahoehoe flow. After various photographs from Wentworth and Macdonald (1953).

stratigraphic sequence (Fig. 13–8). As can be seen in the figure, a vertical section cut transverse to the tongues looks like pillow lava; associated rocks must therefore be examined to determine if the deposits are subaerial or subaqueous (see *Subaqueous flows*, below). Stearns and Macdonald (1946) and Wentworth and Macdonald (1953) have described other useful structures and forms in basalts of Hawaii.

Somewhat more viscous lavas tend to be scoriaceous near their tops and bases. If their upper surfaces are rough and jagged, they are often called *aa* lava. Pahoehoe flows commonly change to this type of lava as they move away from their source, especially where they flow over steep slopes. Some *scoria flows* consist of little else but a jumble of rough lumps, balls, and plates, which can be mistaken for pyroclastic (explosive) deposits. Flows may spread evenly over a plain, but on slopes of a few degrees they move as streams through their older and less mobile parts. These streams can become strongly discordant to levees of solidified lava which form along their borders (Fig. 13–9).

The interiors of most aa flows consist of massive lava, but some have a platy cleavage guided by sheets of thin vesicles. Phenocrysts

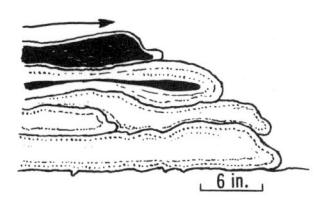

Fig. 13–8. Small pahoehoe tongues ("toes") accumulating to form a body whose transverse cross section (right) has a pillowed appearance. The right view is after a photograph from Stearns (1940, facing p. 88).

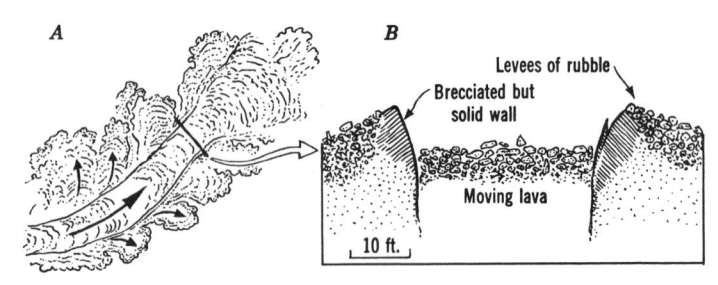

Fig. 13–9. Restricted lava stream in a large flow. (A) Map showing generalized pattern of ridges and lava movements, with major stream cutting through older lobes of the same flow; based on an aerial photograph (Krauskopf, 1948, facing p. 1278). (B) Enlarged section across major stream channel, showing moraine-like levees of scoria and brecciated but solid walls; after Krauskopf (1948, p. 1279).

are generally aligned in flow patterns, but these may be obscure. The more silicic rocks may show a vague banding similar to flow structures in viscous lavas (Section 13–5). Stratified or partially filled amygdales can be used to determine the original attitudes of some flows (Fig. 13–10A).

Columnar joints form in the nonscoriaceous cores of many flows, and they are commonly divided into two tiers (Fig. 13–11A). Some flows, however, have only one tier whereas others have three, so that interflow sediments or soils should be used where possible for counting flow sequences. In determining the dip of jointed flows at isolated outcrops, it cannot be assumed that the columns lie normal to the con-

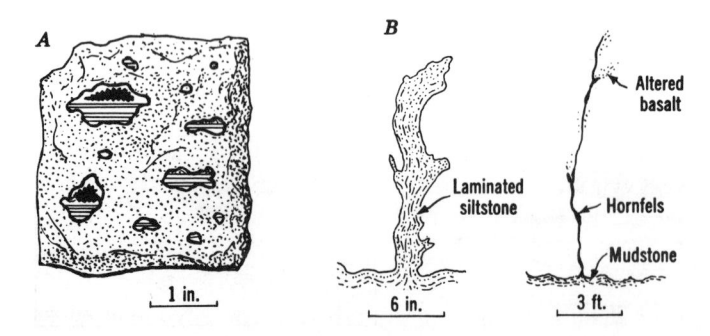

Fig. 13–10. (A) Hand specimen showing amygdales and geodes with stratified fillings of chalcedony and quartz. (B) Spiracles of laminated siltstone (left) and baked mud in basalt flow near Fallon, Nev.

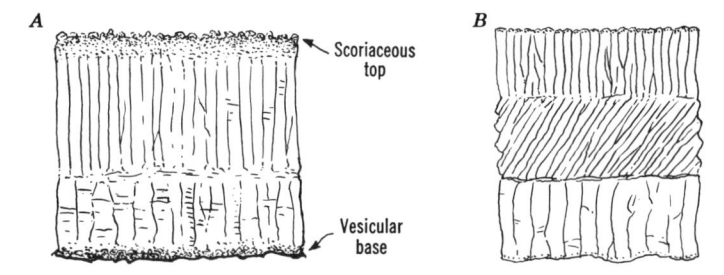

Fig. 13-11. (*A*) Section through a two-tier columnar lava flow. (*B*) Cross section through a columnar flow in which creep of the liquid interior (toward the right) resulted in inclined growth of the central tier of columns; after Waters (1960, p. 352).

tact of the flow, for widely divergent fans of columns are present in some flows. Waters (1960) described a particularly useful relation where columns in the center of flows are tilted in the direction of flow (Fig. 13-11*B*).

Pipe amygdales are uncommon in these more viscous flows, but large *spiracles* form where air, mud, or steam from wet ground rises into flows when they are still fluid (Fig. 13-10*B*). The upper ends of the spiracles are typically overturned in the direction of flow (Waters, 1960). Although spiracles can be used to determine stratigraphic sequence, they must be distinguished from clastic dikes formed by sifting of sediments into cracks in the tops of flows, and from dikes of partially fused sedimentary rocks that may intrude the borders of large sills or dikes (Section 13-7). Large-scale steam explosions at the base of a flow may form pipes that lead upward to surficial cinder cones, structures that might be mistaken for the eruptive vents of the flows themselves.

Subaqueous flows. If lavas are erupted under water or flow into water, *pillows* form where spurts of lava burst upward or outward and expand until chilled (Fig. 13-12). The glassy skins and zones of vesicles shown in the figure are characteristic, though some have hollow cores like the pahoehoe tongues described above (Fig. 13-8). Pillow lavas, however, are generally associated with aqueous sedimentary rocks or with beds of fragmented basaltic glass or palagonite. Palagonite tuffs are dark greenish-gray to black, have a subvitreous to waxy luster, and commonly carry calcite amygdales or cements. They may weather brown or may crystallize to iron-rich clays or chlorites, which give them a gray-green color and more stony appearance.

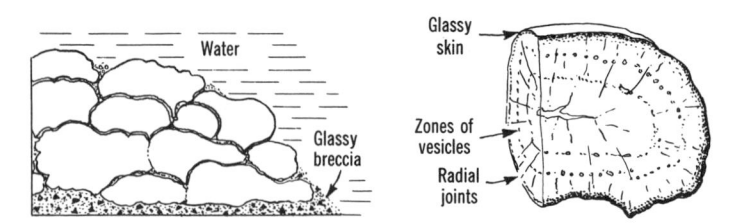

Fig. 13-12. Pillow lava, showing generalized section through deposit (left) and enlarged segment of single pillow. The pillows' rounded tops and cusped bottoms can be used to determine stratigraphic sequence in deformed rocks.

Clastic textures may be difficult to see in recrystallized rocks, but the tuffs can be recognized in some marine sequences by their association with red and green cherts and calcareous rocks. Fuller (1931) and Waters (1960) have described palagonite deposits in which the primary dips of foreset beds record the direction of flow of lava tongues into water.

13-5. Structures of Silicic Lavas

Silicic and moderately silicic (intermediate) lavas may be fluid enough to spread evenly for distances of several miles; many, however, are so viscous they form thick, short tongues or steep-sided domes. The shapes of ancient flows can often be determined by mapping their contacts with adjacent strata, but it is necessary to map flow structures of lavas that are poorly exposed or greatly dissected. Although the flow structures consist in part of oriented phenocrysts, their more obvious components are parallel layers of lighter and darker materials, ranging from 0.01 to 0.5 in. thick. The layers are typically folded and contorted in a way to suggest drag or liquid drift of one layer over another (Fig. 13-13A). The patterns of flow structures shown in Fig. 13-13B can be used to place scattered outcrops into approximate stratigraphic position, although these relations do not occur in all flows. The lighter layers may consist mainly of small spherulites. Some layers separate along sheets of closely spaced vesicles, giving the rocks a platy cleavage. The surfaces of the cleavage plates usually show scratches, furrows, or linear trains and smears of deuteric minerals (hematite, magnetite, tridymite), indicating translation of one layer over another (Fig. 13-13C). The lineated plates apparently form as the lava changes from a viscous fluid to a solid; in some cases the surfaces are shears which cut across earlier flow structures. The platy cleavage is more closely spaced in the lower parts of a flow or along the walls of a lava channel or conduit, where it may dip steeply (Fig. 13–13D). The overall pattern of flow layers and flow lineations in a viscous tongue is shown in Fig. 13–14.

In summary, five types of directional structures may be mapped in viscous flows: (1) planar, and perhaps linear, orientation of phenocrysts, (2) flow layers (generally parallel to phenocrysts), (3) shear surfaces (generally but not always parallel to earlier flow layers), (4) fold-axis lineations formed at right angles to flow, by drag of layers, and (5) lineations on surfaces of cleavage plates, formed parallel to direction of movement.

Domes. Viscous lava commonly accumulates directly over a vent to form a steep-sided *dome*. The lava in most domes is so nearly solid it rises as steeply inclined sheets and prisms, which protrude at the

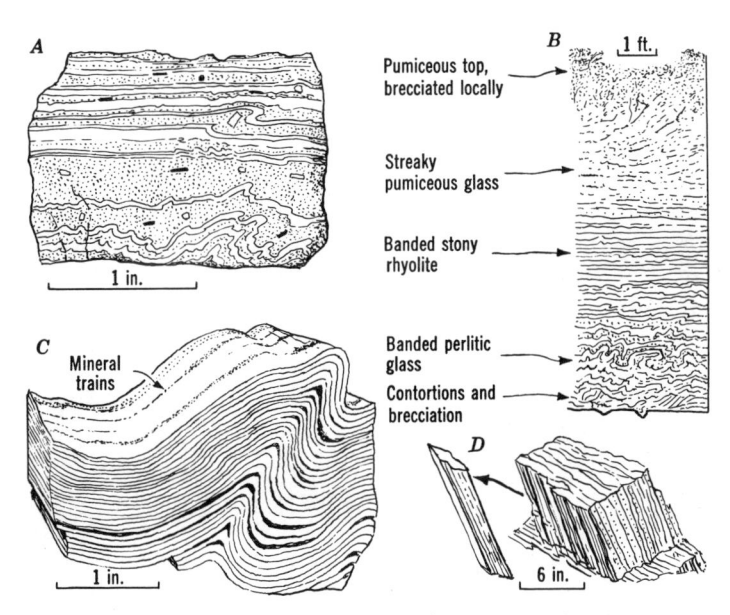

Fig. 13–13. Flow structures in silicic rocks. (A) Typical flow-banded rhyolite. (B) Generalized section through rhyolite flow near Silver Peak, Nev. (C) Folded flow layers that have separated and been coated by cristobalite. Note the two lineations. Santa Rosa Range, Nev. (D) Strongly furrowed flow layers with lineations that plunge steeply, parallel to direction of flow. Rhyolite intrusion southeast of Fallon, Nev.

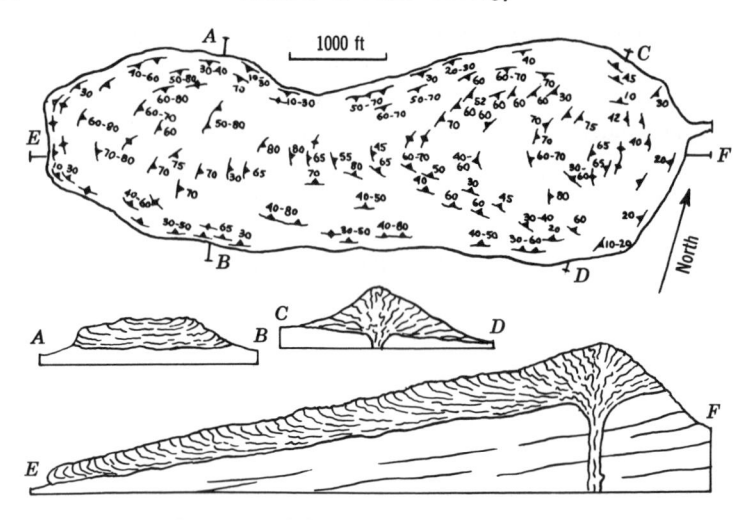

Fig. 13–14. Map and sections of the Watchman dacite flow, Crater Lake National Park, Oregon. After Williams (1942, p. 45). Courtesy of the Carnegie Institution of Washington.

upper surface of the dome as plates and spines. As these masses disintegrate, avalanches of angular fragments cascade down the sides of the growing dome (Fig. 13–15A). Inward-dipping flow layers in the solid interior of the dome are hidden by the apron of fragmental debris except at the summit, where partly broken prisms and plates can be seen. The steeply inclined surfaces of the prisms may be grooved in a way comparable to the shear surfaces described above.

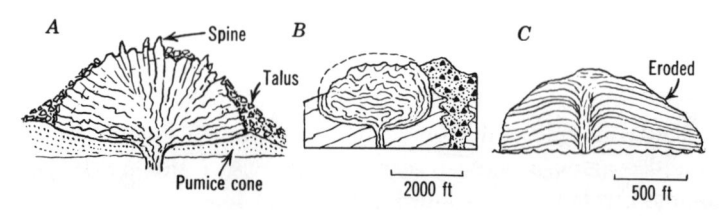

Fig. 13–15. (A) Viscous dome with protruding spines and talus apron. Mainly after examples described by Williams (1932). (B) Dome of Divide (Raker) Peak, Lassen National Park, protruded through flank of andesite cone. After Williams (1932, p. 145). (C) Eroded dome forming a small island off Ischia, Italy; it is cut by a fault so that its internal structures are shown clearly. After Rittmann (1930, p. 43).

Flow structures in some domes are roughly concentric with their gross shapes, but bulbous domes with strongly concentric planar structures are more likely to form as shallow intrusions (Fig. 13–15B). An uncommon but interesting type of extrusive dome is that formed when moderately fluid lavas erupt in rapid succession from a central conduit and pile up in a steep mound (Fig. 13–15C). The positions of vents beneath uneroded domes and viscous flows are indicated by patterns of flow structures and alteration zones. Fluids that pass upward through the vent commonly replace the lavas with tridymite, opal, cristobalite, or high-temperature quartz. The rocks may also be argillized to punky, light-colored masses stained variably by hematite, hydrated ferric oxides, or yellow and green clays.

13–6. Pyroclastic and Closely Related Deposits

Although magmas of all compositions may form pyroclastic deposits, the more viscous (silicic and alkaline) ones are far more frequently fragmented by explosions or frothed by rapid expansion of gases. The characteristics of pyroclastic deposits depend not only on the eruption itself but also on geographic conditions such as rainfall, drainage, and relief. Pyroclastic units may consist of simple, tabular beds or they may have linear or irregular shapes. Their textures and internal structures may be of great variety. The origin of a given deposit is determined by tracing out its shape, examining its composition and small-scale features, and recording lateral variations of these features. Ash-fall beds or distinctive ignimbrite sheets that can be traced laterally are especially valuable because they provide ideal stratigraphic datum planes (see, for example, Mackin, 1960). Useful characteristics of a deposit include the size, sorting, shape, and composition of fragments and the degree of their compaction, welding, or alteration. Brief descriptions of the principal kinds of deposits follow.

1. *Cones of scoria, pumice, or ash.* Formed by mild to moderate explosions from a central conduit; vague to distinct beds of various grain sizes dip outward from conduit; abundant bombs or shaped lapilli and coarse ash indicate nearness of vent; accidental blocks may be common; relation with interbedded lavas likely to be complicated by mixing and channeling; agglutination and gas alteration very close to vent.

2. *Widespread blankets of ash and lapilli.* Formed by major (Vulcanian) explosions; materials driven high in atmosphere so that they are cooled during fall (except very close to vent); differentiation produced

by rates of fall of particles results in concentration of large fragments and crystals, especially mafic ones, near source and concentration of glass, especially pumiceous ash, away from source; individual layers may also be graded from base to top; deposits thicken markedly near volcano; in distal parts, they thin gradually away from volcano; at any one place, thickness is the same on hilltops as in valleys, although this relation may be altered quickly by erosion; delicate pumice lumps of initial deposit not abraded. See also the descriptions by Williams (1942).

3. *Hot flows of ash and larger fragments* (*ignimbrites*). Formed from glowing avalanches or pyroclastic flows, driven mainly by gravity; deposits thick in valleys and thinner over plains, with ridges largely uncovered; deposits unsorted and therefore individually unstratified; some consist of chaotic mixtures of large blocks in a fine-grained matrix, others of vitric lapilli and ash with or without crystal fragments; pumice and scoria fragments typically abraded, and bedrock surface under flow may be abraded and striated; high temperatures indicated by charred wood, reddening of pumice, and, in some cases, welding of vitric materials in the lower part of deposit; hot gases may form spiracles (fossil fumaroles) or may recrystallize and lithify much of deposit; columnar joints common, and larger lithic fragments may show contraction joints (Fig. 13–16). See also the descriptions by Gilbert (1938) and Smith (1960).

4. *Volcanic mudflow deposits* (*lahars*). Formed by rapid mixing of loose pyroclastic debris and abundant water; move down river systems to form thick linear deposits in valleys and broader deposits on plains; mostly chaotic mixtures of large and small fragments in a fine-grained matrix, but contain lenses of sorted sand and gravel where streams reworked the deposits; wood uncharred, unless charred fragments have been incorporated from ignimbrite deposits; may interfinger distally with stream, lake, or marine deposits. See also Anderson (1933) and Fisher (1960).

5. *Water-laid pyroclastic beds.* Formed where pyroclastic materials form in or fall into water, or where loose materials are reworked by water; if reworked, many fragments are abraded, size sorted, and mixed locally with nonvolcanic debris, but single deposits erupted into quiet water may be unstratified and entirely volcanic; parallel orientation of glass chips may give vitric tuffs a striking planar fabric; vitric marine and lacustrine deposits commonly argillized to bentonite beds, or altered diagenetically to cherty keratophyre tuffs or to zeolitic beds;

Fig. 13–16. Andesite block in unsorted tuff-breccia deposit, showing radial joints interpreted to indicate cooling after fragment broke from its source. After Shelton (1955, pl. 4).

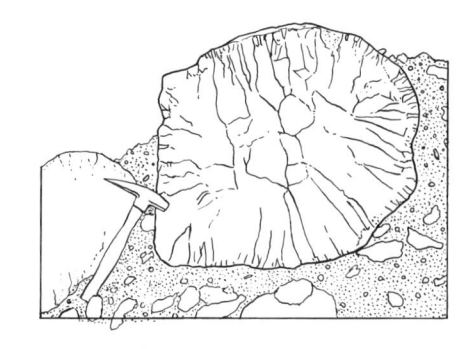

deposits of basic glass and palagonite commonly associated with pillow lavas (Section 13–4).

Various kinds of deposits may be interstratified around one volcano, and some deposits may grade laterally into other types. The upper or distal parts of an ignimbrite deposit, for example, may be incorporated into a volcanic mudflow, or an ash blanket may be overridden by, and incorporated into, a pyroclastic flow. Where the textures and compositions of successive ash-blanket deposits are very similar, the sequences can be subdivided at weathered zones, thin vegetative layers, or layers and lenses of stream deposits or wind-worked materials. Contacts between closely similar ignimbrite sheets can often be drawn at the base of each zone of welded or compacted pumice fragments.

13–7. Volcanic Feeders and Related Intrusions

Volcanic intrusions that cannot be mapped to scale should be plotted by symbols showing the trend of their longest dimension. Central conduits often lie at the focus of radiating fractures or at the intersection of two fractures. Positions of major subjacent intrusions ("magma chambers") can sometimes be found by mapping concentric patterns of dikes or elliptical groups of volcanic pipes. Where the major intrusions are exposed, they are commonly phaneritic and therefore plutonic in appearance.

The rocks of volcanic feeders and intrusions can sometimes be matched with the flows and pyroclastic beds that they feed. Direct connections between feeders and flows are rarely seen; hence most correlations must be made by matching the lithology of the intrusions with that of nearby flows and tuffs, and by determining the sequence of intrusion in composite necks and correlating it with stratigraphic sequences.

Dikes and sills. In mapping dikes and sills, the major difficulty is to distinguish them from associated lavas. Near the surface, steep basaltic feeders may be as fine-grained and vesicular as the lavas they intrude; however, they can be distinguished by their steeply inclined flow structures and steep, chilled contacts. At greater depth, where the dikes are less vesicular, most of them have simple sets of columnar joints lying at right angles to their contacts (Fig. 13–17A). The central parts of more deeply seated basic dikes consist of coarser-grained rock (diabase or dolerite), even though the margins are aphanitic. The central parts of thick basalt flows, however, may also be coarse-grained.

Dikes of silicic rocks are typically fine-grained, even at considerable depths, and are commonly less tabular than dikes formed from more fluid magmas. They generally have flow structures like those of silicic lavas; these structures lie parallel to their steep contacts and may be flow-folded or brecciated along them. At moderate depths, most silicic intrusions have margins of dark glass or very fine-grained spherulitic rock that grade inward to stony, more coarsely spherulitic, or very fine-grained phaneritic rocks. They may also show steeply inclined autobrecciated bands. Pyroclastic rocks may be formed where pyroclastic explosions tap deeply into viscous feeder-dikes.

Sills may be particularly difficult to recognize because their flow structures and contacts lie parallel to those of the lavas and tuffs that they intrude. They can be identified most readily where they cut across flows, send apophyses into overlying beds, or carry inclusions of the overlying rocks. They are also typically less vesicular and have

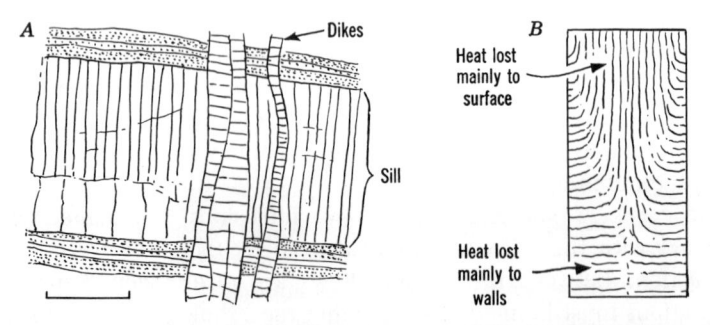

Fig. 13–17. Joint patterns in cross sections of minor intrusions. (A) Dolerite (diabase) sill cut by two dikes; Bradford Bay, Isle of Skye, Scotland. (B) Idealized vertical section through cylindrical volcanic feeder, showing change in joint pattern with depth.

simpler primary joint patterns than lavas with which they are associated, and they lack the highly vesicular or eroded tops that characterize many flows. Thick or more deeply seated sills are generally rather coarse-grained and alter the overlying rocks perceptibly. Many basic sills and dikes weld silicic vitric tuffs for distances of a few inches or feet from their contacts. Some melt siltstones or sandstones which may then be mixed or injected into the intrusions.

Volcanic necks. The necks or conduits of central-type volcanoes are roughly equidimensional in plan view, although they may be connected with complexes of radiating, intersecting, or concentric dikes and sills. Some are only pipelike enlargements or channels along dikes. In deeply eroded areas it can be difficult to determine if an intrusive pipe was ever a volcanic conduit. However, exposures near the original level of eruption show a systematic outward dip of lavas and pyroclastic beds erupted from the vent. High-level exposures of necks are also likely to contain bodies of agglomerate or agglutinated breccias that have slid down from the crater walls. Some may contain fragments of water-laid sediments that were deposited in lakes or ponds in the overlying crater or caldera and have subsided into the neck. The border zones of silicic necks are commonly perlitic or pumiceous near the surface, especially where they intruded water-rich wall rocks. Intense alterations of rocks by escaping volcanic gases are also indicative of exposures near the original vent level (see *Domes,* Section 13–5). Finally, the pattern of columnar joints in otherwise massive necks may indicate the relative depth of exposure (Fig. 13–17B).

Pyroclastic textures alone do not mean that the rocks formed near the level of the original vent, for explosions can reach to considerable depths. Most clastic necks consist of nonvitric breccias or tuff-breccias containing various amounts of broken country rocks. The fragments of country rock may be angular, abraded, or well rounded and in some pipes they constitute almost all of the breccia. In general, the deeper the source of the fragments, the smaller they are, and the more they are rounded (see Williams, 1936, p. 132).

Intrusive sequence in composite or multiple necks is determined by the usual relations of cross-cutting, diking, chilled margins, and inclusions. Where basaltic and silicic magmas are erupted simultaneously into a neck or dike, the basic magma can chill against the silicic one and at the same time be intruded by it (see Bailey and McCallien, 1956). Anomalous fine-grained margins can also be produced by contact-alteration of older rocks by younger magmas.

Superficial intrusions. Intrusive relations produced at or near the surface of a volcanic field may give spurious indications of volcanic feeders. Lava streams, for example, can sink beneath and flow under older lavas, thereby producing local, superficial sills (Krauskopf, 1948, p. 1280). Deep cracks in lavas are commonly filled from above with loose ash, forming pyroclastic dikes that may be very impressive in a single outcrop. Williams and Meyer-Abich (1955, p. 32) described superficial pyroclastic dikes formed where fractures opened during earthquakes and were injected with water-soaked pumice.

References Cited

Anderson, C. A., 1933, The Tuscan formation of northern California: *University of California Publications in Geological Sciences*, v. 23, pp. 215–276.

Bailey, E. B., and W. J. McCallien, 1956, Composite minor intrusions, and the Slieve Gullion complex, Ireland: *Liverpool and Manchester Geological Journal*, v. 1, pp. 466–501.

Fisher, R. V., 1960, Criteria for recognition of laharic breccias, southern Cascade Mountains, Washington: *Bulletin of the Geological Society of America*, v. 71, pp. 127–132.

Fuller, R. E., 1931, The aqueous chilling of basaltic lava on the Columbia River Plateau: *American Journal of Science*, v. 21, pp. 281–300.

Gilbert, C. M., 1938, Welded tuff in eastern California: *Bulletin of the Geological Society of America*, v. 49, pp. 1829–1862.

Krauskopf, K. B., 1948, Lava movement at Parícutin Volcano, Mexico: *Bulletin of the Geological Society of America*, v. 59, pp. 1267–1283.

Mackin, J. H., 1960, Structural significance of Tertiary volcanic rocks in southwestern Utah: *American Journal of Science*, v. 258, pp. 81–131.

Rittmann, A., 1930, *Geologie der Insel Ischia:* Zeitschrift für Vulkanologie, Ergänzungsband 6, 265 pp.

Shelton, J. S., 1955, Glendora volcanic rocks, Los Angeles Basin, California: *Bulletin of the Geological Society of America*, v. 66, pp. 45–90.

Smith, R. L., 1960, Ash flows: *Bulletin of the Geological Society of America*, v. 71, pp. 795–842.

Stearns, H. T., 1940, *Geology and ground-water resources of the islands of Lanai and Kahoolawe, Hawaii:* Hawaii, Div. of Hydrography, Bulletin 6, 177 pp.

Stearns, H. T., and G. A. Macdonald, 1942, *Geology and ground-water resources of the island of Maui, Hawaii:* Hawaii, Div. of Hydrography, Bulletin 7, 344 pp.

———, 1946, *Geology and ground-water resources of the island of Hawaii:* Hawaii, Div. of Hydrography, Bulletin 9, 363 pp.

Waters, A. C., 1955, Geomorphology of south-central Washington, illustrated by the Yakima East quadrangle: *Bulletin of the Geological Society of America*, v. 66, pp. 663–684.

———, 1960, Determining direction of flow in basalts: *American Journal of Science*, Bradley Volume, v. 258-A, pp. 350–366.

Wentworth, C. K., and G. A. Macdonald, 1953, *Structures and forms of basaltic rocks in Hawaii*, U.S. Geological Survey, Bulletin 994, 98 pp.

Wentworth, C. K., and Howel Williams, 1932, *The classification and terminology of the pyroclastic rocks:* National Research Council, Bulletin 89, pp. 19–53.

Williams, Howel, 1932, The history and character of volcanic domes: *University of California Publications in Geological Sciences*, v. 21, pp. 51–146.

———, 1936, Pliocene volcanoes of the Navajo-Hopi country: *Bulletin of the Geological Society of America*, v. 47, pp. 111–171.

———, 1942, *The geology of Crater Lake National Park, Oregon:* Carnegie Institution of Washington, Publication 540, 162 pp.

Williams, Howel, and Helmut Meyer-Abich, 1955, Volcanism in the southern part of El Salvador: *University of California Publications in Geological Sciences*, v. 32, pp. 1–64.

14

Field Work with Igneous and Igneous-Appearing Plutonic Rocks

14-1. Concepts of Plutonic Geology that Apply to Field Studies

The procedures and terminology presented in this chapter relate both to igneous rocks and to metamorphic (typically metasomatic) rocks which have the appearance of igneous rocks. This organization has been used because field studies can rarely be started on the basis of genetic classifications. Similarly, studies involving so-called *plutonic* and *volcanic* rock associations should not be limited by genetic concepts, for coarse-grained ("plutonic") igneous rocks are known to have formed within a few thousand feet of the surface or even as segregations in lavas. Moreover, migmatites, thought generally to indicate deep-lying processes, occur also around some high-level intrusions. Preconceived ideas regarding depths of geologic processes must therefore be applied with caution. Thorough, unbiased studies are generally essential to understanding conditions of igneous activity. In this book, volcanic, plutonic, and metamorphic studies have been separated into three chapters purely for convenience; in actuality, the various features considered in these chapters may often be found inseparably associated.

Whereas many volcanic processes can be observed and measured directly, plutonic processes must be interpreted indirectly. Sequences of plutonic events are imprinted as a series of superimposed structures and textures, and all too commonly the last of these has erased the others. Interpreting plutonic rocks thus requires sifting out evidence by working back in time from the youngest or most obvious features to relics of former events. These events must often be reconstructed from data gathered at widely distributed outcrops, and therefore the events can be understood only after mapping is well advanced. Within a given pluton, there are four general classes of features that can be studied, those indicating: (1) movements of magma or solid rock, (2) sequence of intrusion, (3) rate and sequence of crystallization, and (4) concentration and movement of fluids. Mapping should generally cover more than one of these classes of features. The more important described here are:

1. Rock variants, based on composition or texture.
2. Various kinds of contacts, including contacts within plutons.

3. Various kinds of planar or linear fabrics and structures, especially those related to inclusions.
4. Swarms of parallel dikes, veins, or fractures.
5. Distribution of various kinds of altered rock.

Mapping the country rocks for considerable distances around a pluton may be more useful than detailed studies within the body, as may delineation of metamorphic zones in the surrounding terrain (Section 15–13).

14–2. Cartographic Units of Plutonic Igneous Rocks

Plutonic igneous rocks may be mapped as separate units based on compositions, textures, or small-scale structures. Some nearly identical rocks can be mapped separately on the basis of intrusive (age) relations. Just as in sedimentary terrains, ideal cartographic units have genetic meaning. Because of the lack of simple stratigraphic relations in most plutons, the continuity of plutonic igneous units cannot be predicted at the outset of a survey. A reconnaissance should consequently be made before starting to map a large number of units that may be discontinuous, gradational, or otherwise obscure. Even after units have been selected, several weeks' work may be required to determine the best criteria for mapping them. In general, the more units that can be mapped in a pluton or group of plutons, the better the chances of determining their origin.

Although geographic names need not be used during the field season, they must be selected before a report can be written. Generally speaking, selection of names is based on the principles and rules that guide the naming of sedimentary units (Section 12–2).

Units based on composition. Units based mainly on mineral composition (mode) are used most widely. Although many have been formalized by geographic names (as *Lebanon Granite, Bonsall Tonalite*), formal designations should be deferred until a rock has been mapped over so large an area that *its contact relations to other rocks are known*. Where these relations are unknown, where correlations have not been checked, or where the rock has not been studied petrographically, informal designations, as *granite of Mapes Hill, basic rocks of Selby Valley*, should be used. A system that is useful for composite intrusions is to give a geographic name to the pluton and refer the various rock units to it, as *syenite of Jonas stock, layered gabbros of Hadley pluton*. Although these names are used informally, they should not duplicate existing formation names.

Units based on texture and structure. Units based on textures or structures may have at least as much genetic meaning as units based on composition; however, most of them are not formalized as formations, especially where they are variants in a body with one composition. Because textures and structures within a pluton vary with *conditions* of origin, it is appropriate to use the term *facies* for these synchronous units. Where mapped within one compositional unit, they might be given a name such as *fine-grained facies of Smith granite;* where mapped through a suite of closely related compositional units, a name such as *layered facies of Coldwell pluton* might be used. These units can be based on characteristics such as porphyritic texture, seriate texture, various ranges of grain size, poikilitic grains, grain shapes (as hypidiomorphic versus allotriomorphic), miarolitic cavities, compositional or textural layering, linear structures, various schlieren, protoclastic textures and structures, and, possibly, abundance of inclusions. An example of a project where textural and structural facies were critical to interpreting a pluton has been given by Waters and Krauskopf (1941).

Units related by age or association. Customarily, related plutonic igneous formations have not been assembled into *groups;* instead, the term *series* has been widely used (in some cases implying a sequence of units). Because of the restricted stratigraphic meaning of *Series,* it is preferable to use the term *Group* in formal designations. Terms such as *suite, sequence,* or *association* can be used informally. Inserting the adjectives *igneous, intrusive,* or *metasomatic* will help clarify their meaning, as *Wardlaw intrusive sequence, metasomatic suite of Jacks Valley.*

Units based on alterations. Units based on deuteric or secondary alterations can be utilized to trace movements of fluids in plutons and may be essential to interpreting ore deposits. These units are fundamentally *alteration zones,* superimposed on the original compositional or textural units. If geographic names are needed to designate alteration zones, they should be used informally, as *kaolinized Jonas Granite, uralite zone of Hartford Gabbro,* and *chloritized breccias of Haley fault.*

14–3. Naming Plutonic Rocks

Where the origin of a rock cannot be determined readily, the name used should be based on texture and composition. Names suggested in this section are appropriate for rocks with phaneritic (visible) grains just as those in Section 13–3 apply to rocks that are largely aphanitic.

The name *rhyolite*, for example, would apply to intrusive as well as extrusive rocks of appropriate texture and composition.

Figure 14–1 presents a classification of phaneritic rocks based on their principal minerals. At each triangular corner, the rock would consist only of the mineral shown there; compositions of rocks at intermediate points are proportional to the distances from the appropriate corners, as indicated by the numbers on the diagrams. Adamellites, for example, contain less than 40 percent of mafic minerals and more than 10 percent of quartz, while between one third and two thirds of their total feldspars are plagioclase. The various 10 percent divisions are used instead of 5 percent divisions (as those of Johannsen, 1939) to enlarge the fields of some names and narrow those of others. The adjectives *quartz-bearing* and *nepheline-bearing* are useful for rocks in the fields along the line joining K feldspar and plagioclase (as *quartz-bearing syenite*). The names that follow are based on special textures or compositions:

Porphyritic rocks (as *porphyritic granite*) have very large grains in a phaneritic groundmass (but see *porphyries*, below).

Porphyries (as *granite porphyry*) have phenocrysts in a very fine-grained (less than about 0.5 mm) phaneritic groundmass.

Pegmatites (as *granite pegmatite*) are very coarse-grained and commonly show marked textural variations.

Aplites (as *granite aplite*) have a fine-grained granular texture (with or without phenocrysts) and contain less than a percent or two of mafic minerals.

Lamprophyres are dark porphyries that contain abundant mafic grains both as phenocrysts and in the groundmass. Many varieties have been named, but groundmass minerals may be so difficult to recognize that terms such as *biotite lamprophyre* and *hornblende-augite lamprophyre* must be used at most outcrops.

Many useful textural varieties of plutonic rocks can be named by using adjective modifiers. Traditional classifications of grain size (as *fine-grained*, less than 1 mm; *medium-grained*, 1–5 mm; *coarse-grained*, more than 5 mm) may or may not be useful to a given project. These limits can be adjusted arbitrarily, or, far better, actual grain sizes can be recorded (as *1 mm granite*). The adjective *seriate* may be used for rocks whose phenocyrsts range from large grains to groundmass size. Classification of feldspathic rocks into hypidiomorphic and allotriomorphic varieties may be useful because wholly anhedral feldspars (especially plagioclase) suggest recrystallization in the solid or nearly

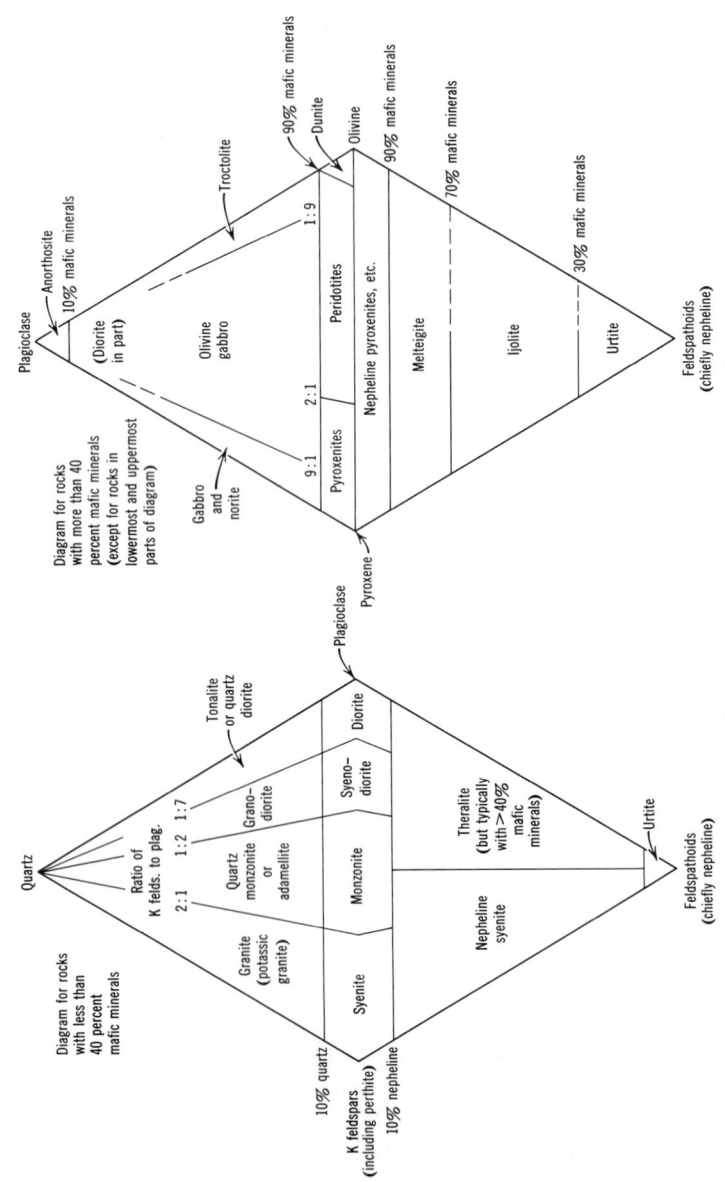

14-1. Classification of igneous and igneous-appearing plutonic rocks.

Field Work with Igneous and Igneous-Appearing Plutonic Rocks

solid state. Oriented fabrics are also important, and they will be considered in Section 14–5.

Color index (percent of mafic grains, by volume) affords a simple means of classifying some rocks. The charts of Appendix 3 can be used to estimate color index, although variations of overall color with grain size must be considered also. In addition to color index, the ratios between distinctive mafic minerals can often be used to identify rocks whose essential minerals are difficult to estimate in hand specimen. Many adamellites, for example, carry only biotite as a mafic accessory, whereas most granodiorites carry both biotite and hornblende.

Specific gravities. Rocks that look alike in hand specimen can sometimes be named and mapped according to their specific gravities. Specific gravity balances are too cumbersome to carry in the field, but they can be used in camp to determine large numbers of specimens. Readings within about 0.03 of absolute specific gravity can be made quite quickly if rocks are fresh and not too porous. This degree of precision is sufficient to distinguish many pairs of similar rocks. Another method, suggested by R. G. Coleman (personal communication), is to carry wide-mouthed bottles of graduated heavy liquids in the field and to drop small pieces of rock into them to see if they sink or float. Three or four $1\frac{1}{2} \times 3$ in. amber bottles should be adequate if the approximate specific gravities of the rocks are known in advance. Mineral chips of suitable specific gravity should be placed in each bottle so that changes in the specific gravities of the liquids can be detected. Because the liquids are volatile and reactive, the bottles should be carried in a strong case, not in a knapsack.

14–4. Contacts of Plutonic Units

Contacts between plutons and country rocks will here be called *external* contacts, and those between units within plutons will be called *internal* contacts. Either type of contact may be sharp or gradational and, in addition, may be described by such terms as concordant, discordant, sinuous, angular, irregular, smooth, blocky, veined, and so forth. Contacts can also be classified on a genetic basis, but usually not until they have been extensively mapped and examined. Geometric relations of dikes, inclusions, and other structures, however, may show *at a particular place* that the contact originated by intrusion, by metasomatism, or by mixing of two magmas.

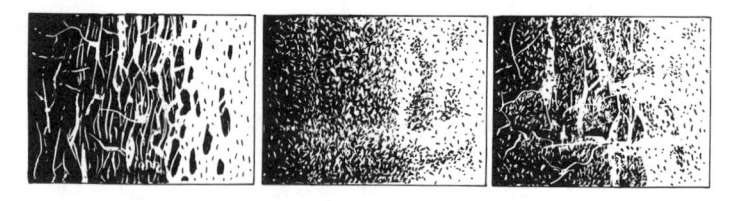

Fig. 14–2. Veined (left), permeated (middle), and veined and permeated gradations between homogenous igneous and metamorphic rocks.

External contacts. Sharp external contacts can be mapped by the procedures described in Section 4–7. They must, however, be walked out completely because they can change locally to broadly gradational zones.

General procedures for mapping gradational contacts are given in Section 4–9. Gradations of plutonic rocks to country rocks may be *veined, permeated* (or *porphyroblastic*), or some mixture of the two (Fig. 14–2). Depending on the nature of the gradations and the degree of exposure, the contacts can be drawn either where the mixture of country rock and plutonic rock consists of approximately equal parts of each, or where the plutonic rock forms a continuous matrix to fragments of country rock. The veined or permeated country rocks may also be mapped as a zone along the contact (Section 15–14). Dikes in the zone of gradation should be examined for geometric proof of dilation, or lack of dilation (Fig. 14–3), for discordant flow structures and foreign inclusions (Fig. 14–4A), and for evidence of forceful movements of magma against or into wall rocks (Fig. 14–4B).

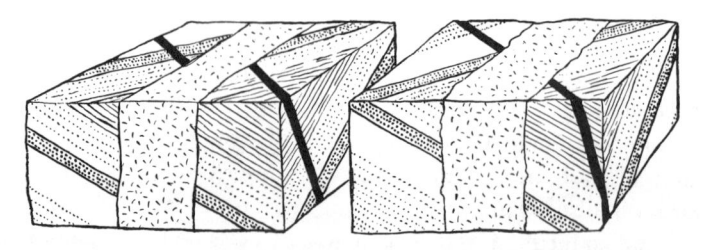

Fig. 14–3. Intrusive (dilative) dike (left) and replacement dike. The beds and black vein in the left block have been separated at right angles to the dike walls, as can be proven by matching their lines of intersection across the dike. In the case of the replacement dike, all structures project through the dike as though it were not there, showing that there has been no dilation.

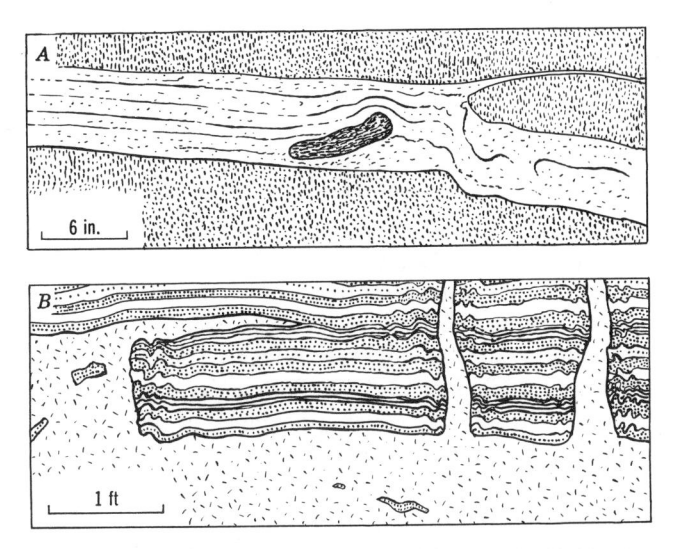

Field Work with Igneous and Igneous-Appearing Plutonic Rocks

Fig. 14–4. Intrusive relations. (A) Horizontal granitic dike with flow structures that are discordant to the fabric of wall rocks, a rounded inclusion of schist that is foreign to the walls, and geometrically matching walls. Near Monadhliath pluton, southwest of Aviemore, Scotland. (B) Wedge-shaped apophyses that compress bedding adjacent to their walls; at contact of granodiorite stock, Santa Rosa Range, Nev.

Interpretations of veins and dikes are discussed further in Section 15–14.

Internal contacts. Sharp internal contacts between contrasting rocks can be mapped readily; however, they may change unpredictably to gradational zones of intermediate rocks. Discontinuous sharp contacts between similar rocks may indicate local erosion or intrusion by the magmatic core of a pluton. Some may be faults that passed from nearly solid rock into mobile magma. Where rocks are similar, minor textural or compositional characteristics must be utilized to distinguish sharp contacts. The contacts are sometimes indicated by discordant planar or linear fabrics, by swarms of inclusions, by abrupt changes in dike or fracture patterns, or by zones of protoclasis or alteration. If the contacts go undetected until mapping reveals major structural discontinuities, outcrops should be revisited to determine age relations. These relations can be determined on the basis of: (1) dikes of the younger rock in the older, (2) inclusions of the older rock in the younger, (3) alteration of the older rock by the younger, and (4) the fact that the younger rock cuts across veins or inclusions in the older.

Gradational internal contacts are typically obscure and therefore difficult to map. Some gradations within plutons are so broad that the term *contact* loses its structural significance. In feldspathic rocks, the gradations can often be traced on the basis of color index, or on the basis of the ratios of critical mafic minerals, as biotite and hornblende in granitic rocks. Although fresh, dark-colored rocks are difficult to map by color index, weathered surfaces often show minerals clearly. Feldspars are generally whitened by weathering, while olivine is altered to limonitic pits and hypersthene is bleached to golden or bronze-colored cleavable grains. Gradational contacts between mafic rocks can also be keyed on the first appearance of distinctive accessory minerals, as biotite, olivine, or quartz.

After the criteria for drawing gradational contacts have been established by studying several complete gradations, rock chips from these gradations can be carried in the field as a standard for plotting other outcrops. The color-spot method is especially useful for this type of mapping (Section 4–10).

Zones of mixed (hybrid) magmas. Some gradational zones suggest that one magma intruded into or against another. If both magmas were largely liquid and had similar physical properties, the only textural evidence of mixing may be corroded grains that have survived interaction in the hybrid magma. If one magma was nearly solid, the hybrid zone can be expected to show partially disintegrated and reacted inclusions of one rock in the other. Where the more crystalline magma was pliable, the inclusions may be spindle shaped or streaked into lines and stratiform mixtures of the two rocks. A striking structural relation is that where one rock occurs both as dikes *and* as inclusions in another. This condition may arise because the two magmas had different temperatures of crystallization. Gabbroic magma, for example, may locally be chilled against granitic magma.

Although the rocks in the hybrid zone may have compositions suitable to simple mixing of the two adjacent magmas, volatile materials can cause strikingly heterogeneous textures. Where basic and granitic magmas have mixed, for example, the basic rocks may have large poikilitic grains of potassium feldspar or biotite, and the granitic rocks may have clots of mafic grains or orthoclase phenocrysts with rims of plagioclase.

Deformed contacts. Convincing evidence of intrusion or metasomatism may be difficult to find where contacts have been deformed

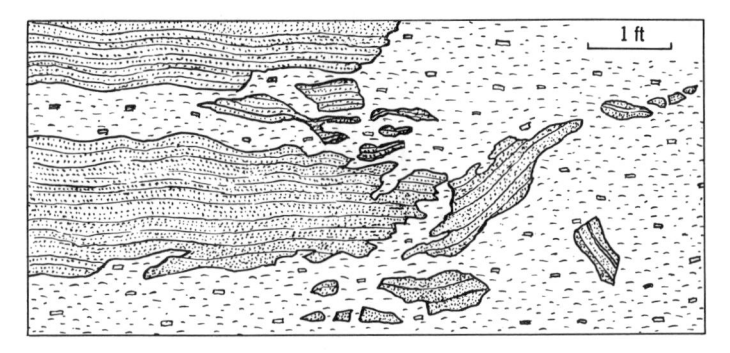

Field Work with Igneous and Igneous-Appearing Plutonic Rocks 281

Fig. 14–5. Granitic rock with a foliation that passes into country rocks without deflection. Contact zone of Flamanville granite at Anse de Sciotot, Normandy, France (see also the figures of Martin, 1953).

during late-magmatic or postmagmatic stages. Structural relations commonly appear contradictory. In Fig. 14–5, for example, the granitic rock is intrusive, but late-magmatic deformation of both the intrusion and its walls has produced a through-going planar fabric that makes the dikes appear metasomatic. The strongly lineated, plicated contact of Fig. 14–6 was probably also formed in this way.

If deformation occurs when the rocks are almost solid but still hot enough to recrystallize moderately, broken or bent grains should be apparent (Fig. 14–7C in Section 14–5). Adjacent country rocks should also be sheared and granulated.

Fig. 14–6. Contact of the Ardclach granite on Hill of Aitnoch, Nairnshire, Scotland. Note that the planar fabric of the granite crosses the contact at various angles but is typically parallel to layering in the adjoining metamorphic rocks. There is a strong lineation in the schists but not in the granite.

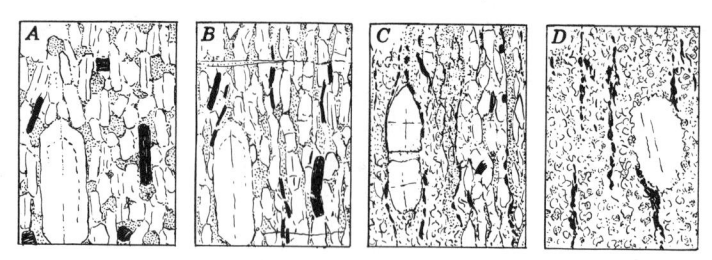

Fig. 14–7. Planar fabrics in plutonic (here, granitic) rocks. Quartz is stippled, feldspar unpatterned, and mafic grains black. (A) Simple flow fabric in hypidiomorphic granular rock. (B) Fabric showing effects of late flow (note change in shapes of quartz grains compared to A). (C) Strongly protoclastic fabric. (D) Granitic gneiss, showing lenses of biotite grains and simple granoblastic texture which have resulted from recrystallization during and following deformation.

Dynamothermal metamorphism typically makes igneous rocks schistose or gneissose (Section 15–3). Although the new foliations commonly parallel contacts for long distances, they cross them locally, either without deviation or with a distinct sigmoidal bend. These relations must be examined over large areas to determine whether the body is a metamorphosed igneous pluton or a metasomatic pluton.

14–5. Planar and Linear Fabrics in Plutonic Rocks

Rocks have a *planar fabric* when platy grains are oriented so as to be more or less parallel and linear grains lie in various directions in this plane. Rocks with a *linear fabric* have elongate grains aligned in a linear sense, as logs in a stack of cord wood. Many plutonic igneous rocks have discernable planar fabrics and less obvious linear fabrics that lie in the plane of the planar fabric. These terms refer to textural aspects of the rocks, that is, to statistical orientations of shaped grains. Fabrics may or may not be associated with planar or linear structures such as compositional layers (or bands), schlieren, and inclusions; furthermore, rock fabrics may or may not be parallel to these structures.

The genetic term *flow structures* should be reserved until the origin of a given fabric has been estimated, for planar and linear fabrics may originate in a number of ways. They may form: (1) by streaming or viscous flow in magma that is largely liquid (hence the term *flow structure*), (2) by rotation, shearing, and recrystallization in magma that contains only moderate or small amounts of liquid, (3) by entirely metamorphic deformation (solid flow), (4) by accumulation and compaction of tabular or elongate crystals and magma on a smooth floor, (5) by outward growth of elongate grains from a solid surface, and (6) by inheritance of metamorphic fabrics, as during metasomatism. The following criteria can be used to interpret textures at any one outcrop.

1. *True igneous flow.* Many grains are euhedral or subhedral (Fig. 14–7A); biotite is often in pseudohexagonal, equidimensional grains that are not well oriented. Tabular minerals, especially feldspars and biotite, are best oriented; quartz is in equidimensional anhedral patches or in euhedra and rounded subhedra; lineation is typically formed by hornblende prisms, elongate feldspar tablets, and pyroxene prisms.

2. *Largely solid igneous flow.* Quartz is anhedral and somewhat lenticular or lineate; biotite typically forms flakes or aggregate masses with grossly linear or planar shapes; feldspars are locally bent or broken; fractures at about right angles to linear and planar fabrics are common (Fig. 14–7B). Mafic aggregates in basic rocks show somewhat elongate or lensoid patterns. *Protoclastic* textures occur where recrystallization is slight, leaving abundant evidence of granulation; characteristically, streaks of biotite and thin sheets of recrystallized quartz are molded around lenticular or rhombic fragments (porphyroclasts) of feldspar (Fig. 14–7C).

3. *Solid (metamorphic) flow.* Fabric is stronger and more persistent than in igneous types, with exception of some protoclastic rocks; feldspars are anhedral and in granoblastic aggregates with quartz (Fig. 14–7D); mafic grains, especially biotite, are in well-oriented individual plates or aggregates, as in semischists and gneisses (Section 15–3); in more schistose rocks, fine-grained oriented muscovite and chlorite can pseudomorph former feldspars and mafic grains.

Sedimentation fabrics may be closely similar to flow fabrics. Inherited metamorphic fabrics in metasomatic rocks should be parallel to and continuous with the fabrics of surrounding metamorphic rocks, unless deformation has taken place during metasomatism.

Fabrics can be interpreted more completely by observing their relations to contacts, inclusions, compositional layering, and fractures. In Fig. 14–4A, for example, the rounded shape of the transported inclusion and the sinuous pattern of the banding indicate that the fabric is a true flow structure. In Fig. 14–5, the continuity of the fabric through inclusions and dike walls shows it formed at a stage

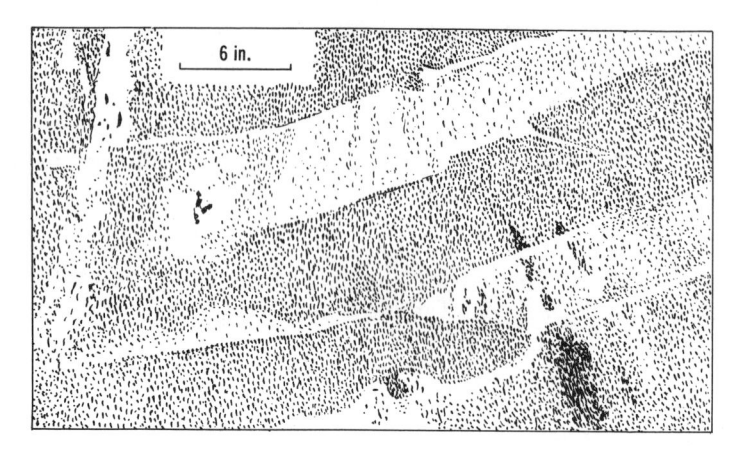

Fig. 14–8. Metasomatic dikes of light-colored granitic rocks in more mafic tonalite, Santa Lucia Range, Calif. Gneiss-like structure of tonalite is in vertical plane perpendicular to page, and passes through dikes. Note (1) less modified septa that cross the dikes, (2) continuation of dark inclusions from tonalite into dikes, and (3) control of the dikes by a set of cross fractures in the tonalite, locally forming sharp contacts.

when the magma was largely solid (type 2, above). Inherited fabrics in metasomatic rocks can also pass through contacts as well as inclusions (Fig. 14–8). Further relations between fabrics and related structures are described in the next three sections.

Methods. Oriented fabrics that are obscure in hand specimens can often be seen in clean outcrops measuring several feet across. Rocks tend to split along the aligned cleavage surfaces of feldspars and micas, and both the shape of outcrops and the grain of the topography may reflect the fabric of the rocks.

The number of readings and symbols needed per unit area will depend on the consistency of the fabric and the purpose of the project. Several readings should be taken at each large outcrop or group of outcrops before plotting a symbol. If all the readings are within 5 or 10° of their average, the symbol may be plotted with a straight strike line. If attitudes are more variable, the outcrops should be examined to trace out the continuity of the fabric. Moderate rolls or irregularities will be found in many cases, and a sinuous strike line can be used for their symbol. When different ages or kinds of fabrics are present, distinctive symbols should be used for each, and the relations between the fabrics should be sketched and described in the notes. Detailed structural analyses may be useful if two or more directions of rolling or folding are suspected (Section 15–11).

Smooth outcrop surfaces lying at an angle to a planar fabric will appear to have a linear fabric. Surfaces exactly parallel to the planar fabric must therefore be used for measuring lineations. Unless lineations are unusually strong and persistent, more three-dimensional outcrops may be needed than can be found in many areas. Detailed studies, however, can often be made at quarries, road cuts, stream gorges, or sea cliffs. Oriented samples are of great value in interpreting the origins of planar and linear fabrics (Section 15–12).

14–6. Inclusions and Related Structures

The term *inclusion* is used here in a descriptive sense, referring to any foreign appearing fragment or circumscribed body that lies in a more homogeneous rock. Most *schlieren* (Section 14–7) are distinguishable from inclusions since their streaky and elongate shapes are not clearly circumscribed at the ends. Inclusions may be classified as follows:

Xenolith. Fragment of foreign rock.
Xenocryst. Mineral grain of foreign origin.
Cognate xenolith. Fragment of igneous rock formed as an earlier part of a composite or multiple intrusion.
Autolith. Early segregation or accretionary body in magma.
Orbicule. Spheroidal body with concentric or radiate variations in composition (autoliths, in part).
Skialith. Relics of country rocks isolated by metasomatic growth of host.

The shapes and fabrics of inclusions can often be used to interpret the fabric of the surrounding rock. They may also serve as a measure of squeezing, fracturing, attrition, replacement, or overgrowth (as of orbicular rims) within the pluton. The cases shown in Fig. 14–9 are based on actual examples but are simplified to permit comparisons. In case A, angularity of inclusions and sweeping patterns of fabric in the host indicate igneous flow; in B, the planar fabric of the host may be a true flow fabric or may have formed by flow and flattening of nearly solid materials; in C, continuity of fabric through inclusions suggests metamorphism or strong late-igneous flattening; D shows a common type, suggesting early orientation of inclusions by flow of liquid magma, followed by crystallization at rest; E suggests late re-

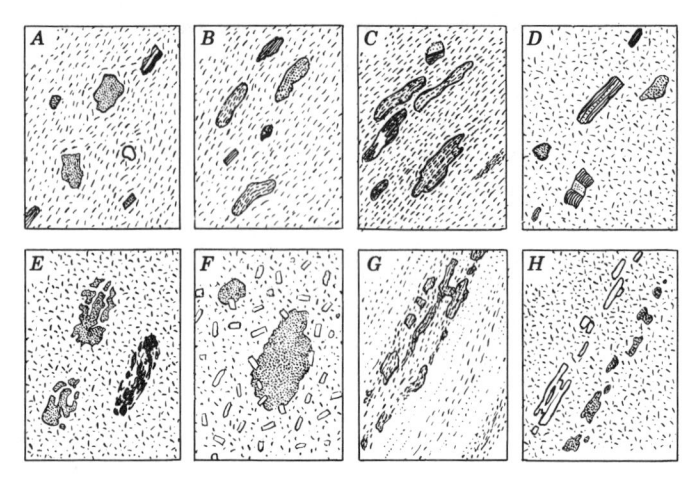

Fig. 14–9. Relations between inclusions and rock fabrics (see text for explanation).

placement of inclusions along internal fractures; in *F*, large feldspar grains (porphyroblasts) have grown in inclusions; the continuity of rock types and structures in cases *G* and *H* point strongly to metasomatism.

Some igneous units can be recognized by the kind or abundance of their inclusions. Moreover, xenoliths of distinctive rocks can sometimes be used to determine the amount of movement or degree of mixing in a magma. If patches or bands within a pluton are characterized by certain kinds of inclusions, their distribution may establish the former continuity of country rock units through the pluton. This would indicate they have been partly replaced or only moderately displaced as the pluton formed. Pitcher (1953) described a detailed field study of this type of relation.

Inclusions showing concentric zones with different textures or compositions provide a record of reactions and exchanges with the host. Inclusions of one original type, as shale or basalt, may be classified according to their degree of reconstitution, and their distribution may indicate movements and reactions in the pluton.

14–7. Layers and Schlieren in Plutonic Rocks

Layers in plutonic bodies are stratiform features that differ from one another in composition, texture, or both; they are often called *bands* because of their appearance on outcrop surfaces. These features are analogous to sedimentary beds, and, like beds, they may be thick or thin, obvious or obscure, and tabular for long distances or lenticular. They may also be folded, broken, or deformed into various shapes. The German term *Schlieren* is sometimes used for these layers, but it is more appropriately applied to streaky, wisplike, or nonstratiform concentrations of dark or light minerals. Alternately, schlieren may be called *streaks, wisps,* or other descriptive names. Most layers and schlieren can be measured and plotted as planar structures; some have linear shapes or associated linear structures that can also be measured and plotted.

Layers and schlieren can form in different ways, and it is possible for several types to occur in a given pluton. It is therefore necessary to classify them in a useful descriptive way that will serve, if possible, to distinguish them genetically. The foliation symbol used for each type should be designated by a small letter or other suitable mark.

Layers and schlieren are here organized, for convenience, into five genetic classes. Although the list is tentative and some of the structures described are problematical, all known possibilities should be considered in the field; otherwise large-scale relations might be grossly misinterpreted.

a. Layers formed by accumulation of crystals on a floor.

1. *Gravity sorting of light and heavy grains from a magma current.* Layers originate essentially as sedimentary beds, and therefore show features such as density grading (with heavier minerals concentrated at base), cut-and-fill relations, cross-bedding, slump-folds and faults, intraformational breccias, and channel-like structures (Fig. 14-10A). Layers may be superimposed rhythmically or intercalated with layers of homogeneous rock that formed between pulses of currents. Indications of tops and bottoms of layers must be consistent. Planar and linear fabrics when present are parallel to layers.

2. *Differentiation after separate injections of magma.* Layering requires repeated injections of fresh magma, each followed by a period of quiescence. Minerals at base of layers are those that formed first, as magnetite, chromite, and olivine in basic rocks; remainder of layers may or may not show discernable variations. Layers are likely to be thicker (tens or hundreds of feet) than those of 1, above, and may show local current effects. Planar fabric may be strong, moderate, or absent; linear fabric should be rare or absent.

b. Layers formed by growth of minerals from a planar interface.

1. *Changes in magmatic conditions may lead to repeated layers.* Minerals tend to have comb fabric at large angles to interface, or may

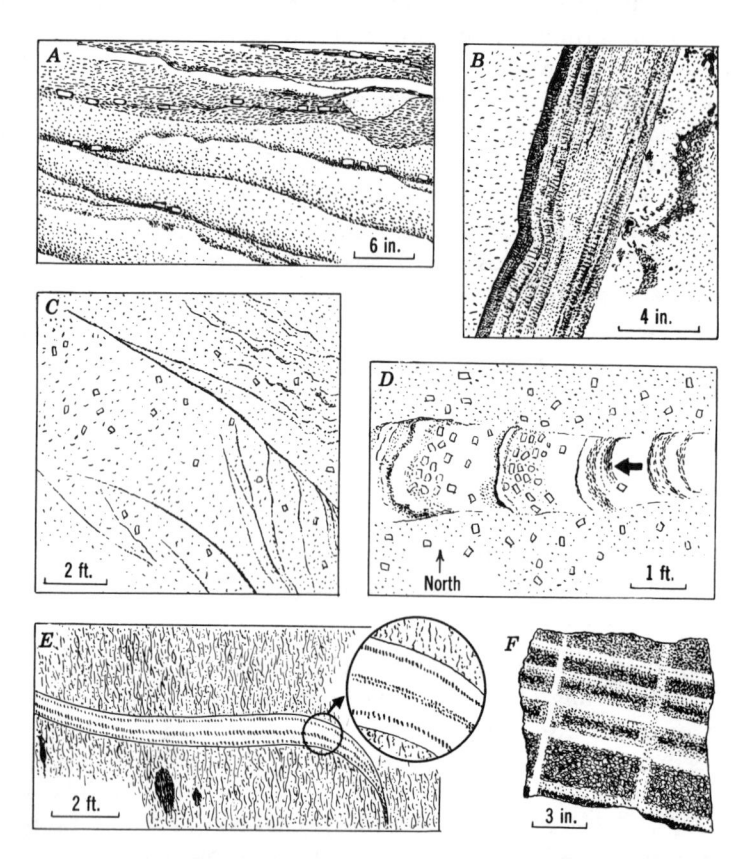

Fig. 14-10. Layers and schlieren in plutonic rocks. All views are essentially vertical sections except *B* and *D*, which are horizontal. (*A*) Systematically graded and eroded(?) layers in tonalite; Tioga region, Sierra Nevada, Calif. (*B*) Comb-fabric in layered pyroxene diorite at contact with earlier tonalite (left); Lake Tahoe, Calif. Courtesy of A. A. Loomis. (*C*) Flow layers and schlieren in granodiorite; Mt. Conness, Sierra Nevada, Calif. (*D*) Dike showing mafic minerals and large feldspar tablets concentrated into layers and schlieren oblique to walls. Arrow indicates sense of late flow. Near Mt. Conness, Sierra Nevada, Calif. (*E*) Thin mafic layers in dike, showing fabrics that lie oblique to contacts of dike but parallel to fabric of the surrounding Aar granite. South of Goschenen, Switzerland. (*F*) Hornblende gabbro, showing two sets of feldspathic "layers." Near Enterprise, Sierra Nevada, Calif.

show budlike or branching shapes (Fig. 14–10B). Probably similar structures can form late in history of pluton by fillings in tension fractures (comb veins).

c. Layers and schlieren formed by various kinds of flow.

1. *Flow of heterogeneous magma.* Original heterogeneities may be caused by contamination, differentiation, mixing of magmas, partial fusion of heterogeneous rock, or unequal distribution of volatiles, and subsequent flow layers may thereby differ in composition, texture, or both. If pseudosedimentary structures are present, they will not give a consistent sense of tops and bottoms (Fig. 14–10C). Primary flow lineation is usually present and should lie in plane of layers. Swirled flow-folds like those of lavas occur locally (Fig. 14–4A).

2. *Extension of heterogeneous magma by late, largely solid flow.* Flattened inclusions and schlieren lie at right angles to principal flow direction (Fig. 14–10D). New fabric commonly crosses primary layering and contacts. Planar fabric is typically of type 2 described in Section 14–5, and it may locally lie obliquely to earlier fabric.

3. *Flow of initially homogeneous magma.* Balk (1932, pp. 24–26) suggested that different mineral species might be sorted into layers because they offered different frictional resistance to moving magma; this hypothesis provides one explanation of graded layers that lie against steep contacts.

d. Metamorphism (especially metasomatism) of bedded rocks.

1. Pseudomorphs of all kinds of sedimentary structures can be formed, causing similarities with type a-1 (p. 287); however, some layers should have compositions that are generally nonigneous, as quartzite, marble, or calc-silicate rocks.

e. Late-magmatic and secondary layers and schlieren.

1. *Metasomatism on fractures subparallel to earlier planar fabric.* This is probably caused by concentration of fluids on planar surfaces or on shears that are subparallel to the planar fabric. Their secondary nature is indicated where the layers cross one another (Fig. 14–10F).

2. *Fillings in earliest tension fractures deformed by late flow.* Veins or dikes emplaced along swarms of tension fractures can be stretched or smeared by late flow so as to look like primary flow layers or schlieren. Here and there, however, they should cross, or they may be traced to less deformed structures.

Layers and schlieren of one type may be modified at later stages. Late flow commonly orients their fabric at a considerable angle to their gross shape (Fig. 14–10E). The orientations and shapes of inclusions can also be used to determine the stage when certain layer-structures formed (Section 14–6).

14–8. Fractures and Related Structures in Plutons

Maps of fractures can be used to interpret late deformation and alterations within plutons, for the maps show the overall pattern of fractures relative to fabric and other features. Moreover, the patterns can be interpreted relative to the shape of a pluton and the structures in the country rocks around it. Some fracture systems can be seen to be closely related to intrusive movements of the rocks; others may be unrelated to intrusion. Balk (1937) described principles and methods for mapping fractures, along with summaries of a number of field studies. Examples of detailed studies are those given by Cloos (1925), Martin (1953), and Hutchinson (1956). The variety of situations found by these investigators demonstrates that each pluton must be examined critically in its own light. It is especially important that fractures be classified as completely as possible at the outcrop and plotted with distinctive symbols.

Geometric relations to fabric. In the widely used system introduced by Hans Cloos, fractures are classified by their relation to primary (flow) *lineations* in the rock. *Cross* (*Q*) *fractures* lie at approximately 90° to the lineation; *longitudinal* (*S*) *fractures* strike parallel to the trend of the lineation and dip steeply, and *primary flat* (*L*) *fractures* dip gently, if at all, and include the lineation (Fig. 14–11). If linear fabrics and schlieren cannot be found, fractures can be related to planar fabrics. Useful and common fractures are those that (1) lie at approximately 90° to the planar fabric, (2) are parallel to the fabric, and (3) form conjugate (complementary) systems that strike either with the fabric or at large angles to it. It may be convenient

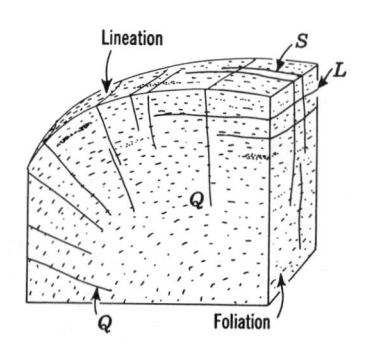

Fig. 14–11. Diagram illustrating terminology of fractures that can be related to linear fabrics or linear schlieren. The curved surface represents the contact of the pluton. After Balk (1937).

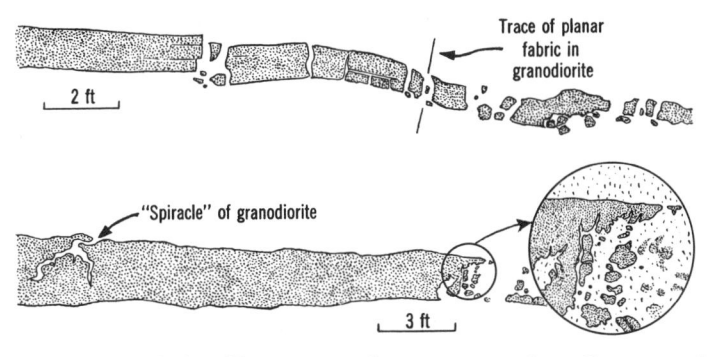

Fig. 14–12. Dark tabular dikes in granodiorite, passing laterally into inclusion swarms. Note intrusive relation of granodiorite to dike at lower left. French Lake, Sierra Nevada, Calif.

to give arbitrary letter designations to these fractures in the notes, but this need not be done on the map if fabrics and primary structures are plotted.

Age classifications. The relative ages of fractures should be determined, for fracture systems of different ages could give a misleading composite map pattern. Early-formed tension fractures are generally filled with fine-grained rocks similar to those of the main pluton, or by pegmatite, aplite, or dark porphyritic rocks. The earliest dikes may be fractured, strewn out, or partially assimilated or granitized by the host rock (Fig. 14–12). These relations show that the pluton solidified both slowly and spasmodically, for late flow followed fracturing. Somewhat younger fractures carry deposits of deuteric minerals such as quartz, epidote, chlorite, actinolite, serpentine minerals, carbonates, hematite, and zeolites. Fracture walls are commonly altered to these minerals. The relative ages of fractures can often be determined where two kinds of veins or dikes cross, or where one deposit has been reopened along a medial or marginal fracture to admit another kind of deposit (Fig. 14–13).

Early shear fractures are characterized by protoclastic zones or bands in which crushed minerals have recrystallized to gneissose or schistose aggregates. Later shears typically bear aphanitic smears of mylonite or gouge.

The youngest fractures in plutons are usually clean, unaltered joints that either formed in the latest stage of solidification or are entirely secondary. The latest secondary joints in many areas lie approximately

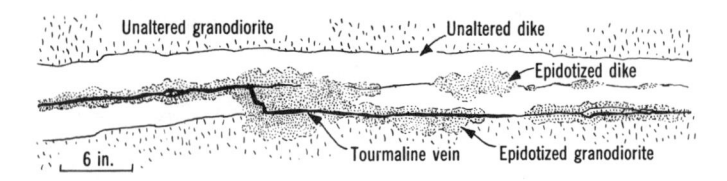

Fig. 14–13. Quartzofeldspathic dike (unpatterned) cut along medial and marginal fractures by tourmaline veins (black). Near English Mt., Sierra Nevada, Calif.

parallel to the topographic surface, suggesting that they formed by erosional unloading.

Genetic classifications. Many fractures in plutons can be classified as either tension fractures or shear fractures. Most tension fractures dilate and fill with dike or vein materials. If they carry no deposits, their walls are commonly altered by fluids that passed along them, yielding anything from sharply demarcated metasomatic dikes (especially of pegmatite and aplite) to vague zones of bleached primary minerals (Fig. 14–8). Shear fractures, on the other hand, are characterized by granulation and smearing of minerals, by abrupt changes in the orientation of rock fabrics at their walls (Fig. 14–14A), by slickensides and grooves, and by offsets of inclusions, layers, schlieren, dikes, and veins. Feather (tension) joints indicate the direction of shear along a principal fracture (Fig. 14–14B). Shears may be classified as normal faults (hanging wall down), reverse faults (hanging wall up, dip more than 45°), thrusts (hanging wall up, dip less than 45°), and strike-slip faults (movement mainly in the direction of strike). The direction and amount of net slip can be determined where a linear feature or two divergently dipping planar features are cut by shear fractures (Fig. 14–14C).

Tabular dikes and veins are commonly granulated or slickensided along their walls or on medial fractures, providing evidence of two stages of movement on one fracture set.

Methods. Except for the most detailed mapping, tension fractures are generally plotted only where they occur in sets or parallel swarms. Individual shears, however, should be plotted as completely as space allows. Symbols for dikes and veins can be distinguished from those for joints by drawing a heavy strike line or a double strike line (Appendix 4). An accessory letter can be used to indicate the type of deposit or alteration. Shears should be plotted by a fault symbol with a short strike line; a number can be used to show the number of shears in a set (Fig. 14–14C). A single symbol is generally used for the aver-

Field Work with Igneous and Igneous-Appearing Plutonic Rocks 293

age of readings ranging through 10° or so within a set, but it should not be used for the average attitude of widely divergent fractures. Several prominent fracture-directions can be plotted as a group of symbols whose strike lines cross at the locality where the readings were taken. Stereographic projections can be used to plot and analyze fractures where cross-cutting relations indicate complex movements. These plots will be especially valuable if the fractures can be related to primary fabrics and structures.

14–9. Alterations of Plutonic Rocks

Field studies of rock alteration should include (1) mapping zones of mineral alteration and crushing, (2) examining and plotting small-scale structures along which alteration has taken place, and (3) sampling

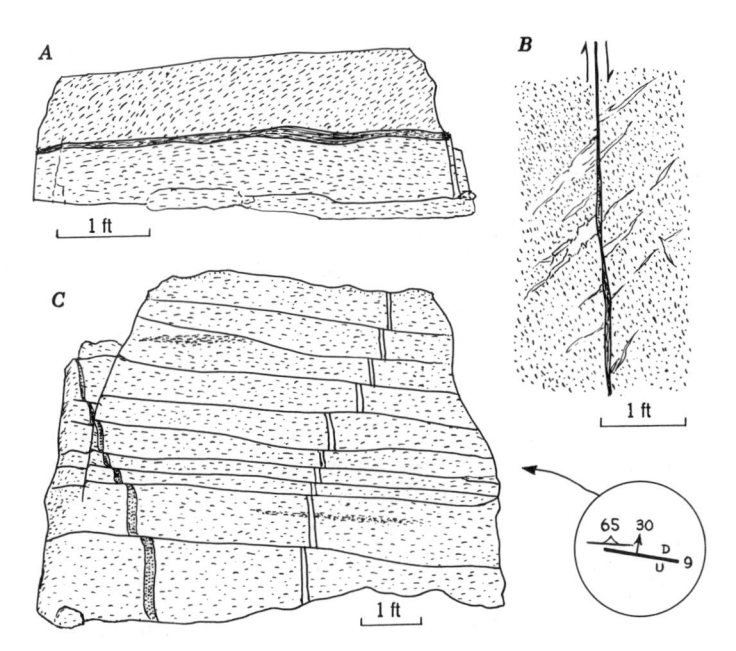

Fig. 14–14. (A) Sheet of protoclastic granodiorite cutting planar fabric; South Fork of Feather River, Calif. (B) Feather joints and bleached rock related to shear in granodiorite; west of Donner Summit, Calif. (C) Large outcrop surface that dips steeply toward the observer, showing set of nine late shears in granodiorite. The two dikes have a component of dip toward one another, thereby recording a systematic sense of offset on the shears. Fabric and schlieren dip steeply into the outcrop, as shown by the map symbols on the right. Near Enterprise, Sierra Nevada, Calif.

where altered and unaltered rocks can be related to the mapped features. Alteration zones should be mapped as an overprint on primary units and structures, as by placing transparent overlays on field sheets and plotting color spots or patterns for a given alteration. These maps can show that the alterations are connected with certain rocks or structures in a pluton, or that they are related to structures and units in the surrounding terrain. Alteration minerals should be matched with fracture sets or systems so that age relations between crossing fractures can be used to date alterations. Some hydrothermal alterations are closely connected with contact metamorphism, and the two should therefore be examined concurrently (see Section 15–13). Examples of detailed studies relating various alterations to fracturing and igneous processes have been given by Gilluly (1946) and Mackin (1947, 1954).

A large number of deuteric and secondary minerals can be used as a basis for mapping. Some occur in a variety of rocks; others are restricted to certain types. Sericitization of plagioclase and chloritization of biotite take place in most feldspathic rocks, especially in granitic rocks. Silicification typically forms fine-grained tough rocks with blurred to obscure relict textures, while argillization (production of clays from silicate minerals) yields softer rocks that generally show clear textural relics. Thoroughly albitized feldspathic rocks may be nearly white, light green (colored by epidote and chlorite), or light red (colored by hematite or hydrated ferric oxides). Other minerals that may characterize various alterations of feldspathic plutonic rocks are zoisite, carbonates, barite, fluorite, zeolites, alunite, pyrite, and various ore minerals, especially chalcopyrite, galena, and sphalerite. Hot fluids in particular can alter granitic rocks to coarse-grained quartz, tourmaline, and muscovite, locally with scheelite, cassiterite, magnetite, pyrrhotite, wolframite, or topaz. Common alterations in basic and ultrabasic rocks are olivine and enstatite to serpentine minerals, pyroxenes to uralite (aggregates of acicular actinolite) and chlorite, and calcic plagioclase to whitish or greenish aggregates of fine-grained albite, epidote, zoisite, sericite, and calcite. Other important secondary minerals in basic rocks are dolomite, ankerite, magnesite, siderite, magnetite, pyrite, prehnite, talc, and various base-metal sulfides. The nepheline in feldspathoidal rocks is commonly altered to cancrinite or to a variety of zeolites or clays.

References Cited

Balk, Robert, 1932, *Geology of the Newcomb quadrangle:* New York State Museum Bulletin 290, 106 pp.

Balk, Robert, 1937, *Structural behavior of igneous rocks:* Geological Society of America, Memoir 5 (reprinted 1959), 177 pp.

Cloos, Hans, 1925, *Einführung in die tektonische Behandlung magmatischer Erscheinungen (Granittektonik)*: Berlin, Gebrüder Borntraeger, 194 pp.

Gilluly, James, 1946, *The Ajo mining district, Arizona:* U.S. Geological Survey Professional Paper 209, 112 pp., with geological maps (1947) and supplement (1949).

Hutchinson, R. M., 1956, Structure and petrology of Enchanted Rock batholith, Llano and Gillespie Counties, Texas: *Bulletin of the Geological Society of America*, v. 67, pp. 763–806.

Johannsen, Albert, 1939, *A descriptive petrography of the igneous rocks, vol 1: Introduction, textures, classifications, and glossary:* Chicago, University of Chicago Press, 318 pp.

Mackin, J. H., 1947, *Some structural features of the intrusions in the Iron Springs district:* Utah Geological Society Guidebook no. 2, 62 pp.

———, 1954, *Geology and iron ore deposits of the Granite Mountain area, Iron County, Utah:* U.S. Geological Survey Mineral Inv. Field Studies Map MF-14 (with cross sections and text).

Martin, N. R., 1953, The structure of the granite massif of Flamanville, Manche, North-West France: *Quarterly Journal of the Geological Society of London*, v. 108, pp. 311–341.

Pitcher, W. S., 1953, The migmatitic Older Granodiorite of Thorr district, County Donegal: *Quarterly Journal of the Geological Society of London*, v. 108, pp. 413–446.

Waters, A. C., and Konrad Krauskopf, 1941, Protoclastic border of the Colville batholith: *Bulletin of the Geological Society of America*, v. 52, pp. 1355–1418.

15

Field Work with Metamorphic Rocks

15–1. Studies of Metamorphic Rocks

Metamorphic rocks indicate sequence and nature of processes within the earth just as sedimentary and volcanic rocks indicate sequence and nature of processes at and near the earth's surface. Studies of metamorphic rocks not only advance our knowledge of earth processes and history but also may be crucial in exploring for mineral deposits or planning engineering projects. Often, these studies must be thorough enough to evaluate sequences of new features, for a given metamorphism is typically the composite effect of several processes and events. The following should be done if possible:

1. Determine the premetamorphic lithology and stratigraphic sequence.
2. Map the premetamorphic rock units and structures.
3. Classify and map the metamorphic rocks and structures.
4. Determine the amounts and kinds of deformation.
5. Map zones of metamorphic minerals or textures.

The relative importance of these studies and the order in which they are undertaken will depend on the individual case. Typically, several can be done simultaneously, and each kind of study will contribute information to the others. Determining the original lithology and distribution of rock units is generally very important because metamorphic changes can be guided or limited by premetamorphic features. Primary textures and structures, for example, can control metamorphic deformation considerably, and primary minerals (or minerals formed during an early stage of metamorphism) often limit those that can form subsequently.

The depth and therefore the pressure at which a given metamorphism took place can sometimes be approximated by broad stratigraphic and structural studies. Zone mapping may indicate directly whether metamorphism was caused by igneous intrusion, by deep burial, or by deformation. Metamorphic temperatures and pressures can sometimes be estimated by comparing mineral assemblages with laboratory data on the synthesis of minerals. In order to do this reliably, the total metamorphic history of a suite of rocks must first be determined by combined field and textural studies. In interpreting an assemblage of hornfelses, for example, it may be important to recognize that some minerals are relict to regional metamorphism, others formed during thermal metamorphism, and still others formed during later stages of hydrothermal metamorphism. This genetic sequence might be established by tracing out rocks and structures relative to intrusions and late vein systems. Because metamorphic age relations are so crucial in these studies, outcrops that show superimposed structures or textures should be studied and sampled with particular care.

15–2. Cartographic Units of Metamorphic Rocks

As with sedimentary and igneous rocks, any physically distinct body of metamorphic rocks that can be plotted on a map can serve as a cartographic unit. There are two kinds of metamorphic units: (1) metamorphosed sedimentary or igneous formations and (2) metamorphic zones. The first kind of unit is based on premetamorphic lithology, the second on the kind or degree of metamorphic changes in the rocks. Because of their fundamental nature, metamorphic zones must be superimposed on metamorphosed formations. The two kinds of units must therefore be mapped separately and distinctly.

Units based on original lithology are so important that all possible methods should be used to recognize and trace them. Ideally, unmetamorphosed formations are traced to the area being studied. Their contacts should be mapped through the metamorphic terrain regardless of how the metamorphic grade or appearance of the rocks may change. These units should be named as outside the area of metamorphism; however, their metamorphic nature should be indicated in some way (as *metamorphosed Errin Conglomerate, slates of the Sheldon Formation, garnet metamorphic zone of the York Shale*).

Although most metamorphic units cannot be traced to less altered formations, their original lithology and sequence can often be interpreted from relict textures and structures. These units should be defined and given geographic names as described in Section 12–2. It is useful if the lithologic part of the name connotes both the original and the metamorphic lithology of the unit (as *Manley Slate, Smith Metabasalt, Lorin Marble*). The description of the unit should cover both the original nature and the metamorphic nature of the rocks.

Units of highly deformed or recrystallized rocks have customarily been formalized by geographic names, even though their original lithology and sequence cannot be determined. In many cases this creates no problem; however, such units may be metamorphic variants of less altered formations named differently elsewhere.

Where unusually strong deformation has affected large rock bodies, with the result that individual rock types can be plotted only as lenticular or irregular relics, informal names should be used for both the collective and individual units (as *Weltner metamorphic complex, amphibolites of Jakes Ridge*). Informal names should also be used where metamorphic units are indistinct because of poor exposure or incomplete study.

Metamorphic zones. A metamorphic zone is the outcrop of a rock mass characterized by certain metamorphic minerals, textures, or structures. Such a zone may be defined on any useful basis as long as it is secondary to the original rock. Because zones show the distribution of certain kinds or degrees of metamorphism, they are essential to understanding the progress and causes of a given metamorphism. Methods used in mapping metamorphic zones are described in Section 15–13.

15–3. Naming Metamorphic Rocks

Rock names used in the field should be based on characteristics that can be recognized at single outcrops and can be correlated with features observed under the microscope. Field classifications must generally be descriptive rather than genetic, but their defining characteristics should be such as to suggest genetic relations as the mapping proceeds. In the system suggested here, the main rock name is based on the texture of the rock and the principal mineral constituents are added as modifying nouns, for example, *hornblende-quartz schist, andalusite-biotite hornfels*. Both parts of these names are thought to carry considerable genetic meaning.

A name must be used consistently, regardless of where a given rock occurs. A schistose rock, for example, should not be called a hornfels because it is found in a contact aureole. Mixed or superimposed textures should be indicated in names or described in notes because they may indicate two or more periods of metamorphism.

The textures that form the basis of the classification are:

Schistose. Platy or elongate grains oriented in a parallel to subparallel way so that the rock cleaves readily; may be classified further as *foliated* or *lepidoblastic* where the preferred orientation and cleavage are planar, and *lineated* or *nematoblastic* where they are linear.

Granoblastic. Grains approximately equidimensional; platy and linear grains oriented at random or so subordinate that cleavage is not developed.

Hornfelsic. Grains irregular and interincluded but generally too small to be seen clearly; rocks recognized in the field by unusual toughness, ring to hammer blow (if fresh), and hackly fracture, commonly at all angles to bedding or foliations. Under hand lens, freshly broken surfaces show a sugary coating that will not rub off (formed by rending of interlocking grains).

Semischistose. Intermediate between schistose and nonschistose textures and useful where platy or linear grains are so few or so unevenly distributed that the rock has only a crude cleavage; especially common in partially metamorphosed granular rocks, as sandstones and igneous rocks.

Cataclastic. Clastic textures resulting from breaking and grinding of solid rock materials; characterized by angular, lensoid, or rounded fragments (porphyroclasts) in a very fine-grained and typically streaked or layered groundmass. The term *mortar structure* can be used for completely unoriented arrangements, whereas the terms *phacoidal, flaser,* and *augen* are useful for rocks with lenticular relics (Fig. 15–1). These textures and structures are especially obvious where cataclastic sheets cut through massive or bedded rocks.

Grain-size terms. Actual sizes should be recorded where possible; the terms *fine-grained, medium-grained,* and *coarse-grained* might well be adapted to the rocks being studied because a single set of dimensions cannot be applied with equal meaning to all rock groups (an unusually coarse-grained hornfels, for example, might be finer grained

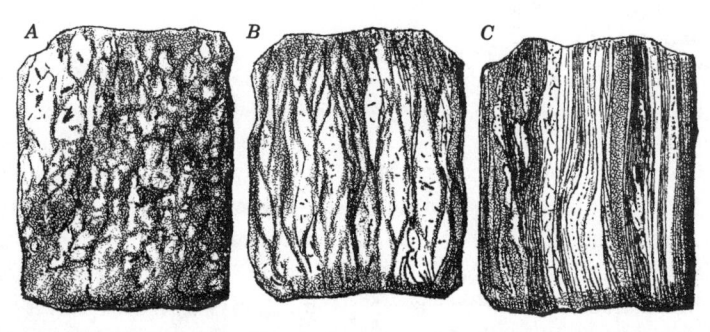

Fig. 15–1. Cataclastic rocks. (*A*) Mortar structure in cataclastic granite. (*B*) Phacoidal metadiorite. (*C*) Mylonitized granite. All samples are 3 in. across.

than a typical medium-grained schist). The adjective *porphyroblastic* applies to any rock where large grains have grown *during metamorphism* in a finer-grained groundmass.

The various rocks can be classified as follows:

a. Schistose rocks

1. *Schist.* Grains can be seen without using a microscope.
2. *Phyllite.* All (or almost all) grains of schistose groundmass too small to be seen without a microscope, but cleavage surfaces have sheen caused by reflections from oriented minerals.
3. *Slate.* Very cleavable but surfaces dull; tougher than shales and with cleavage commonly oblique to bedding.
4. *Phyllonite.* Appearance like phyllite, but cataclastic origin indicated by relict rock slices, slip folds, and porphyroclasts.

b. Granoblastic rocks

1. *Granulite* or *granofels.* Granoblastic rocks, irrespective of mineral composition; because *granulite* can connote special compositions and conditions of origin, the name *granofels* may be preferred (Goldsmith, 1959).
2. *Quartzite, marble,* and *amphibolite.* Compositional names that generally connote granoblastic texture; exceptions should be modified where clarity requires, as *schistose amphibolite, hornblende-plagioclase hornfels, hornblende-plagioclase schist.*
3. *Tactite (skarn).* A useful name for heterogeneous calc-silicate granulites and related metasomatic rocks of typically uneven grain.

c. Hornfelsic rocks

1. All called *hornfels,* or, if relict features are prominent, the adjective *hornfelsic* may be used with the original rock name (as *hornfelsic andesite, hornfelsic conglomerate*).

d. Semischistose rocks

1. *Semischist.* Fine grained (typically less than ¼ mm) so that individual platy or lineate grains are indistinct; relict features often common.
2. *Gneiss.* Generally coarser than ½ mm with small aggregates of platy or lineate grains in separate lenses, blades, or streaks in an otherwise granoblastic rock (Fig. 14–7D); platy or lineate structures may be distributed evenly through the rock or may be concentrated locally as layers or lenses (*banded gneiss*); some layers or lenses may be entirely granoblastic or schistose.

e. Cataclastic rocks

1. Where original nature of rock is still apparent, the adjective *cataclastic* can be used, as *cataclastic granite*, or special structural terms may be used, as *flaser gabbro, phacoidal rhyolite*.
2. *Mylonite.* Crushing so thorough that rock is largely aphanitic and generally dark colored; may be layered and crudely fissile but not schistose like phyllonite; porphyroclasts commonly rounded or lenticular.
3. *Ultramylonite, pseudotachylite.* Aphanitic to nearly vitreous-appearing dark rock typically injected as dikes into adjoining rocks.

The textures of many low-grade metamorphic rocks are so dominantly relict that names such as *massive metabasalt* and *semischistose metadacite* are most appropriate. Where metamorphism (typically hydrothermal) has produced one or two prominent minerals, rocks with dominant relict textures can be given names such as *chloritized diorite* and *sericitized granite*. Strongly metasomatized rocks with coarse or bizarre textures may require such special names as *greisen, quartz-schorl rock,* and *corundum-mica rock*.

Polymetamorphic rocks. Polymetamorphic rocks are those that have been affected by more than one period or kind of metamorphism, and they should be given suitably compounded names wherever possible. The commonest types form where late shearing or hydrothermal fluids cause retrograde metamorphism (diaphthoresis) of higher grade rocks. Names such as *sheared gneiss* and *phyllonitized granulite* can be used where the changes are mainly cataclastic, and names such as *chloritized garnet-biotite schist* and *albitized amphibolite* can be used where the changes are mainly hydrothermal. Another common type of polymetamorphism is produced where igneous bodies intrude regionally metamorphosed rocks, commonly forming crudely fissile hornfels from phyllite. Magmas and associated fluids may also inject or alter adjoining rocks to produce *veined schists, feldspathized granulites, permeated schists,* and so on (see Section 15–14).

Superposition of two schistose or semischistose fabrics is another common type of polymetamorphism. It may be indicated by folded or kinked platy or linear grains, by crinkled, knobby, or offset cleavage surfaces, or by thin bands of new schist cutting obliquely across an older schistosity. Names like *crumpled* or *plicated schist,* or *sheared* or *phyllonitized schist* can be used for these rocks. Because crumpled or sheared schists or cleavages can be produced during one period of metamorphism, the foliations should be studied relative to folds and associated structures (Section 15-8).

15-4. Premetamorphic Lithology and Sequence

The successful interpretation of metamorphism depends considerably on determining the original lithology of a given suite of rocks. Not only must the original rock types be determined, but also their textures, compositions, and small-scale structures. An argillaceous sandstone, for example, is likely to react differently to metamorphic processes than a well-sorted quartzose sandstone, yet each may bear relict sand grains after metamorphism. Similarly, thermal metamorphism of phyllite may produce a decidedly different product than thermal metamorphism of a clay shale. These differences are caused in part by compositions (including volatile substances), in part by mechanical response to deformation, and in part by the effects of permeability and grain size on the rates of mineral reactions.

Some mineral assemblages are directly indicative of original rock types. Almost all quartzites, quartz-rich schists, marbles, and calc-silicate rocks are of sedimentary origin. They should be examined for relics of coarse textures (as in metaconglomerates and biohermal limestones), sand-size layers, or thinly laminated beds that suggest fine-grained sediments. Micaceous rocks with grains of kyanite, andalusite, or sillimanite form from claystones rich in illite or kaolinite, while schists with staurolite, chloritoid, or abundant pink garnets form from more iron-rich (chloritic or montmorillonitic) claystones or siltstones. Ultrabasic rocks yield such unique rocks as talc schists and anthophyllitic schists, but few other meta-igneous rocks can be recognized on the basis of composition alone. Amphibolites, for example, can form from basalt, diabase, gabbro, basic tuffs, tuffaceous sediments, mafic graywackes, or dolomitic mudstones. At least as great a variety of rocks can yield metamorphic rocks consisting of micas, feldspars, and quartz. Relict textures and structures must therefore be sought carefully in these rocks. The brief suggestions that follow can be supplemented by the descriptions and figures in the three preceding chapters.

Metasedimentary rocks are most easily recognized by their relict bedding, especially where some layers or lenses are calcite bearing or quartz rich. Amphibolites of sedimentary origin are commonly intercalated with gray-green calcareous layers rich in diopside or scapolite or brownish layers containing much biotite. Associated micaceous rocks often contain ellipsoidal zoned bodies of calc-silicate minerals that formed from calcareous or dolomitic concretions.

Metalavas typically form thick massive layers which are so resistant to dynamic metamorphism that porphyritic textures and igneous structures remain locally, except in the most altered rocks. Pillows and large blocks or bombs are often recognizable, and smaller pyroclastic fragments can occasionally be seen on smooth wet surfaces.

Intrusive bodies, even those of small size, tend to retain their granular or porphyritic textures, even after strong dynamothermal metamorphism. These textures may be more obvious in slightly weathered outcrops than under the microscope. Small bodies may show dark, fine-grained (originally chilled) margins, apophyses, and inclusions; however, similar features can be formed by cataclasis, and therefore the textures of the rocks must be examined carefully. Larger bodies should be mapped to locate diagnostic cross-cutting relations. Metamorphosed contact aureoles can be recognized around some intrusive bodies by noting the destruction of typical contact-metamorphic minerals and the partial conversion of hornfels to schist.

Stratigraphic sequence. The most valuable implement for unraveling metamorphic history is a firm stratigraphic sequence of premetamorphic units. The surest way of determining sequence is to trace lithologic units to (or from) less metamorphosed terrains. Where this is not possible, sequence can be determined by (1) selecting and mapping cartographic units that are based on original lithology and (2) finding fossils or relict structures that indicate tops of beds or flows. Mapping formations establishes the pattern of folds and faults and thereby determines the *geometric sequence* of the rock units. This sequence, however, may be completely inverted in strongly deformed rocks, so that the true sequence has to be established by stratigraphic methods. When comparing metamorphosed and unmetamorphosed sequences, possible changes in thicknesses should be considered. It will commonly be found that metamorphosed units have been thinned or thickened along slip surfaces that cut the bedding at a low angle. Incompetent units are characteristically thinned on fold limbs and thickened at fold hinges.

Where fossils are present, they provide exceedingly valuable evidence of sequence and may permit correlations with unmetamorphosed formations. They occur most frequently where metamorphic cleavages lie parallel to bedding (usually on the limbs of folds), or in blocks or slices of marble or quartzite that have escaped strong dynamic metamorphism. In metavolcanic sequences, calcareous beds, lenses, and pockets in pillowed lavas and associated tuffs can yield well-preserved fossils. Although massive fossils would seem to be the only ones that would be preserved, delicate fossils can be found here and there on cleavage surfaces of slates, phyllites, or associated calcareous rocks. They will usually be distorted, but a paleontologist familiar with the original forms can frequently recognize them.

Stratigraphic sequence can also be determined from relict small-scale structures such as those described and illustrated in Chapters 12 and 13. These structures can be preserved nearly perfectly or can be distorted or blurred to various degrees. Cross-bedding in quartzites is frequently recognizable, even after rather high-grade metamorphism (Fig. 15–2A). Faulting on arcuate surfaces, however, may produce somewhat similar patterns in schists and gneisses (Fig. 15–2B). Metamorphosed size-graded beds commonly show gradations in color or composition; aluminous silicates may cluster in their more argilla-

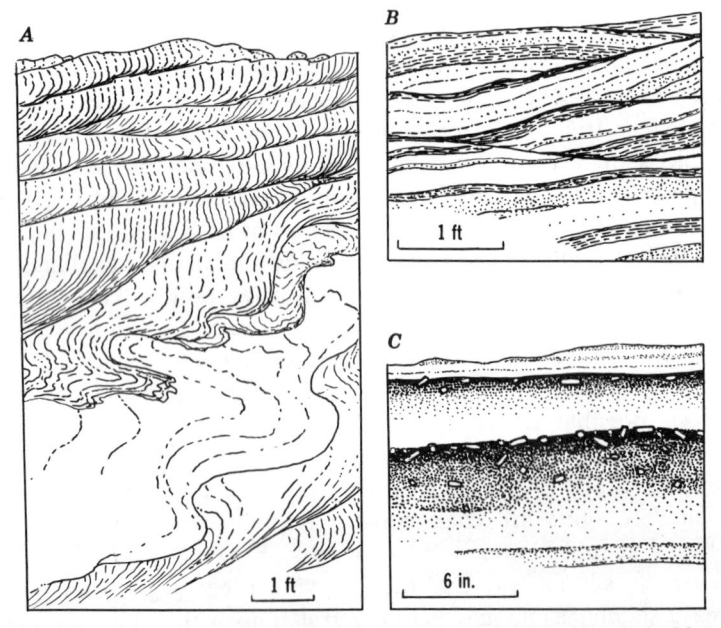

Fig. 15–2. Structures in metasedimentary rocks. (A) Relict cross-bedding in Moine epidote-mica quartzites, distorted by flowage of unknown age. Face of quarry at Loch Shinn, Sutherland, Scotland. (B) Pseudo-cross-bedding in schists and quartzites; Santa Lucia Range, Calif. (C) Andalusite porphyroblasts at tops of graded beds, aureole of Ardara pluton, County Donegal, Ireland.

ceous (upper) parts (Fig. 15–2C). The larger of the various sole markings can be found in many hornfelses and low-grade schists. The tops of beds can also be determined from the geometric relations between folded beds and secondary cleavages (Section 15–8).

15–5. Studying Metamorphic Deformation

Metamorphism almost always results in deformation. Studies of metamorphism should ascertain the nature of a given deformation, its directions and magnitude, and its age relative to recrystallization. These studies must often cover a host of structural details; however, they should be planned so that basic mapping precedes local detailed analyses. One way of organizing these studies is as follows:

1. Make a general geologic study of the area by mapping units based on premetamorphic lithology, by determining sequence from relict structures, and by plotting principal foliations and lineations.

2. Re-study major structures (as folds shown by distribution of lithologic units on map), carefully noting the relation of minor (outcrop-size) structures to them. At the same time observe relations of foliations and lineations to both minor and major structures. Microscopic study of oriented samples will help in interpreting foliations and lineations.

3. Where structural relations are unsolved (perhaps because of two or more periods of deformation), analyze the best-known parts of the major structures by plotting various foliations and lineations on spherical projections. This constitutes a statistical comparison between the various minor structures and the major ones.

4. If deformation is still unsolved, especially where two or more deformations must be interpreted in granular rocks, make a thorough microscopic study of oriented samples, using petrofabric (statistical) methods when necessary.

If formational units cannot be mapped as in step 1, it may be possible to map the rocks in zones characterized by certain textures and structures (Section 15–13). The remaining studies could then be applied to each of these zones.

15–6. Foliations and Lineations

Metamorphic foliations and lineations can form in a number of different ways, some of which are not well understood. Although they should be interpreted as fully as possible in the field, mapping must generally be started by classifying foliations and lineations on a descriptive basis.

Foliations. Foliation can be applied as a general term to metamorphic structures with platy, layered, or planar aspects. These structures can be named as follows.

Layering. Assemblages of parallel tabular bodies differing in composition or texture (essentially like bedding in sedimentary rocks); often called *banding* because of appearance on two-dimensional outcrops.

Schistosity. Platy or elongate grains oriented in a parallel to subparallel way, so that rock cleaves readily; *slaty cleavage* in very fine-grained rocks is probably analogous.

Gneissose structure. Small aggregates of platy or elongate grains in separate lenses, blades, or streaks; grain sizes typically greater than 0.5 mm.

Slip cleavage. Closely but finitely spaced subparallel surfaces of cleavage along which slip has taken place.

Fracture cleavage. Closely but finitely spaced subparallel surfaces of cleavage along which no slip has taken place.

Secondary cleavage, oblique cleavage. Useful substitutes when the foregoing terms cannot be applied confidently.

The maximum dimension limiting the terms *closely spaced* is generally about ¼ in. in fine-grained schistose rocks and about 2 in. in coarse-grained schists and massive rocks.

Where it is possible to discern two or more foliations of one kind but of different ages, the sequence can be numbered, as *slip cleavage 1 and 2*. Notations can also be simplified by assembling foliations into groups related by age or geometry. This is commonly done by using the letter S with a subscript number; for example, compositional layering and a parallel schistosity could be labeled S_1, slip cleavage and a parallel schistosity S_2, and so forth.

Lineations. A *lineation* is a preferred (subparallel) orientation of one or more kinds of linear features. A great variety of lineations may be studied and mapped. Some give direct clues as to the amount, kind, and direction of deformation; others may be problematical until the map is complete and their patterns can be compared to all other deformational features. If the scale of the map is small, lineations can be plotted and compared on spherical projections. Where lineations can be related to distorted primary structures, their genetic implications will be particularly clear.

Ideally, all kinds and degrees of lineations should be mapped, but limitations of time and scale usually make it necessary to select only a few. Especially useful are axes of folds whose closure can be observed at large outcrops or closely spaced groups of outcrops. Smaller-scale lineations associated with folds include crenulations in schists, lines of intersection of bedding and secondary cleavages, intersections of two or more cleavages or other foliations, and such related structures as boudins and rods. Relict structures that can be deformed into linear bodies include pebbles, fossils, inclusions, oöids (oöliths), pellets, amygdales, lapilli, and volcanic blocks or bombs. If the original shapes of these structures are known, their new lineate shapes can be used to determine the direction and amount of deformation. Ernst Cloos (1947) has described a study of this type.

Slip surfaces may have linear mineral smears, striations, or grooves indicating direction of movement. Lineations formed by elongate mineral grains or by elongate aggregates of grains are commonly formed where rock bodies have recrystallized during or after deformation.

A small letter can be placed at the origin of each lineation arrow to designate the kind of lineation, and a small number can be used to signify the relative age of the lineation. Linear structures that have formed at the same time, or by closely related movements, can be grouped into families, as *lineations 1, lineations 2*. These designations help keep data meaningful, especially when all readings cannot be plotted on a map.

15-7. Geometric Styles of Folding

Not only can folds be plotted as linear elements, but their general shapes or geometric styles can be used to assess the distribution and intensity of deformation. Marked variations in fold style over an area should therefore be described or mapped. At the same time, the relations of folds to foliations and lineations should be noted, as described in Section 15-8. These data will be particularly useful in determining the distribution of two or more periods of folding.

The following suggestions regarding terminology are illustrated by Fig. 15-3. An important aspect of fold style is the degree of divergence of the limbs and their symmetry relative to the axial plane (*A*, *B*, and *C* in the figure). Terms relating to the degree of curvature at the hinge line, to the degree of similarity between the folds in successive layers, and to the attenuation and slipping out of limbs are shown by examples D through J. The shapes of axial planes and the alignment of axes give a measure of distribution and regularity of movement, with variations shown by examples K through P. Cases M, N, and P typically occur in soft rocks or in deeply buried rocks, in which rapid plastic deformation is possible. Figures 15–15 and 15–16 show other examples of "flowage" folding in mixtures of metamorphic and granitic materials.

Refolding. Complex patterns like those of Fig. 15–3P may indicate refolding. The critical relations in outcrops are a persistent geometric arrangement of culminations and depressions formed by folding of

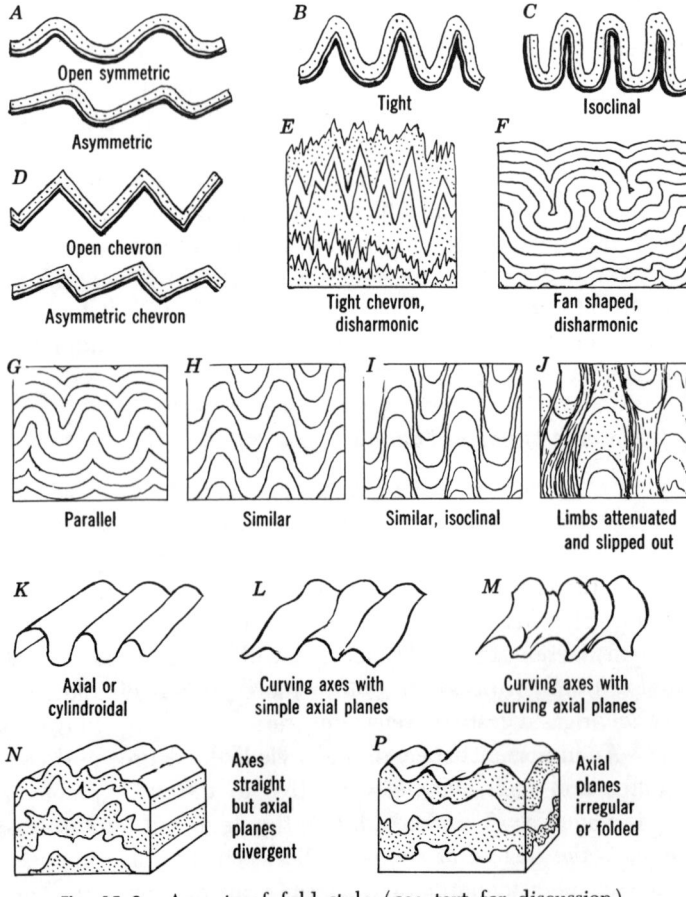

Fig. 15–3. Aspects of fold style (see text for discussion).

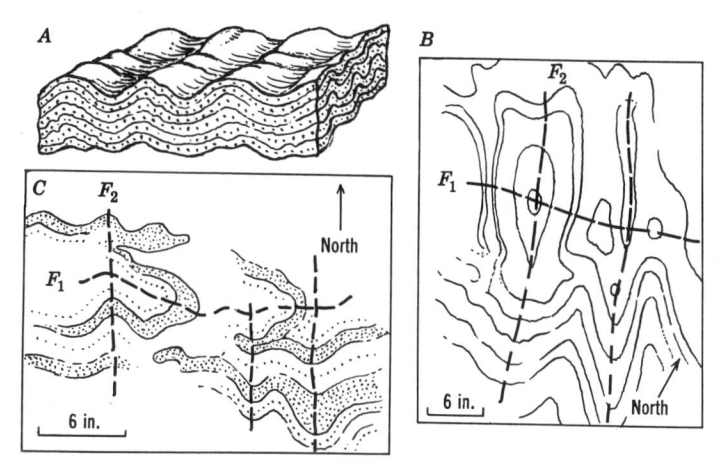

Fig. 15-4. Refolded rocks. (A) Hand specimen of knobby schist from an outcrop where two axes of folding can be seen clearly. West of Healdsburg, Calif. (B) Horizontal glaciated surface in Strath Conon, Scotland, showing the approximate axial traces of older (F_1) and younger (F_2) sets of folds. (C) Inclined outcrop surface near locality of B, showing refolded isoclinal folds. These outcrops were kindly shown to the writer by Michael J. Fleuty.

old limbs and axes (Fig. 15-4A and B), or the outright evidence shown by folded isoclinal folds (Fig. 15-4C). Hinges of the earlier folds are likely to be destroyed in most places, but crudely quadrate or rhomboid patterns may still be seen on smooth surfaces.

15-8. Foliations and Lineations Related to Folds

When geometric styles of folds are noted and described, foliations and lineations associated with them should be studied, measured, and plotted. If there is not enough space for all these features on a map, the axes of individual folds and perhaps one or two prominent foliations can be plotted. Their relations to other foliations and lineations can then be sketched and described in the notes. Drawings are most useful if made when looking parallel to the fold axis, for most linear structures related to folding trend either parallel to the fold axis or at right angles to it, presumably because of rotation, slipping, and flowage around the axis during metamorphism. Aberrant lineations and foliations may have been caused by irregularities in the original rock bodies, by geometric variations as the folds grew, or by a second period of folding.

The general size of the folds and the strength of the rock materials should be noted as these various features are studied. A large body of incompetent beds, for example, is likely to fold less regularly than a few beds within the body. The relative amounts of flexure, solid flow, and slip should be assessed at each group of outcrops. Time relations between deformation and recrystallization are also important, as illustrated by the examples described in the next paragraph.

Relations between older and younger foliations and lineations can be seen most easily in medium to coarse-grained schists, gneisses, and granulites. Figure 15–5A shows a case where schistosity is exactly parallel to the layering (probably relict bedding) in parallel-folded rocks. It is almost certain that this schistosity is older than the folds. In case B, the rocks were also schistose before they were folded, but

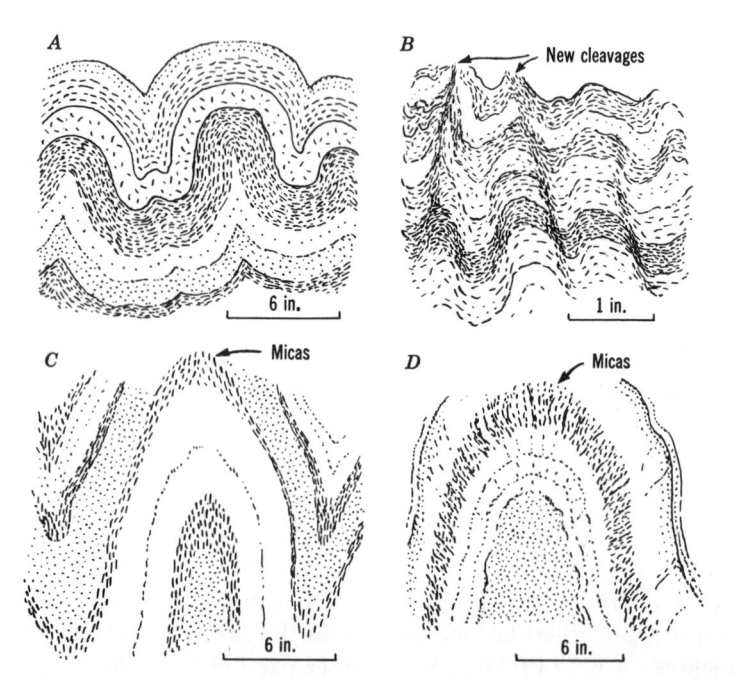

Fig. 15–5. Relations between foliations and folds in schists and related rocks. (A) Folded Moine(?) schists and quartzites with parallel granitic sheet; 300 ft south of Monadhliath pluton, near Aviemore, Scotland. (B) Crumpled schists from Strath Oykell, Sutherland, Scotland. (C) Schist and feldspathic quartzites; Valley of Kinzig River, Black Forest, Germany. (D) Schist and feldspathic quartzites; north of Scourie, Sutherland, Scotland.

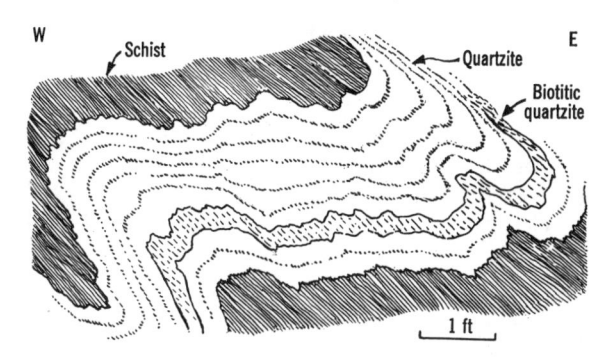

Fig. 15–6. Slip-folded Moine(?) schists and quartzofeldspathic rocks, west of Aviemore, Scotland. Note that micas (short dashes) in the thin partings between quartz-rich layers are oriented parallel to the cleavage in adjoining schist.

a new foliation had begun to form along some limbs and axial planes. At the outcrop, the rocks cleave almost as readily in this new direction as along the older schistosity. In case C, the rocks recrystallized considerably during folding, for the micas at fold hinges have been pressed or sheared so as to grow parallel to the axial planes. The mica-rich layer in case D has also recrystallized during folding, but here the new orientation is similar to that of secondary cleavages in many finer-grained rocks (compare Fig. 15–7). Finally, the case shown in Fig. 15–6 strongly suggests slip folding (translation of thin sheets of rock parallel to the new schistosity). The critical relations are: (1) the schistosity is approximately parallel to the axial planes of the folds, regardless of the compositions of the layers, (2) the folding is similar (harmonic) from one layer to the next, and (3) the distance subtended by any one cleavage surface on a given layer is approximately constant, regardless of where it occurs in the fold (this results in an apparent attenuation of layers on fold limbs).

Cleavages in fine-grained rocks. Individual grains cannot be seen in slates and phyllites, but relations between cleavages and bedding can be used to interpret deformation. Slaty cleavage commonly lies parallel to the axial planes of argillaceous beds that were folded tightly as they were metamorphosed. Figure 15–7A shows this relation diagrammatically; the rocks cleave on an infinite number of surfaces parallel to the lined pattern of the drawing. This cleavage can form when compression normal to the axial planes flattens the rocks as they recrystallize. It can also form in consequence of shearing on

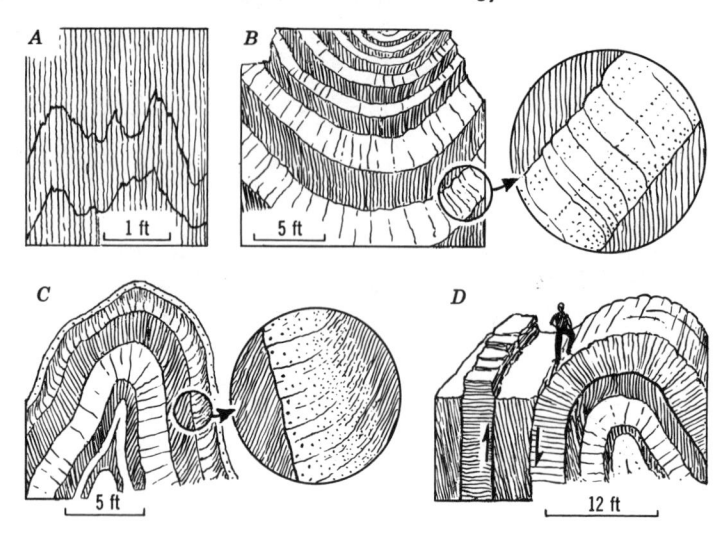

Fig. 15–7. Secondary cleavages in fine-grained rocks. (A) Simple axial plane cleavage in slate. (B), (C) and (D) Secondary cleavages in folded phyllites and quartzites of Santa Rosa Range, Humboldt Co., Nev.

closely spaced surfaces that parallel the axial planes. The origin of the cleavage can sometimes be determined if deformed relict fossils, oöids, or sand grains can be observed. Oriented samples should therefore be collected for petrographic studies (Section 15–12).

Where competent and incompetent beds are intercalated, secondary cleavages are commonly oblique to axial planes, except at fold hinges. Figures 15–7B, C, and D show several cleavage patterns in folded quartzites and phyllites. The cleavages illustrated are either fracture cleavages or slip cleavages with displacements of 1 mm or less. Regardless of how they form (this is much argued), oblique cleavages such as these are valuable for determining tops of beds at isolated outcrops and for estimating the position of an outcrop relative to the overall geometric form of a fold. The angle formed where cleavage surfaces pass from phyllite layers into quartzite layers is especially useful. Outcrops at the crests or troughs of folds show cleavages that pass from one bed to another without deflection, whereas outcrops on limbs of folds show a distinct angle in the cleavage at the contacts between beds. The tops of the beds can be determined by the sigmoidal curve of cleavages in micaceous or feldspathic quartzite beds (Fig. 15–7B) or by the unilateral curve in graded beds (Fig. 15–7C). A useful empirical rule for determining tops of obliquely

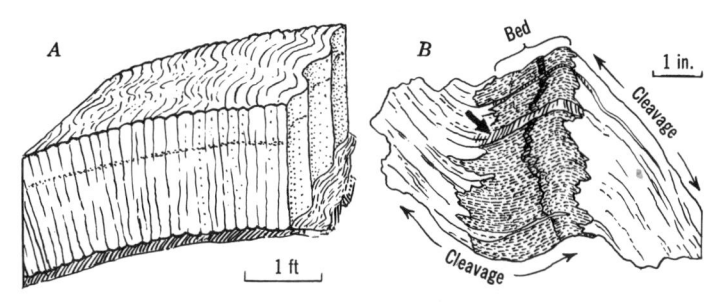

Fig. 15–8. Folded secondary cleavages. (A) Gently dipping quartzite bed showing vertical fracture cleavage deformed into steeply plunging minor folds; Santa Rosa Range, Nev. (B) Older cleavage (thin arrows) folded and cut locally by a younger cleavage (heavy arrow). The patterned bed has been offset into a sawtooth shape along the earlier cleavage. From Shenandoah River Valley, south of Harpers Ferry. After Nickelsen (1956, p. 262).

cleaved beds is that the acute angle between the cleavage and a bedding surface points in the direction the adjoining bed slipped toward the hinge of an adjacent fold (Fig. 15–7D). These determinations of sequence should be tested by reference to relict sedimentary structures because an entire sequence may have been overturned before it was folded and metamorphosed.

Examples of field studies in which cleavages and related lineations were used to interpret larger structures are given by Wilson (1951) and Nickelsen (1956).

Folded cleavages. Folded cleavages provide evidence of a second folding of beds (Fig. 15–8A). If a second set of cleavages formed with the younger folds, however, the beds themselves may be difficult to recognize (Fig. 15–8B). Statistical methods may be helpful in sorting out the structures formed during each deformation (Section 15–11).

Boudins and rods. Thin beds, veins, or dikes enclosed in incompetent materials are frequently broken, rotated, and squeezed into linear bodies during metamorphism and folding. The commonest of these are *boudins* (literally, sausages); two types of these linear structures are shown in Figs. 15–9A and 15–16D. *Rods* or *rodding structure* are chiefly quartz or quartzofeldspathic bodies that have been folded, dismembered, or rolled out at right angles to their lengths (Fig. 15–9B). Judged by their streamlined profiles, these bodies are typically deformed as they are emplaced; many appear to be segregated metamorphically from the surrounding rock (Gilbert, 1953).

Boudins and rods should plunge parallel to the fold with which

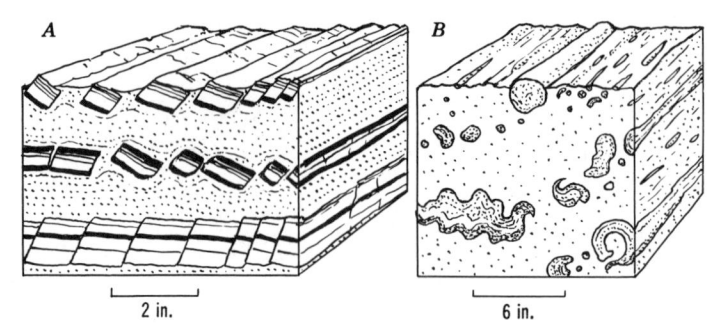

Fig. 15–9. (*A*) Cleaved siliceous beds in limestone, partially disrupted into linear rhombs; Palmetto Mts., Nev. (*B*) Rods and related linear relics of silicate veins in marble; Santa Lucia Range, Calif.

they are associated. They must, however, be examined carefully to make sure they are not relict sedimentary structures (Fig. 12–9*C*).

Drag folds. Although true drag folds can form during flexure-slip folding, asymmetric folds in metamorphosed thin-bedded rocks can also form by slip or solid flow oblique to bedding (Fig. 15–10). In either case, these minor folds should plunge parallel to the main fold on which they formed, and their asymmetry should indicate their rela-

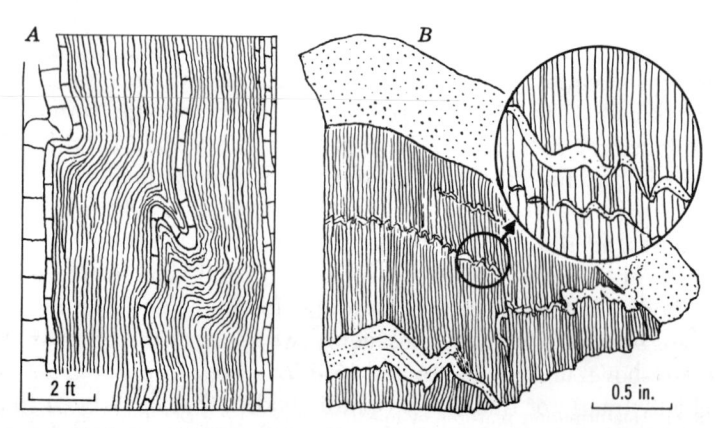

Fig. 15–10. (*A*) Drag fold in shale, formed by bedding-plane slip; an anticlinal crest would lie to the observer's right. Baltimore and Ohio Railroad cut, Pinto, Md. After Gair (1950, p. 871). (*B*) Part of small fold in phyllite and quartzite, showing asymmetric crinkles of siliceous laminae and vein in cleaved phyllite. Santa Rosa Range, Nev.

tive position on a major fold. They can also be used to determine stratigraphic sequence, provided they formed during the major folding. A later slip folding, for example, could form minor folds with the opposite sense of asymmetry. These various minor folds must also be distinguished from relict sedimentary slump or flow folds (Fig. 12–14).

15–9. Deformation Structures in Massive Rocks

Foliations and lineations in massive rocks are less striking than those in bedded rocks, but they can indicate local stress-strain relations more reliably because of the homogeneity of the rocks. Large intrusive bodies are ideal units, and thick lava flows, tuff-breccias, quartzites, metaconglomerates, or marble beds can also be useful.

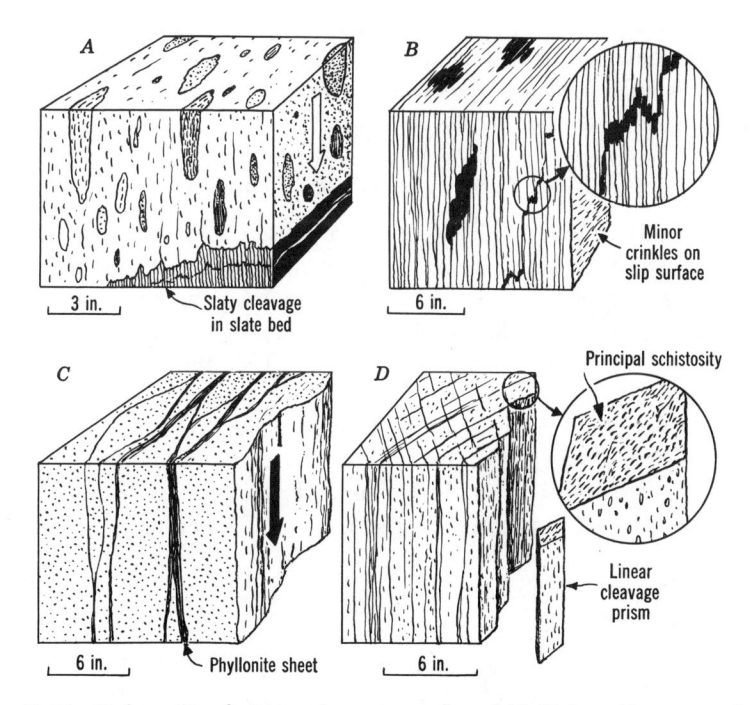

Fig. 15–11. Deformation features of massive rocks. (*A*) Deformed metaconglomerate near Mariposa, Calif. (*B*) Slipped-out quartz vein and inclusions in metamorphosed dike; near Bidwell Bar, Calif. (*C*) Metabasalt cut into lineate forms by sheets of phyllonite; heavy arrow indicates plunge. Near Oroville, Calif. (*D*) Semischistose meta-andesite cut into rhombic prisms by intersecting cleavage surfaces; near Bidwell Bar, Calif.

Metamorphic solid flow resulting in flattening and rotation of grains typically forms a simple schistosity. If the rocks are extended most in one direction, equidimensional bodies will become linear (triaxial) ellipsoids whose longest axes lie in the plane of the schistosity (Fig. 15–11A). A massive rock can also be made schistose if sheared on closely spaced surfaces. Shearing in a systematic sense would deform equidimensional bodies into ellipsoids whose longest axis would lie oblique to the schistosity (Fig. 15–11B). Veins or other tabular structures would be expected to show slip-folded forms. The slip surfaces can often be recognized by their thin layers of phyllonite and by the way they part and join to form linear bodies (Fig. 15–11C). Mineral and smear lineations on the slip surfaces may indicate the last direction of slip.

Slip sometimes takes place on two complementary (conjugate) sets of surfaces, cutting the rock into prisms with rhombic cross sections. In some cases, shearing of this type takes place intermittently, or contemporaneously, with flattening and rotation of grains. The rock may then show various slip surfaces and schistose fabrics that intersect in a pronounced lineation (Fig. 15–11D).

15–10. Joint and Vein Patterns in Deformed Rocks

Almost all dynamothermally metamorphosed rocks are cut by one or more sets of joints or veins. The commonest set lies at approximately right angles (nearer 80°) to prominent lineations; other fractures strike parallel to the lineations (Fig. 15–12A). Complementary (conjugate) sets such as those in Fig. 15–12B and C may also be associated with folded rocks. Whether they are true shears or not, if they are oriented consistently to fold axes, they can be used to trace folds through obscure outcrops.

Veins or fractures with drusy mineral coatings or altered walls should be mapped in preference to clean joints, which may be younger

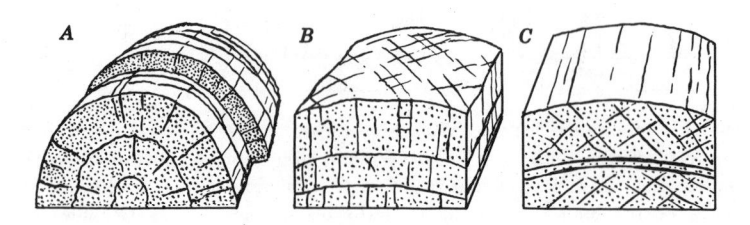

Fig. 15–12. Common fracture patterns in deformed rocks.

than the metamorphism. Segregations of quartz, epidote, and carbonates are especially common on fractures formed during low-grade metamorphism. Fractures in higher grade rocks are likely to bear quartz, alkali feldspars, and various metamorphic minerals. Sequences of veins can be determined in some areas, while folding or slicing out of veins may indicate the relative ages of two deformations.

15–11. Analyses on Spherical Projections

Structures can be analyzed statistically on spherical projections when they cannot be mapped to scale. These analyses have the ultimate purpose of interpreting deformational movements, although their more immediate use may be to detect obscure folds or to determine whether there has been more than one direction of folding. In any case, it is important to plot structures whose sense of displacement can be seen at the outcrop. The analyses are generally made on Lambert equal-area (Schmidt) projections, as described by Phillips (1954) and Badgley (1959).

A general and often preliminary analysis can be made when the general mapping is completed (step 1, Section 15–5). A transparent overlay is attached to a net and each bedding or foliation symbol is plotted as a point that represents the pole (line) perpendicular to the structure. Linear structures are also plotted as points, but with different symbols or colors. If the overlay becomes too crowded, different kinds of structures can be plotted on separate transparent sheets. These sheets can be superimposed to compare the geometric relations between the structures. If points of lineations define a distinct axis around which the points of planar structures form a girdlelike band, a strong case for simple (cylindroidal) folding has been made (Fig. 15–13). If the plot shows considerable divergence of some structures, the outcrops should be re-examined. If the divergence is still unexplained, or if the plot shows only vaguely preferred trends, the area may be analyzed further, as follows:

1. Divide the area into subareas, each consisting of a geometric part of a major fold, as a hinge or limb, or a major segment of a structural zone (Section 15–13).

2. Construct a plot for each subarea by measuring many attitudes; the readings should be distributed as evenly over the subarea as possible.

3. Compare the diagrams of subareas to determine whether the spread of points on the general plot was caused by geographic varia-

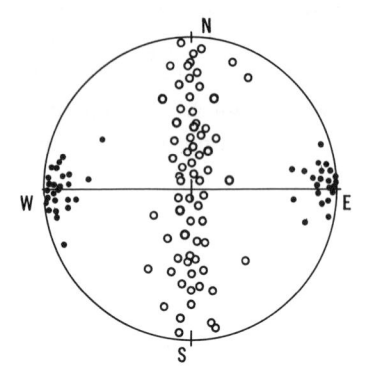

Fig. 15–13. Plot of east-west fold axes (solid dots) and poles to bedding and cleavage planes (circles), based on an equal-area net.

tions from one subarea to the next; perhaps it was caused because the structures in one or more of the subareas formed at different times.

4. Re-examine outcrops to determine the relative ages of structures.

5. If there were two ages of folding, it may be possible to unfold (undeform) structures to determine the patterns of earlier folds (see Phillips, 1954, p. 28; Badgley, 1959, p. 195).

Depending on the area concerned, many useful variants of these methods can be used. Weiss and McIntyre (1957) and Ramsay (1958) have described studies that combined mapping and structural analyses.

15–12. Oriented Samples for Microscopic Studies

Oriented rock samples provide a means of interpreting structures plotted on the map. They are especially useful in identifying cleavages and lineations in fine-grained rocks. Although laboratory procedures are not dealt with in this book, it should be emphasized that they need not consist of petrofabric (statistical) studies alone. The microscope can be used in simpler ways to determine movements on foliation surfaces, relative ages of foliations, and age relations between minerals and structures.

Samples must be marked so they can be oriented relative to the geologic map. For rocks with a planar foliation, the strike and dip of the foliation can be marked as in Fig. 15–14A. The attitude of the plane is then measured and recorded along with the description of the specimen. *The relations of this surface to bedding and metamorphic features must be sketched and described in the notes.* The rock is marked either before being knocked from the outcrop, or else after being fitted exactly back in place on the outcrop. If there is no planar foliation, a strip of tape placed parallel to the horizon with the word *level* on it will record part of the orientation; a second piece of tape on a nearly horizontal surface is then marked with a north arrow (Fig. 15–14B). Alternately, pieces of tape marked with strike and dip symbols can be placed on any two planar surfaces of the rock; the attitudes of the two surfaces are then measured and recorded in the notes. A felt-tip pen or colored pencils can be used instead of tape.

If petrofabric studies will be made, literature on this subject should be read before collecting samples. Samples for petrofabric studies should be chosen after mapping is well advanced and their role in solving problems is clear.

15–13. Mapping Metamorphic Zones

Rocks subjected to a variety of metamorphic conditions can be mapped profitably as a series of zones based on key minerals, textures, or structures. This procedure is founded on the principal that rocks with the same initial composition and structural properties should develop characteristics that depend only on conditions of metamorphism. Each set of characteristics affords a basis for mapping a zone, and the progression of zones gives unique evidence of the kinds and intensities of metamorphism. The pattern of zones, in turn, may indicate the cause of metamorphism, for example, where zones are concentric around an intrusion.

Zone mapping requires a reconnaissance and preferably a complete mapping of premetamorphic units. The criteria for the zones should be selected during this mapping. These criteria should be based on one premetamorphic rock type, and it must be widely distributed,

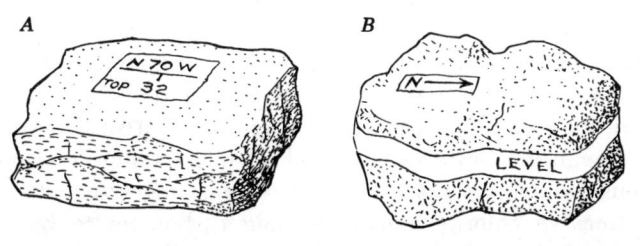

Fig. 15–14. Oriented rock samples, marked by tape labels.

readily recognizable, and sensitive to variations in metamorphic conditions. Zones can be mapped by plotting color spots or letter symbols for outcrops with suitable characteristics. Because many metamorphic changes are broadly gradational, the first appearance or disappearance of a key mineral, texture, or structure must often be used to map zone contacts (Section 4-9). Reversals and other complications should be described in the notes. New key minerals, for example, may be concentrated locally on shear zones or near veins, indicating localized effects of granulation and fluids. Relics of low-grade rocks in high-grade zones can indicate that *the high-grade rock passed through the states of lower grade zones.*

Because of variations in rock properties and metamorphic histories, methods that succeed in one area may fail in another. Each area must be considered individually; the suggestions that follow present only a few possibilities.

Mineral zones in regional terrains. The most widely used mineral zones in regional metamorphic terrains are based on the first appearance of certain *index* minerals in schistose metashales. Slates containing clays and clastic chlorite and micas are first recrystallized to phyllites or schists with new metamorphic chlorite and muscovite; then, the following minerals appear with increasing degrees of metamorphism: red-brown biotite, almandine garnet, staurolite (if the rocks are rich in iron), kyanite, and sillimanite. This well-known sequence is particularly useful because the minerals are commonly distinctive in hand specimen. Several samples should be examined at each outcrop because only a few layers may have suitable chemical compositions or be coarse-grained enough to show index minerals. Age relations between minerals must also be noted, for chlorite and muscovite can form by retrograde reactions in high-grade rocks.

Basic rocks, as metabasalts, can be zoned by the following criteria: (1) disappearance of chlorite (to form hornblende), (2) disappearance of epidote (to form plagioclase and hornblende), (3) appearance of red garnet, (4) change of green hornblende to brown hornblende, and (5) appearance of pyroxenes, especially hypersthene. Retrograde metamorphism must again be considered; for example, epidote and chlorite in amphibolites may be retrograde.

Mineral zones can be mapped in impure calcareous or dolomitic rocks, although moderate variations of chemical composition may result in misleading variants. The following changes may be useful: (1) reactions of chlorite, muscovite, and carbonates to biotite and epidote (or zoisite), (2) appearance of hornblende, (3) disappearance of epidote (to form plagioclase and hornblende), and (4) appearance of pyroxene. Garnet is commonly present in all but the lowest grade zones.

Regional zones of textural reconstitution. Metamorphic rocks that have similar mineral assemblages over broad areas may be zoned on textural changes. Grain sizes, for example, can be used to subdivide low-grade metashales into zones of slates, phyllites, and schists. Zones can also be based on the progressive granulation and concurrent recrystallization of coarser-grained granular rocks. An example of such a project is the division of chlorite-zone rocks made by Hutton and Turner (Hutton and Turner, 1936; Turner, 1948, p. 38).

Structural zones. Some rocks that were originally homogeneous over a given area can be zoned on the basis of progressive structural changes. Zones can be based on geometric forms of folds (Section 15-7), sizes of folds, degree of shearing or brecciation of primary features, or spacing of shear surfaces in massive rocks. Assemblages of several structural and textural characteristics can often be used effectively. Sutton and Watson (1951) have described an example of this type of mapping.

Zones of contact metamorphism. Contact metamorphism of fine-grained rocks may produce textural and mineralogical changes that can be mapped in zones. The first appearance of small porphyroblasts or aggregate mineral knots can be used to draw the outer contact of an aureole in slates or phyllites. An inner and an outer zone may be based on the change of the spotted rocks to tough hornfelses or mica-rich schists. In some cases, an innermost zone of feldspathic veining or permeation can be plotted. The zones can also be based on the appearance or disappearance of minerals; some possibilities in metashales are: chlorite, muscovite, red-brown biotite, andalusite, garnet, staurolite, cordierite, and sillimanite.

Contact aureoles in basic rocks can be mapped as either textural or mineralogical zones, or both. The outer contact of an aureole can often be drawn where low-grade schists and related rocks have been converted to tough, fine-grained hornfelses. An inner zone can sometimes be based on increased grain size, as the first appearance of megascopic hornblende and plagioclase, or the appearance of feldspathic veins. The mineralogic changes described for regionally metamorphosed rocks can also be used to map contact zones.

Contact zones can be based on calcareous and dolomitic rocks, although the appearance of key minerals may depend on small differences in composition. Some possible mineral appearances are: wollastonite in sandy or cherty limestones, tremolite and then diopside in siliceous dolomites, epidote and then lime-rich garnets in argillaceous limestones, and phlogopite in argillaceous dolomites. Zone patterns can be complicated greatly by the introduction of substances along fractures or crush zones. An example of the use of mineral zones to develop a pattern of progressive metasomatism is given by Burnham (1959), who based three zones on the successive appearances of monticellite, idocrase, and garnet.

15-14. Migmatites and Related Rocks

Various mixtures of granitic materials and metamorphic rocks have been classed as *migmatite* (literally, *mixture-rock*). Some geologists have used the term broadly while others have restricted it considerably; it is a term that has been defined on both descriptive and on genetic grounds. Unless the term is defined in a restricted sense for a particular project, migmatites might be considered a broad class of rocks that can be subdivided and given more specific names. Metamorphic rocks that carry macroscopic veins, pods, or other small bodies of granitic rock can be called *veined* rocks (*veined schist, veined granulite*). Metamorphic rocks peppered with grains of feldspars and quartz can be called *permeated* or *feldspathized* rocks (*permeated schist, feldspathized quartzite*). Intermediate cases can be given such names as *veined permeated schist* (Fig. 14-2). Mixtures consisting of a granitic rock host with fragments or relics of metamorphic rocks can be mapped as granitic rocks, perhaps as an inclusion-rich or contaminated facies of a given pluton. Genetic terms such as *injected schist, zone of metasomatic veins* can be used where evidence is ample. Many migmatites, however, have formed by complex combinations of the following processes: injections of magma, partial melting or solution of country rocks, local metamorphic segregation, granitization by extraneous fluids, granitization by diffusion of ions, and gross movements of mixed materials.

Migmatites can be related to intrusions, deformations, and other events by mapping them as overprints on basic lithologic units and structures. Mapping should be based primarily on amounts of granitic materials, although symbols and patterns can also show structural types (Fig. 15-15). The compositions of veins can also provide a means of classifying different kinds or ages of migmatites. The granitic materials can sometimes be classified genetically. Injection, for example, produces such effects as dilation of older structures, matching vein

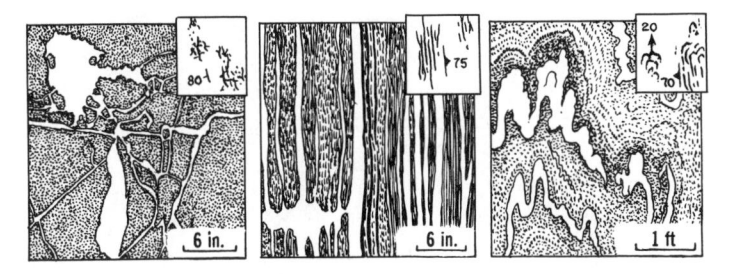

Fig. 15–15. Structural varieties of migmatites, with inserts showing patterns that may be plotted in color on field sheets. The left-hand and middle examples are from the Feather River region, Calif., and the other is from Nairnshire, Scotland.

walls, bending and breaking of septa in veins, moved inclusions, foreign inclusions, and planar fabrics that are parallel to vein walls but not to the foliation of the adjoining rocks (Figs. 14–3, 14–4, and 14–9). Metasomatism may be indicated by proof of lack of dilation of walls, inclusions (skialiths) in place, relics of former textures and structures, gradational contacts, and mafic segregations at or near the contacts of veins (Figs. 14–3, 14–8, 14–9E, and 14–9F). Metasomatism is indicated strongly where vein materials are closely related to the compositions of the rocks in which they occur. Veins in metashales, for example, might contain abundant potassium feldspar, mica, and garnet, whereas those in adjoining amphibolite layers consist mainly of plagioclase, hornblende, and pyroxene.

Observations should also be made on deformation structures associated with migmatites. The contrast between the angular appearance of some veins and the complexly folded, plastic appearance of others indicate important differences in the histories of rock masses (Figs. 15–15 and 15–16A). The irregular rotation ("swimming") of schist relics may suggest deformation of soft, perhaps partially molten, granitic materials (Fig. 15–16B). In contrast, deformation of solid veins tends to produce slipping-out, boudinage, and rodding (Fig. 15–16C and D). Extreme post-vein deformation may yield porphyroclastic schists or phyllonites. Sequences of structural types can indicate that deformation went on concurrently with vein formation. Maps of deformation structures may show that the migmatites moved as gross plutonic bodies, perhaps intruding less mobile parts of a metamorphic terrain.

The original definitions and descriptions of migmatites given by Sederholm (1907, 1926) are particularly instructive. Read (1931)

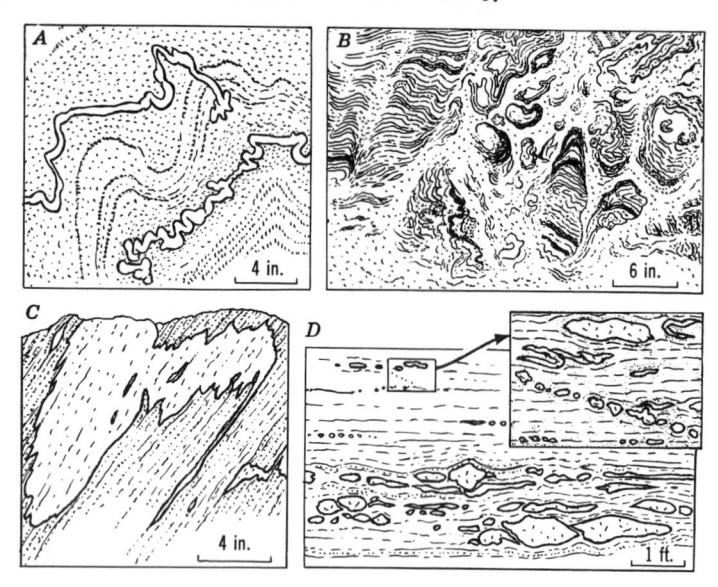

Fig. 15–16. Deformed granitic veins in migmatites. (*A*) Entrail-like (ptygmatic) folds in veins that injected folded biotitic granulites; near Inverness, Scotland. (*B*) Balls and swirls of complexly deformed schist and heterogeneous granite; northeast shore of Loch Ness, Scotland. (*C*) Slip-folded granitic sheet in biotitic granulites; near Bettyhill, Sutherland, Scotland. (*D*) Boudins of granite in biotitic granulites, with insert showing "string-of-beads" (small rods) lying across foliation. Same area as *C*.

and Crowder (1959) have described terrains where migmatites were treated as cartographic units.

References Cited

Badgley, P. C., 1959, *Structural methods for the exploration geologist*, New York, Harper and Brothers, 280 pp.

Burnham, C. W., 1959, Contact metamorphism of magnesian limestones at Crestmore, California: *Bulletin of the Geological Society of America*, v. 70, pp. 879–920.

Cloos, Ernst, 1947, Oölite deformation in the South Mountain fold, Maryland: *Bulletin of the Geological Society of America*, v. 58, pp. 843–918.

Crowder, D. F., 1959, Granitization, migmatization, and fusion in the northern Entiat Mountains, Washington: *Bulletin of the Geological Society of America*, v. 70, pp. 827–878.

Gair, J. E., 1950, Some effects of deformation in the central Appalachians: *Bulletin of the Geological Society of America*, v. 61, pp. 857–876.

Goldsmith, Richard, 1959, Granofels, a new metamorphic rock name: *Journal of Geology*, v. 67, pp. 109–110.

Hutton, C. O., and F. J. Turner, 1936, Metamorphic zones in north-west Otago: *Royal Society of New Zealand, Transactions*, v. 65, pp. 405–406.

Nickelsen, R. P., 1956, Geology of the Blue Ridge near Harpers Ferry, West Virginia: *Bulletin of the Geological Society of America*, v. 67, pp. 239–270.

Phillips, F. C., 1954, *The use of stereographic projection in structural geology*: London, Edward Arnold Ltd., 86 pp.

Ramsay, J. G., 1958, Superimposed folding at Loch Monar, Inverness-shire and Ross-shire: *Quarterly Journal of the Geological Society of London*, v. 113, pp. 271–307.

Read, H. H., 1931, *The geology of central Sutherland* (explanation of Sheets 108 and 109): Memoirs of the Geological Survey, Scotland, 238 pp.

Sederholm, J. J., 1907, *Om granit och gneis, deras uppkomst, uppträdande och utbredning inom urberget i Fennoscandia* (English summary: *On granite and gneiss, their origin, relations and occurrence in the pre-Cambrian complex of Fenno-scandia*, pp. 91–110): Commission Géologique de Findlande, Bulletin 23, 110 pp.

———, 1926, *On migmatites and associated pre-Cambrian rocks of southwestern Finland* (Part II): Commission Géologique de Findlande, Bulletin 77, 143 pp.

Sutton, John, and Janet Watson, 1951, The pre-Torridonian metamorphic history of the Loch Torridon and Scourie areas in the North-West Highlands, and its bearing on the chronological classification of the Lewisian: *Quarterly Journal of the Geological Society of London*, v. 106, pp. 241–307.

Turner, F. J., 1948, *Mineralogical and structural evolution of the metamorphic rocks:* Geological Society of America, Memoir 30, 342 pp.

Weiss, L. E., and D. B. McIntyre, 1957, Structural geometry of Dalradian rocks at Loch Leven, Scottish Highlands: *Journal of Geology*, v. 65, pp. 575–602.

Wilson, Gilbert, 1951, The tectonics of the Tintagel area, North Cornwall: *Quarterly Journal of the Geological Society of London*, v. 106, pp. 393–432.

———, 1953, Mullion and rodding structures in the Moine series of Scotland: *Proceedings of the Geologists' Association*, v. 64, pp. 118–151.

Appendixes

APPENDIX 1. CHECK LISTS OF EQUIPMENT AND SUPPLIES

A. FOR GENERAL GEOLOGIC WORK

Adhesive tape for labeling
Aerial photographs and indexes
Barometer
Base maps
Camera, film, and accessories
Canteen
Cement, cellulose
Chemicals, as required for staining and other tests
Cold chisels
Color pencils, waterproof
Colored tape or paint for marking outcrops
Compass, Brunton or other
Clinometer (or Brunton compass)
Drawing board
Erasers for pencils
Erasers, soft and hard, for drawing
Field case for maps and photographs
Field glasses
Field notebook or note paper
Grain-size cards (Section 12-3)
Hammer, small, for trimming fossils
Hammer, geologist's
Hand lens
Hand level (or Brunton compass)
Hydrochloric acid, dilute
Ink, black, waterproof
Ink, colored (red, green, blue, brown)
Inks, colored, opaque, for marking photographs (Section 5-10)
Knapsack
Latex, liquid, for fossil casts
Lettering set
Looseleaf folder
Magnet, pocket
Microscope, binocular
Mineral hardness set

Needles, for marking photographs
Paper, cross-ruled
Paper, scratch
Paper, soft, for wrapping fossils, etc.
Paper, tracing
Peep-sight alidade and traverse board
Pen, ballpoint
Pen, contour
Pen, drop circle
Pen holders
Pen points, assorted
Pen, ruling
Pencils, assorted, 3B to 9H
Pencil pocket-clips
Pencil pointer (file or sandpaper)
Plastic, liquid (for fossils, etc.)
Pocket knife
Proportional dividers
Protractors
Reference library
Sample bags
Scale, plotting, 6 in.
Scale, triangular, 12 to 24 in.
Slide rule
Snake bite kit
Specific gravity balance
Specific gravity liquids (Section 14-3)
Stereographic projection net
Stereoscopes
Straightedge, steel
Tables, mathematical
Tally counter
Tracing sheets (as for photo overlays)
Triangles, drawing
T-square
Typewriter
Watch

B. FOR PLANE TABLE SURVEYING

Alidade with adjusting pins, soft brush, and screw driver
Beam compass
Dark glasses
Erasers
Field glasses
Field notebooks
Flagging for signals and point markers
Keel (crayons)
Lumber for signals
Manufacturer's instruction book
Nails, assorted
Needles for marking points
Paper, brown, for covering sheet
Pencils, 2H to 9H
Pencil pointer (file or sandpaper)
Plane table board in case
Plane table sheets
Protractor, large
Reading glass (about 3×)
Rod, stadia
Scales, plotting
Stadia tables or slide rule
Straightedge, steel
Tables, mathematical
Tape, 6 ft
Tape, steel, 100 ft
Tracing paper for 3-point locations
Tripod, Johnson, with cap
Wire for signals

C. FOR CONTROL SURVEYING WITH TRANSIT AND TAPE

Aerial photographs
Ephemeris, solar
Erasers
Field glasses
Field notebooks
Flagging for signals
Hand level (or Brunton compass)
Hatchet
Keel (crayons), red and blue
Logarithms, 7-place
Lumber for signals and stakes
Nails, assorted
Paint for signals
Paint brushes
Paper for computations
Pencils, 1H to 9H
Pliers with wire cutters
Plumb bobs for taping
Reading glass (about 3×)
Rod, Philadelphia
Slide rule
Tape, steel, 100 to 300 ft
Taping pins
Taping scale
Taping thermometer
Transit with tripod and plumb bob
Wire for signals

APPENDIX 2. ABBREVIATIONS FOR FIELD NOTES

The abbreviations suggested here are only a few of many possibilities. Most words of six letters or less are omitted because little is gained in abbreviating them. Meanings can be clarified in some cases by capitalizing the first letter of nouns. A more complete list has been given by J. G. Mitchell and J. C. Maher (1957, Suggested abbreviations for lithologic descriptions: *Bulletin of the American Association of Petroleum Geologists*, v. 41, pp. 2103–2107).

abundant	abnt	crystal	Xl
acicular	acic	crystalline	xln
aggregate	Aggr	diameter	Diam
amorphous	amor	different	diff
amount	Amt	disseminated	dissem
angle	∠	dolomite	Dol
angular	ang	elevation	Elev
andesite	And	equivalent	equiv
anhedral	anhed	evaporitic	evap
anhydrite	Anhy	exposure	Exp
approximate	approx	feldspathic	feld
arenaceous	aren	foliated	fol
argillaceous	arg	foraminifera	Foram
arkosic	ark	formation	Fm
asphaltic	asph	fracture	Frac
average	Ave	fragmental	frag
bedded	bdd	glauconitic	glauc
bedding	Bdg	granite	Gr
bentonite	Bent	granodiorite	Grd
biotite	Bio	granular	gran
bituminous	bit	graptolite	Grap
boulder	Bldr	graywacke	Gwke
brachiopod	Brach	greenstone	Grnst
calcareous	calc	gypsiferous	gyp
carbonaceous	carb	hematitic	hem
cavernous	cav	horizontal	horiz
cemented	cmt	hornblende	Hbl
chalcedony	Chal	hornfels	Hfls
claystone	Clst	igneous	ign
cobble	Cbl	inclusion	Incl
conglomerate	Cgl	interbedded	intbdd
contact	Ctc	intrusion	Intr
cross-bedded	xbdd	irregular	ireg
cross-bedding	Xbdg	joint	Jnt
cross-laminated	xlam	laminated	lam

limestone	Ls	regular	reg
maximum	max	rhyolite	Rhy
member	Mbr	rocks	Rx
metamorphic	met	rounded	rnd
mudstone	Mdst	sandstone	Ss
muscovite	Musc	saturated	sat
nepheline	Neph	secondary	sec
nodular	nod	sedimentary	sed
oölitic	oöl	shale	Sh
orthoclase	Orth	serpentine	Sp
outcrop	Otcp	siliceous	sil
pebble	Pbl	siltstone	Sltst
peridotite	Perid	soluble	sol
permeability	Perm	station	Sta
phenocryst	Phen	structure	Struc
phosphatic	phos	tabular	tab
plagioclase	Plag	temperature	T
point	Pt	unconformity	Unconf
porphyritic	porph	variegated	vrgt
probable	prob	vegetation	Veg
pyritized	py	vertebrate	Vrtb
quartz	Qz	volcanic	volc
quartzite	Qzt	xenolith	Xen

APPENDIX 3. CHARTS FOR ESTIMATING PERCENTAGE COMPOSITION OF ROCKS AND SEDIMENTS

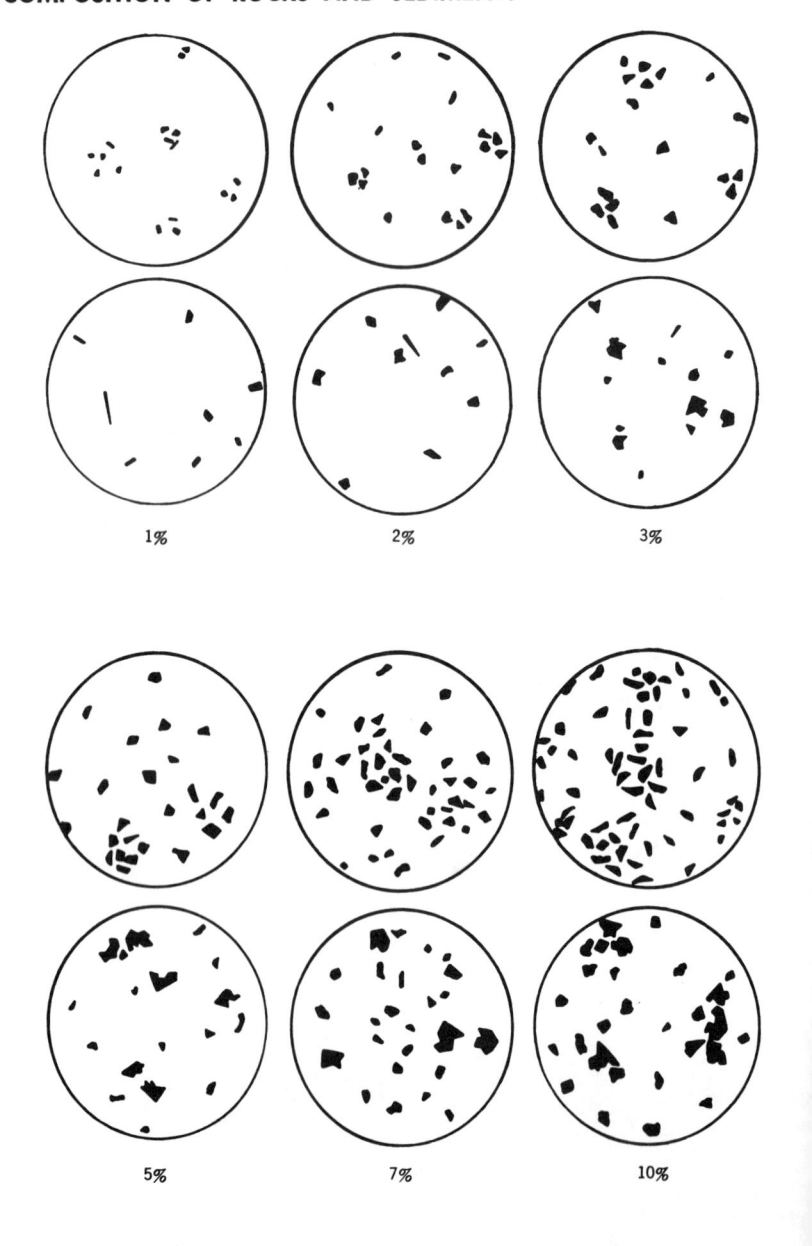

1%   2%   3%

5%   7%   10%

Prepared by R. D. Terry and G. V. Chilingar for *Journal of Sedimentary Petrology* (v. 25, pp. 229–234, 1955); reprinted as *Data Sheet 6* of *Geotimes*, available from the American Geological Institute

APPENDIX 3 (Continued)

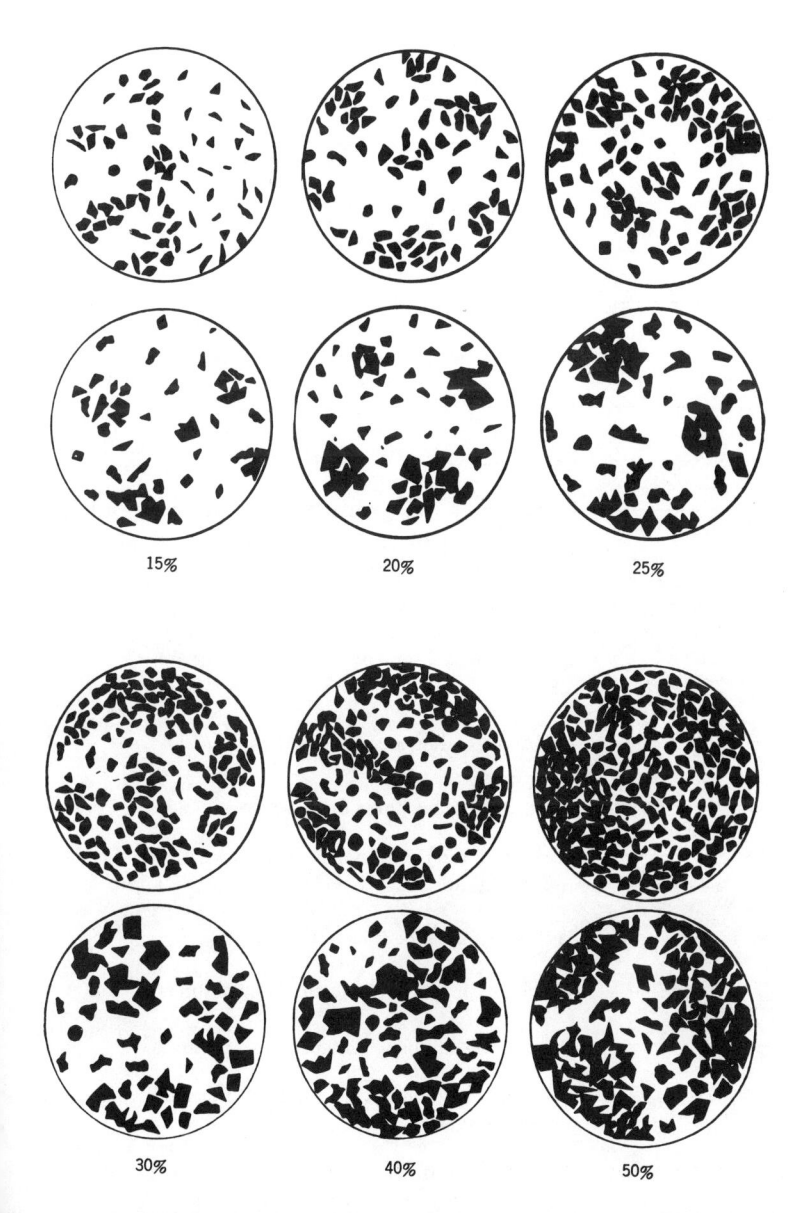

15% 20% 25%

30% 40% 50%

101 Constitution Ave., N.W., Washington 25, D.C. Reprinted here by permission of the authors and the Society of Economic Paleontologists and Mineralogists.

APPENDIX 4. SYMBOLS FOR GEOLOGIC MAPS

1. Contact, showing dip
2. Contact, vertical (left) and overturned
3. Contact, located approximately (give limits)
4. Contact, located very approximately
5. Gradational contact (a new symbol)
6. Contact, projected beneath mapped units
7. Fault, showing dips
8. Fault, located approximately (give limits)
9. Fault, existence uncertain
10. Fault, projected beneath mapped units
11. Possible fault (as located from aerial photographs)
12. Fault, showing trend and plunge of linear features (D, downthrown side; U, upthrown side)
13. Fault, showing relative horizontal movement
14. Thrust faults; T or sawteeth in upper plate
15. Fault zones, showing average dips
16. Normal fault; hachures on downthrown side
17. Anticline (top) and syncline, showing trace of axial plane and plunge of axis; dashed where located approximately
18. Anticline, existence uncertain
19. Anticline, projected beneath mapped units
20. Asymmetric anticline; steeper limb to south
21. Overturned anticline (top) and syncline, showing trend and plunge of axis
22. Overturned anticline, showing dip of axial plane
23. Doubly plunging anticline, showing culmination
24. Vertically plunging anticline

Appendix 4

25 Inverted (synformal) anticline

26 Monocline or flexure in homocline

27 Axial trend of small anticline (left) and syncline

28 Axial trend of folds that are too small to plot individually; patterns show general shapes of folds in profile

29 Strike and dip of bedding

30 Strike and dip of overturned bedding

31 Strike and dip of bedding where tops of beds are shown by primary features

32 Strike of vertical bedding; stratigraphic tops to north

33 Horizontal bedding

34 Undulatory or crumpled beds

35 Strike and dip of bedding, uncertain

36 Strike of bedding certain but dips uncertain

37 Strike and dip of foliations

38 Strike of vertical foliations

39 Horizontal foliations

40 Strike and dip where bedding parallels foliation

41 Strike and dip of joints (left) and veins or dikes

42 Strike of vertical joints (left) and veins or dikes

43 Horizontal joints (left) and veins or dikes

44 Trace (left) and mapped shape of ore vein

45 Body of high-grade ore, with stipples showing wall-rock alteration

46 Body of low-grade ore

47	→30	Trend and plunge of lineation
48	⋄⁹⁰	Vertical lineation
49	↔	Trend of horizontal lineation
50	⤻	Trend of intersection of cleavage and bedding
51	x→ ◊→	Trends of intersections of two cleavages
52	o→ □→ hbl→	Trends of pebble, mineral, etc., lineations
53	⤻ →→	Trends of lineations lying in planes of foliations
54	⇇ ⇉	Trends of horizontal lineations lying in planes of foliations
55	⩳⁹⁰	Vertical lineation and foliation

ACCESSORY SYMBOLS FOR SMALL-SCALE MAPS

56	▫ ▫-	Shafts, vertical (left) and inclined
57	>— >+—	Adits, open (left) and inaccessible
58	>—< X	Trench (left) and prospect
59	✵	Mine, quarry, or glory hole
60	✕	Sand, gravel, or clay pit
61	● -✦-	Oil well (left) and gas well
62	-✦-	Well drilled for oil or gas, dry
63	◐ -✦-	Wells with shows of oil (left) and gas
64	⌽ -O-	Oil or gas well, abandoned (left) and shut in
65	● o ⌽	Water wells; flowing (left), nonflowing, and dry (right)

ACCESSORY SYMBOLS FOR LARGE-SCALE MAPS (plotted to scale)

66	⬭	Glory hole, open pit, or quarry
67	⬬ ⬭ ⊙	Trench (left), open cut, and pit (right)
68	⇉ ⇉	Portal of tunnel or adit, that on right with open cut

Appendix 4

69 — Dump, showing track

70 — Shafts at surface, vertical (left) and inclined

71 — Shaft extending through a level (left), and bottom of shaft

72 — Inclined shaft, with chevrons pointing down

73 — Raise or winze, head (left) and foot

74 — Raise or winze extending through a level

75 — Level working, showing ore chute (left) and inaccessible area

76 — Lagging or cribbing along working, with filled area to right

77 — Drill holes, horizontal (left) and inclined at 30° (showing horizontal projection of end of hole)

Note: map symbols used for topographic, hydrographic, and cultural features are explained and illustrated in color on the folded sheet "Topographic Maps," free on application to the Director, U.S. Geological Survey, Washington 25, D.C.

APPENDIX 5. LITHOLOGIC SYMBOLS FOR CROSS SECTIONS AND COLUMNAR SECTIONS

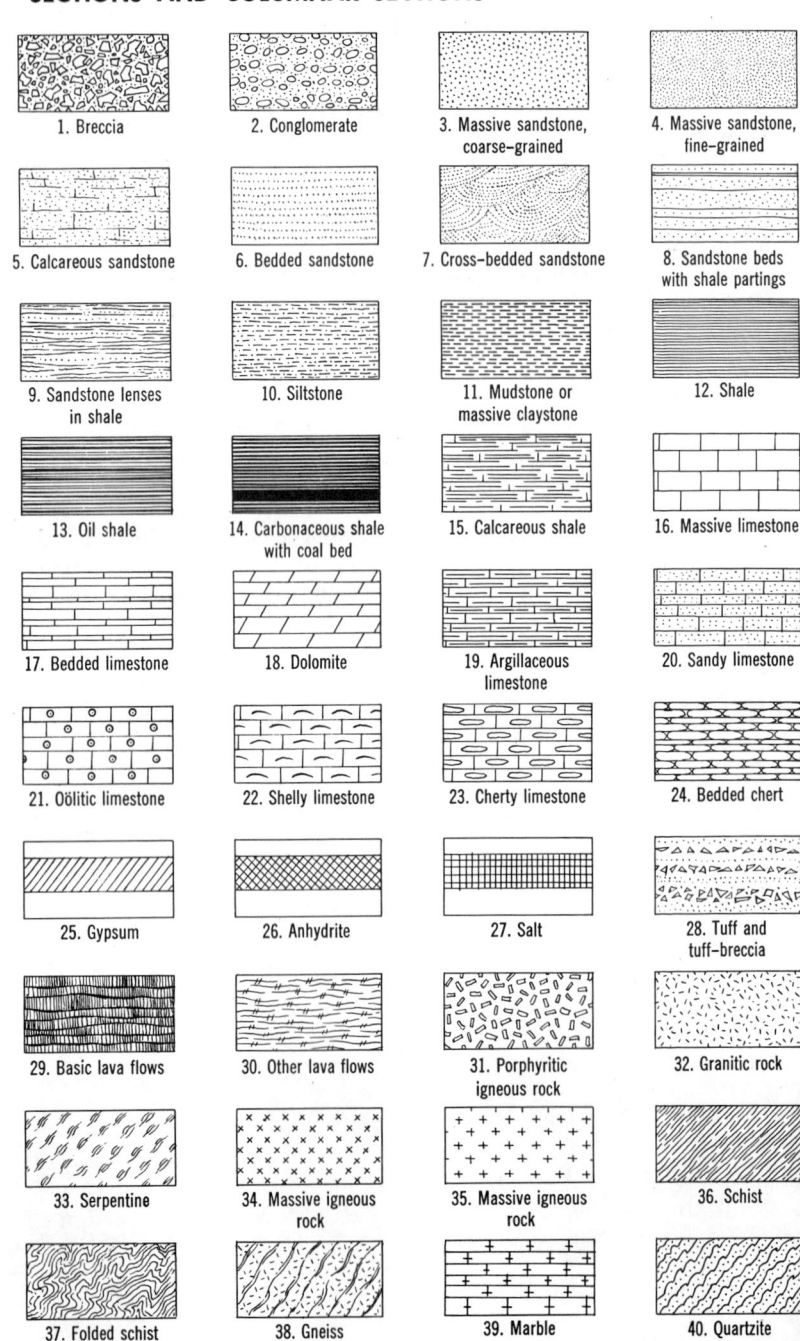

1. Breccia
2. Conglomerate
3. Massive sandstone, coarse-grained
4. Massive sandstone, fine-grained
5. Calcareous sandstone
6. Bedded sandstone
7. Cross-bedded sandstone
8. Sandstone beds with shale partings
9. Sandstone lenses in shale
10. Siltstone
11. Mudstone or massive claystone
12. Shale
13. Oil shale
14. Carbonaceous shale with coal bed
15. Calcareous shale
16. Massive limestone
17. Bedded limestone
18. Dolomite
19. Argillaceous limestone
20. Sandy limestone
21. Oölitic limestone
22. Shelly limestone
23. Cherty limestone
24. Bedded chert
25. Gypsum
26. Anhydrite
27. Salt
28. Tuff and tuff-breccia
29. Basic lava flows
30. Other lava flows
31. Porphyritic igneous rock
32. Granitic rock
33. Serpentine
34. Massive igneous rock
35. Massive igneous rock
36. Schist
37. Folded schist
38. Gneiss
39. Marble
40. Quartzite

APPENDIX 6. ISOGONIC CHART OF THE UNITED STATES AND NEIGHBORING AREAS

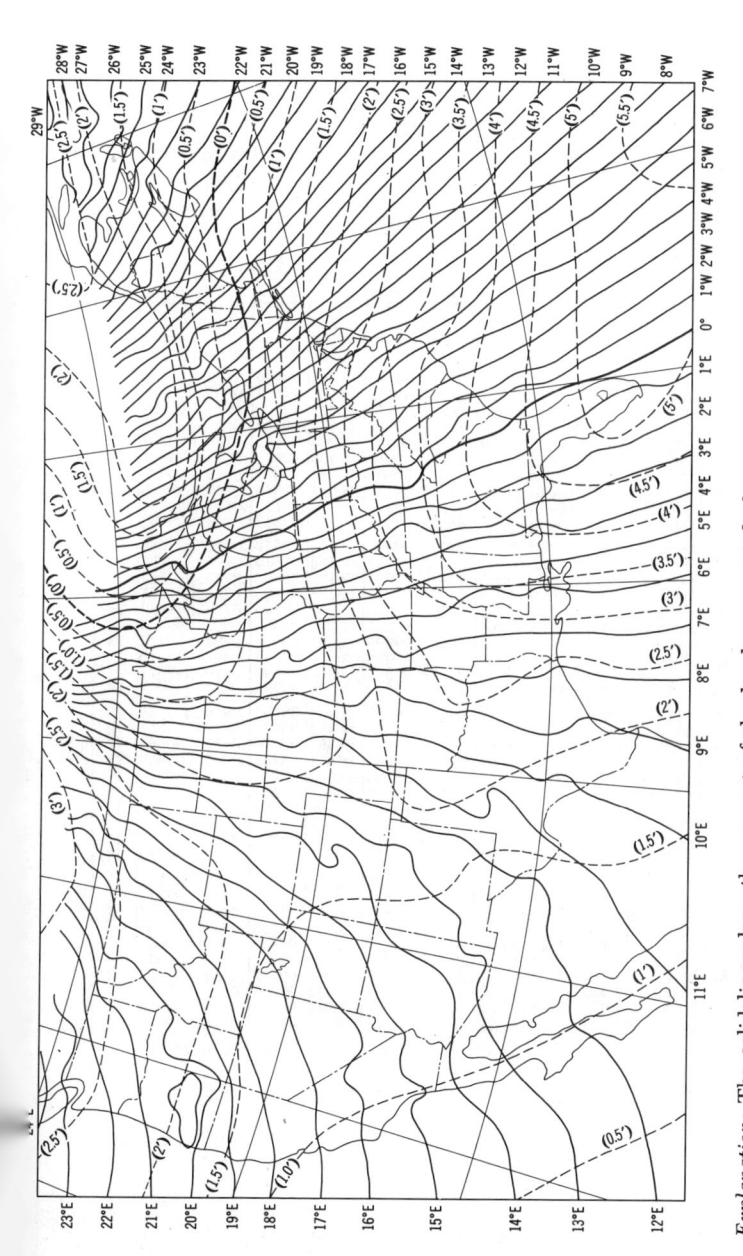

Explanation: The solid lines show the amount of the local magnetic declination in 1960; the dashed lines show the probable change in declination for each year. At the southern tip of Texas, for example, the declination in 1960 was 8° 50' east of true North; in 1970 it will be 8° 50' − (10 × 02'), or 8° 30' E. *Source: Isogonic chart of the United States,* U.S. Coast and Geodetic Survey Chart 3077, 1960.

APPENDIX 7. TABLES OF NATURAL TRIGONOMETRIC FUNCTIONS

′	0° Sine	0° Cosin	1° Sine	1° Cosin	2° Sine	2° Cosin	3° Sine	3° Cosin	4° Sine	4° Cosin	′
0	.00000	One.	.01745	.99985	.03490	.99939	.05234	.99863	.06976	.99756	60
1	.00029	One.	.01774	.99984	.03519	.99938	.05263	.99861	.07005	.99754	59
2	.00058	One.	.01803	.99984	.03548	.99937	.05292	.99860	.07034	.99752	58
3	.00087	One.	.01832	.99983	.03577	.99936	.05321	.99858	.07063	.99750	57
4	.00116	One.	.01862	.99983	.03606	.99935	.05350	.99857	.07092	.99748	56
5	.00145	One.	.01891	.99982	.03635	.99934	.05379	.99855	.07121	.99746	55
6	.00175	One.	.01920	.99982	.03664	.99933	.05408	.99854	.07150	.99744	54
7	.00204	One.	.01949	.99981	.03693	.99932	.05437	.99852	.07179	.99742	53
8	.00233	One.	.01978	.99980	.03723	.99931	.05466	.99851	.07208	.99740	52
9	.00262	One.	.02007	.99980	.03752	.99930	.05495	.99849	.07237	.99738	51
10	.00291	One.	.02036	.99979	.03781	.99929	.05524	.99847	.07266	.99736	50
11	.00320	.99999	.02065	.99979	.03810	.99927	.05553	.99846	.07295	.99734	49
12	.00349	.99999	.02094	.99978	.03839	.99926	.05582	.99844	.07324	.99731	48
13	.00378	.99999	.02123	.99977	.03868	.99925	.05611	.99842	.07353	.99729	47
14	.00407	.99999	.02152	.99977	.03897	.99924	.05640	.99841	.07382	.99727	46
15	.00436	.99999	.02181	.99976	.03926	.99923	.05669	.99839	.07411	.99725	45
16	.00465	.99999	.02211	.99976	.03955	.99922	.05698	.99838	.07440	.99723	44
17	.00495	.99999	.02240	.99975	.03984	.99921	.05727	.99836	.07469	.99721	43
18	.00524	.99999	.02269	.99974	.04013	.99919	.05756	.99834	.07498	.99719	42
19	.00553	.99998	.02298	.99974	.04042	.99918	.05785	.99833	.07527	.99716	41
20	.00582	.99998	.02327	.99973	.04071	.99917	.05814	.99831	.07556	.99714	40
21	.00611	.99998	.02356	.99972	.04100	.99916	.05844	.99829	.07585	.99712	39
22	.00640	.99998	.02385	.99972	.04129	.99915	.05873	.99827	.07614	.99710	38
23	.00669	.99998	.02414	.99971	.04159	.99913	.05902	.99826	.07643	.99708	37
24	.00698	.99998	.02443	.99970	.04188	.99912	.05931	.99824	.07672	.99705	36
25	.00727	.99997	.02472	.99969	.04217	.99911	.05960	.99822	.07701	.99703	35
26	.00756	.99997	.02501	.99969	.04246	.99910	.05989	.99821	.07730	.99701	34
27	.00785	.99997	.02530	.99968	.04275	.99909	.06018	.99819	.07759	.99699	33
28	.00814	.99997	.02560	.99967	.04304	.99907	.06047	.99817	.07788	.99696	32
29	.00844	.99996	.02589	.99966	.04333	.99906	.06076	.99815	.07817	.99694	31
30	.00873	.99996	.02618	.99966	.04362	.99905	.06105	.99813	.07846	.99692	30
31	.00902	.99996	.02647	.99965	.04391	.99904	.06134	.99812	.07875	.99689	29
32	.00931	.99996	.02676	.99964	.04420	.99902	.06163	.99810	.07904	.99687	28
33	.00960	.99995	.02705	.99963	.04449	.99901	.06192	.99808	.07933	.99685	27
34	.00989	.99995	.02734	.99963	.04478	.99900	.06221	.99806	.07962	.99683	26
35	.01018	.99995	.02763	.99962	.04507	.99898	.06250	.99804	.07991	.99680	25
36	.01047	.99995	.02792	.99961	.04536	.99897	.06279	.99803	.08020	.99678	24
37	.01076	.99994	.02821	.99960	.04565	.99896	.06308	.99801	.08049	.99676	23
38	.01105	.99994	.02850	.99959	.04594	.99894	.06337	.99799	.08078	.99673	22
39	.01134	.99994	.02879	.99959	.04623	.99893	.06366	.99797	.08107	.99671	21
40	.01164	.99993	.02908	.99958	.04653	.99892	.06395	.99795	.08136	.99668	20
41	.01193	.99993	.02938	.99957	.04682	.99890	.06424	.99793	.08165	.99666	19
42	.01222	.99993	.02967	.99956	.04711	.99889	.06453	.99792	.08194	.99664	18
43	.01251	.99992	.02996	.99955	.04740	.99888	.06482	.99790	.08223	.99661	17
44	.01280	.99992	.03025	.99954	.04769	.99886	.06511	.99788	.08252	.99659	16
45	.01309	.99991	.03054	.99953	.04798	.99885	.06540	.99786	.08281	.99657	15
46	.01338	.99991	.03083	.99952	.04827	.99883	.06569	.99784	.08310	.99654	14
47	.01367	.99991	.03112	.99952	.04856	.99882	.06598	.99782	.08339	.99652	13
48	.01396	.99990	.03141	.99951	.04885	.99881	.06627	.99780	.08368	.99649	12
49	.01425	.99990	.03170	.99950	.04914	.99879	.06656	.99778	.08397	.99647	11
50	.01454	.99989	.03199	.99949	.04943	.99878	.06685	.99776	.08426	.99644	10
51	.01483	.99989	.03228	.99948	.04972	.99876	.06714	.99774	.08455	.99642	9
52	.01513	.99989	.03257	.99947	.05001	.99875	.06743	.99772	.08484	.99639	8
53	.01542	.99988	.03286	.99946	.05030	.99873	.06773	.99770	.08513	.99637	7
54	.01571	.99988	.03316	.99945	.05059	.99872	.06802	.99768	.08542	.99635	6
55	.01600	.99987	.03345	.99944	.05088	.99870	.06831	.99766	.08571	.99632	5
56	.01629	.99987	.03374	.99943	.05117	.99869	.06860	.99764	.08600	.99630	4
57	.01658	.99986	.03403	.99942	.05146	.99867	.06889	.99762	.08629	.99627	3
58	.01687	.99986	.03432	.99941	.05175	.99866	.06918	.99760	.08658	.99625	2
59	.01716	.99985	.03461	.99940	.05205	.99864	.06947	.99758	.08687	.99622	1
60	.01745	.99985	.03490	.99939	.05234	.99863	.06976	.99619	.08716	.99619	0
′	Cosin	Sine	Cosin	Sine	Cosin	Sine	Cosin	Sine	Cosin	Sine	′
	89°		88°		87°		86°		85°		

Appendix 7

′	5° Sine	5° Cosin	6° Sine	6° Cosin	7° Sine	7° Cosin	8° Sine	8° Cosin	9° Sine	9° Cosin	′
0	.08716	.99619	.10453	.99452	.12187	.99255	.13917	.99027	.15643	.98769	60
1	.08745	.99617	.10482	.99449	.12216	.99251	.13946	.99023	.15672	.98764	59
2	.08774	.99614	.10511	.99446	.12245	.99248	.13975	.99019	.15701	.98760	58
3	.08803	.99612	.10540	.99443	.12274	.99244	.14004	.99015	.15730	.98755	57
4	.08831	.99609	.10569	.99440	.12302	.99240	.14033	.99011	.15758	.98751	56
5	.08860	.99607	.10597	.99437	.12331	.99237	.14061	.99006	.15787	.98746	55
6	.08889	.99604	.10626	.99434	.12360	.99233	.14090	.99002	.15816	.98741	54
7	.08918	.99602	.10655	.99431	.12389	.99230	.14119	.98998	.15845	.98737	53
8	.08947	.99599	.10684	.99428	.12418	.99226	.14148	.98994	.15873	.98732	52
9	.08976	.99596	.10713	.99424	.12447	.99222	.14177	.98990	.15902	.98728	51
10	.09005	.99594	.10742	.99421	.12476	.99219	.14205	.98986	.15931	.98723	50
11	.09034	.99591	.10771	.99418	.12504	.99215	.14234	.98982	.15959	.98718	49
12	.09063	.99588	.10800	.99415	.12533	.99211	.14263	.98978	.15988	.98714	48
13	.09092	.99586	.10829	.99412	.12562	.99208	.14292	.98973	.16017	.98709	47
14	.09121	.99583	.10858	.99409	.12591	.99204	.14320	.98969	.16046	.98704	46
15	.09150	.99580	.10887	.99406	.12620	.99200	.14349	.98965	.16074	.98700	45
16	.09179	.99578	.10916	.99402	.12649	.99197	.14378	.98961	.16103	.98695	44
17	.09208	.99575	.10945	.99399	.12678	.99193	.14407	.98957	.16132	.98690	43
18	.09237	.99572	.10973	.99396	.12706	.99189	.14436	.98953	.16160	.98686	42
19	.09266	.99570	.11002	.99393	.12735	.99186	.14464	.98948	.16189	.98681	41
20	.09295	.99567	.11031	.99390	.12764	.99182	.14493	.98944	.16218	.98676	40
21	.09324	.99564	.11060	.99386	.12793	.99178	.14522	.98940	.16246	.98671	39
22	.09353	.99562	.11089	.99383	.12822	.99175	.14551	.98936	.16275	.98667	38
23	.09382	.99559	.11118	.99380	.12851	.99171	.14580	.98931	.16304	.98662	37
24	.09411	.99556	.11147	.99377	.12880	.99167	.14608	.98927	.16333	.98657	36
25	.09440	.99553	.11176	.99374	.12908	.99163	.14637	.98923	.16361	.98652	35
26	.09469	.99551	.11205	.99370	.12937	.99160	.14666	.98919	.16390	.98648	34
27	.09498	.99548	.11234	.99367	.12966	.99156	.14695	.98914	.16419	.98643	33
28	.09527	.99545	.11263	.99364	.12995	.99152	.14723	.98910	.16447	.98638	32
29	.09556	.99542	.11291	.99360	.13024	.99148	.14752	.98906	.16476	.98633	31
30	.09585	.99540	.11320	.99357	.13053	.99144	.14781	.98902	.16505	.98629	30
31	.09614	.99537	.11349	.99354	.13081	.99141	.14810	.98897	.16533	.98624	29
32	.09642	.99534	.11378	.99351	.13110	.99137	.14838	.98893	.16562	.98619	28
33	.09671	.99531	.11407	.99347	.13139	.99133	.14867	.98889	.16591	.98614	27
34	.09700	.99528	.11436	.99344	.13168	.99129	.14896	.98884	.16620	.98609	26
35	.09729	.99526	.11465	.99341	.13197	.99125	.14925	.98880	.16648	.98604	25
36	.09758	.99523	.11494	.99337	.13226	.99122	.14954	.98876	.16677	.98600	24
37	.09787	.99520	.11523	.99334	.13254	.99118	.14982	.98871	.16706	.98595	23
38	.09816	.99517	.11552	.99331	.13283	.99114	.15011	.98867	.16734	.98590	22
39	.09845	.99514	.11580	.99327	.13312	.99110	.15040	.98863	.16763	.98585	21
40	.09874	.99511	.11609	.99324	.13341	.99106	.15069	.98858	.16792	.98580	20
41	.09903	.99508	.11638	.99320	.13370	.99102	.15097	.98854	.16820	.98575	19
42	.09932	.99506	.11667	.99317	.13399	.99098	.15126	.98849	.16849	.98570	18
43	.09961	.99503	.11696	.99314	.13427	.99094	.15155	.98845	.16878	.98565	17
44	.09990	.99500	.11725	.99310	.13456	.99091	.15184	.98841	.16906	.98561	16
45	.10019	.99497	.11754	.99307	.13485	.99087	.15212	.98836	.16935	.98556	15
46	.10048	.99494	.11783	.99303	.13514	.99083	.15241	.98832	.16964	.98551	14
47	.10077	.99491	.11812	.99300	.13543	.99079	.15270	.98827	.16992	.98546	13
48	.10106	.99488	.11840	.99297	.13572	.99075	.15299	.98823	.17021	.98541	12
49	.10135	.99485	.11869	.99293	.13600	.99071	.15327	.98818	.17050	.98536	11
50	.10164	.99482	.11898	.99290	.13629	.99067	.15356	.98814	.17078	.98531	10
51	.10192	.99479	.11927	.99286	.13658	.99063	.15385	.98809	.17107	.98526	9
52	.10221	.99476	.11956	.99283	.13687	.99059	.15414	.98805	.17136	.98521	8
53	.10250	.99473	.11985	.99279	.13716	.99055	.15442	.98800	.17164	.98516	7
54	.10279	.99470	.12014	.99276	.13744	.99051	.15471	.98796	.17193	.98511	6
55	.10308	.99467	.12043	.99272	.13773	.99047	.15500	.98791	.17222	.98506	5
56	.10337	.99464	.12071	.99269	.13802	.99043	.15529	.98787	.17250	.98501	4
57	.10366	.99461	.12100	.99265	.13831	.99039	.15557	.98782	.17279	.98496	3
58	.10395	.99458	.12129	.99262	.13860	.99035	.15586	.98778	.17308	.98491	2
59	.10424	.99455	.12158	.99258	.13889	.99031	.15615	.98773	.17336	.98486	1
60	.10453	.99452	.12187	.99255	.13917	.99027	.15643	.98769	.17365	.98481	0
′	Cosin	Sine	Cosin	Sine	Cosin	Sine	Cosin	Sine	Cosin	Sine	′
	84°		83°		82°		81°		80°		

′	10° Sine	10° Cosin	11° Sine	11° Cosin	12° Sine	12° Cosin	13° Sine	13° Cosin	14° Sine	14° Cosin	′
0	.17365	.98481	.19081	.98163	.20791	.97815	.22495	.97437	.24192	.97030	60
1	.17393	.98476	.19109	.98157	.20820	.97809	.22523	.97430	.24220	.97023	59
2	.17422	.98471	.19138	.98152	.20848	.97803	.22552	.97424	.24249	.97015	58
3	.17451	.98466	.19167	.98146	.20877	.97797	.22580	.97417	.24277	.97008	57
4	.17479	.98461	.19195	.98140	.20905	.97791	.22608	.97411	.24305	.97001	56
5	.17508	.98455	.19224	.98135	.20933	.97784	.22637	.97404	.24333	.96994	55
6	.17537	.98450	.19252	.98129	.20962	.97778	.22665	.97398	.24362	.96987	54
7	.17565	.98445	.19281	.98124	.20990	.97772	.22693	.97391	.24390	.96980	53
8	.17594	.98440	.19309	.98118	.21019	.97766	.22722	.97384	.24418	.96973	52
9	.17623	.98435	.19338	.98112	.21047	.97760	.22750	.97378	.24446	.96966	51
10	.17651	.98430	.19366	.98107	.21076	.97754	.22778	.97371	.24474	.96959	50
11	.17680	.98425	.19395	.98101	.21104	.97748	.22807	.97365	.24503	.96952	49
12	.17708	.98420	.19423	.98096	.21132	.97742	.22835	.97358	.24531	.96945	48
13	.17737	.98414	.19452	.98090	.21161	.97735	.22863	.97351	.24559	.96937	47
14	.17766	.98409	.19481	.98084	.21189	.97729	.22892	.97345	.24587	.96930	46
15	.17794	.98404	.19509	.98079	.21218	.97723	.22920	.97338	.24615	.96923	45
16	.17823	.98399	.19538	.98073	.21246	.97717	.22948	.97331	.24644	.96916	44
17	.17852	.98394	.19566	.98067	.21275	.97711	.22977	.97325	.24672	.96909	43
18	.17880	.98389	.19595	.98061	.21303	.97705	.23005	.97318	.24700	.96902	42
19	.17909	.98383	.19623	.98056	.21331	.97698	.23033	.97311	.24728	.96894	41
20	.17937	.98378	.19652	.98050	.21360	.97692	.23062	.97304	.24756	.96887	40
21	.17966	.98373	.19680	.98044	.21388	.97686	.23090	.97298	.24784	.96880	39
22	.17995	.98368	.19709	.98039	.21417	.97680	.23118	.97291	.24813	.96873	38
23	.18023	.98362	.19737	.98033	.21445	.97673	.23146	.97284	.24841	.96866	37
24	.18052	.98357	.19766	.98027	.21474	.97667	.23175	.97278	.24869	.96858	36
25	.18081	.98352	.19794	.98021	.21502	.97661	.23203	.97271	.24897	.96851	35
26	.18109	.98347	.19823	.98016	.21530	.97655	.23231	.97264	.24925	.96844	34
27	.18138	.98341	.19851	.98010	.21559	.97648	.23260	.97257	.24954	.96837	33
28	.18166	.98336	.19880	.98004	.21587	.97642	.23288	.97251	.24982	.96829	32
29	.18195	.98331	.19908	.97998	.21616	.97636	.23316	.97244	.25010	.96822	31
30	.18224	.98325	.19937	.97992	.21644	.97630	.23345	.97237	.25038	.96815	30
31	.18252	.98320	.19965	.97987	.21672	.97623	.23373	.97230	.25066	.96807	29
32	.18281	.98315	.19994	.97981	.21701	.97617	.23401	.97223	.25094	.96800	28
33	.18309	.98310	.20022	.97975	.21729	.97611	.23429	.97217	.25122	.96793	27
34	.18338	.98304	.20051	.97969	.21758	.97604	.23458	.97210	.25151	.96786	26
35	.18367	.98299	.20079	.97963	.21786	.97598	.23486	.97203	.25179	.96778	25
36	.18395	.98294	.20108	.97958	.21814	.97592	.23514	.97196	.25207	.96771	24
37	.18424	.98288	.20136	.97952	.21843	.97585	.23542	.97189	.25235	.96764	23
38	.18452	.98283	.20165	.97946	.21871	.97579	.23571	.97182	.25263	.96756	22
39	.18481	.98277	.20193	.97940	.21899	.97573	.23599	.97176	.25291	.96749	21
40	.18509	.98272	.20222	.97934	.21928	.97566	.23627	.97169	.25320	.96742	20
41	.18538	.98267	.20250	.97928	.21956	.97560	.23656	.97162	.25348	.96734	19
42	.18567	.98261	.20279	.97922	.21985	.97553	.23684	.97155	.25376	.96727	18
43	.18595	.98256	.20307	.97916	.22013	.97547	.23712	.97148	.25404	.96719	17
44	.18624	.98250	.20336	.97910	.22041	.97541	.23740	.97141	.25432	.96712	16
45	.18652	.98245	.20364	.97905	.22070	.97534	.23769	.97134	.25460	.96705	15
46	.18681	.98240	.20393	.97899	.22098	.97528	.23797	.97127	.25488	.96697	14
47	.18710	.98234	.20421	.97893	.22126	.97521	.23825	.97120	.25516	.96690	13
48	.18738	.98229	.20450	.97887	.22155	.97515	.23853	.97113	.25545	.96682	12
49	.18767	.98223	.20478	.97881	.22183	.97508	.23882	.97106	.25573	.96675	11
50	.18795	.98218	.20507	.97875	.22212	.97502	.23910	.97100	.25601	.96667	10
51	.18824	.98212	.20535	.97869	.22240	.97496	.23938	.97093	.25629	.96660	9
52	.18852	.98207	.20563	.97863	.22268	.97489	.23966	.97086	.25657	.96653	8
53	.18881	.98201	.20592	.97857	.22297	.97483	.23995	.97079	.25685	.96645	7
54	.18910	.98196	.20620	.97851	.22325	.97476	.24023	.97072	.25713	.96638	6
55	.18938	.98190	.20649	.97845	.22353	.97470	.24051	.97065	.25741	.96630	5
56	.18967	.98185	.20677	.97839	.22382	.97463	.24079	.97058	.25769	.96623	4
57	.18995	.98179	.20706	.97833	.22410	.97457	.24108	.97051	.25798	.96615	3
58	.19024	.98174	.20734	.97827	.22438	.97450	.24136	.97044	.25826	.96608	2
59	.19052	.98168	.20763	.97821	.22467	.97444	.24164	.97037	.25854	.96600	1
60	.19081	.98163	.20791	.97815	.22495	.97437	.24192	.97030	.25882	.96593	0
′	Cosin	Sine	Cosin	Sine	Cosin	Sine	Cosin	Sine	Cosin	Sine	′
	79°		78°		77°		76°		75°		

Appendix 7

′	15° Sine	15° Cosin	16° Sine	16° Cosin	17° Sine	17° Cosin	18° Sine	18° Cosin	19° Sine	19° Cosin	′
0	.25882	.96593	.27564	.96126	.29237	.95630	.30902	.95106	.32557	.94552	60
1	.25910	.96585	.27592	.96118	.29265	.95622	.30929	.95097	.32584	.94542	59
2	.25938	.96578	.27620	.96110	.29293	.95613	.30957	.95088	.32612	.94533	58
3	.25966	.96570	.27648	.96102	.29321	.95605	.30985	.95079	.32639	.94523	57
4	.25994	.96562	.27676	.96094	.29348	.95596	.31012	.95070	.32667	.94514	56
5	.26022	.96555	.27704	.96086	.29376	.95588	.31040	.95061	.32694	.94504	55
6	.26050	.96547	.27731	.96078	.29404	.95579	.31068	.95052	.32722	.94495	54
7	.26079	.96540	.27759	.96070	.29432	.95571	.31095	.95043	.32749	.94485	53
8	.26107	.96532	.27787	.96062	.29460	.95562	.31123	.95033	.32777	.94476	52
9	.26135	.96524	.27815	.96054	.29487	.95554	.31151	.95024	.32804	.94466	51
10	.26163	.96517	.27843	.96046	.29515	.95545	.31178	.95015	.32832	.94457	50
11	.26191	.96509	.27871	.96037	.29543	.95536	.31206	.95006	.32859	.94447	49
12	.26219	.96502	.27899	.96029	.29571	.95528	.31233	.94997	.32887	.94438	48
13	.26247	.96494	.27927	.96021	.29599	.95519	.31261	.94988	.32914	.94428	47
14	.26275	.96486	.27955	.96013	.29626	.95511	.31289	.94979	.32942	.94418	46
15	.26303	.96479	.27983	.96005	.29654	.95502	.31316	.94970	.32969	.94409	45
16	.26331	.96471	.28011	.95997	.29682	.95493	.31344	.94961	.32997	.94399	44
17	.26359	.96463	.28039	.95989	.29710	.95485	.31372	.94952	.33024	.94390	43
18	.26387	.96456	.28067	.95981	.29737	.95476	.31399	.94943	.33051	.94380	42
19	.26415	.96448	.28095	.95972	.29765	.95467	.31427	.94933	.33079	.94370	41
20	.26443	.96440	.28123	.95964	.29793	.95459	.31454	.94924	.33106	.94361	40
21	.26471	.96433	.28150	.95956	.29821	.95450	.31482	.94915	.33134	.94351	39
22	.26500	.96425	.28178	.95948	.29849	.95441	.31510	.94906	.33161	.94342	38
23	.26528	.96417	.28206	.95940	.29876	.95433	.31537	.94897	.33189	.94332	37
24	.26556	.96410	.28234	.95931	.29904	.95424	.31565	.94888	.33216	.94322	36
25	.26584	.96402	.28262	.95923	.29932	.95415	.31593	.94878	.33244	.94313	35
26	.26612	.96394	.28290	.95915	.29960	.95407	.31620	.94869	.33271	.94303	34
27	.26640	.96386	.28318	.95907	.29987	.95398	.31648	.94860	.33298	.94293	33
28	.26668	.96379	.28346	.95898	.30015	.95389	.31675	.94851	.33326	.94284	32
29	.26696	.96371	.28374	.95890	.30043	.95380	.31703	.94842	.33353	.94274	31
30	.26724	.96363	.28402	.95882	.30071	.95372	.31730	.94832	.33381	.94264	30
31	.26752	.96355	.28429	.95874	.30098	.95363	.31758	.94823	.33408	.94254	29
32	.26780	.96347	.28457	.95865	.30126	.95354	.31786	.94814	.33436	.94245	28
33	.26808	.96340	.28485	.95857	.30154	.95345	.31813	.94805	.33463	.94235	27
34	.26836	.96332	.28513	.95849	.30182	.95337	.31841	.94795	.33490	.94225	26
35	.26864	.96324	.28541	.95841	.30209	.95328	.31868	.94786	.33518	.94215	25
36	.26892	.96316	.28569	.95832	.30237	.95319	.31896	.94777	.33545	.94206	24
37	.26920	.96308	.28597	.95824	.30265	.95310	.31923	.94768	.33573	.94196	23
38	.26948	.96301	.28625	.95816	.30292	.95301	.31951	.94758	.33600	.94186	22
39	.26976	.96293	.28652	.95807	.30320	.95293	.31979	.94749	.33627	.94176	21
40	.27004	.96285	.28680	.95799	.30348	.95284	.32006	.94740	.33655	.94167	20
41	.27032	.96277	.28708	.95791	.30376	.95275	.32034	.94730	.33682	.94157	19
42	.27060	.96269	.28736	.95782	.30403	.95266	.32061	.94721	.33710	.94147	18
43	.27088	.96261	.28764	.95774	.30431	.95257	.32089	.94712	.33737	.94137	17
44	.27116	.96253	.28792	.95766	.30459	.95248	.32116	.94702	.33764	.94127	16
45	.27144	.96246	.28820	.95757	.30486	.95240	.32144	.94693	.33792	.94118	15
46	.27172	.96238	.28847	.95749	.30514	.95231	.32171	.94684	.33819	.94108	14
47	.27200	.96230	.28875	.95740	.30542	.95222	.32199	.94674	.33846	.94098	13
48	.27228	.96222	.28903	.95732	.30570	.95213	.32227	.94665	.33874	.94088	12
49	.27256	.96214	.28931	.95724	.30597	.95204	.32254	.94656	.33901	.94078	11
50	.27284	.96206	.28959	.95715	.30625	.95195	.32282	.94646	.33929	.94068	10
51	.27312	.96198	.28987	.95707	.30653	.95186	.32309	.94637	.33956	.94058	9
52	.27340	.96190	.29015	.95698	.30680	.95177	.32337	.94627	.33983	.94049	8
53	.27368	.96182	.29042	.95690	.30708	.95168	.32364	.94618	.34011	.94039	7
54	.27396	.96174	.29070	.95681	.30736	.95159	.32392	.94609	.34038	.94029	6
55	.27424	.96166	.29098	.95673	.30763	.95150	.32419	.94599	.34065	.94019	5
56	.27452	.96158	.29126	.95664	.30791	.95142	.32447	.94590	.34093	.94009	4
57	.27480	.96150	.29154	.95656	.30819	.95133	.32474	.94580	.34120	.93999	3
58	.27508	.96142	.29182	.95647	.30846	.95124	.32502	.94571	.34147	.93989	2
59	.27536	.96134	.29209	.95639	.30874	.95115	.32529	.94561	.34175	.93979	1
60	.27564	.96126	.29237	.95630	.30902	.95106	.32557	.94552	.34202	.93969	0
′	Cosin 74°	Sine 74°	Cosin 73°	Sine 73°	Cosin 72°	Sine 72°	Cosin 71°	Sine 71°	Cosin 70°	Sine 70°	′

344 Manual of Field Geology

′	20° Sine	20° Cosin	21° Sine	21° Cosin	22° Sine	22° Cosin	23° Sine	23° Cosin	24° Sine	24° Cosin	′
0	.34202	.93969	.35837	.93358	.37461	.92718	.39073	.92050	.40674	.91355	60
1	.34229	.93959	.35864	.93348	.37488	.92707	.39100	.92039	.40700	.91343	59
2	.34257	.93949	.35891	.93337	.37515	.92697	.39127	.92028	.40727	.91331	58
3	.34284	.93939	.35918	.93327	.37542	.92686	.39153	.92016	.40753	.91319	57
4	.34311	.93929	.35945	.93316	.37569	.92675	.39180	.92005	.40780	.91307	56
5	.34339	.93919	.35973	.93306	.37595	.92664	.39207	.91994	.40806	.91295	55
6	.34366	.93909	.36000	.93295	.37622	.92653	.39234	.91982	.40833	.91283	54
7	.34393	.93899	.36027	.93285	.37649	.92642	.39260	.91971	.40860	.91272	53
8	.34421	.93889	.36054	.93274	.37676	.92631	.39287	.91959	.40886	.91260	52
9	.34448	.93879	.36081	.93264	.37703	.92620	.39314	.91948	.40913	.91248	51
10	.34475	.93869	.36108	.93253	.37730	.92609	.39341	.91936	.40939	.91236	50
11	.34503	.93859	.36135	.93243	.37757	.92598	.39367	.91925	.40966	.91224	49
12	.34530	.93849	.36162	.93232	.37784	.92587	.39394	.91914	.40992	.91212	48
13	.34557	.93839	.36190	.93222	.37811	.92576	.39421	.91902	.41019	.91200	47
14	.34584	.93829	.36217	.93211	.37838	.92565	.39448	.91891	.41045	.91188	46
15	.34612	.93819	.36244	.93201	.37865	.92554	.39474	.91879	.41072	.91176	45
16	.34639	.93809	.36271	.93190	.37892	.92543	.39501	.91868	.41098	.91164	44
17	.34666	.93799	.36298	.93180	.37919	.92532	.39528	.91856	.41125	.91152	43
18	.34694	.93789	.36325	.93169	.37946	.92521	.39555	.91845	.41151	.91140	42
19	.34721	.93779	.36352	.93159	.37973	.92510	.39581	.91833	.41178	.91128	41
20	.34748	.93769	.36379	.93148	.37999	.92499	.39608	.91822	.41204	.91116	40
21	.34775	.93759	.36406	.93137	.38026	.92488	.39635	.91810	.41231	.91104	39
22	.34803	.93748	.36434	.93127	.38053	.92477	.39661	.91799	.41257	.91092	38
23	.34830	.93738	.36461	.93116	.38080	.92466	.39688	.91787	.41284	.91080	37
24	.34857	.93728	.36488	.93106	.38107	.92455	.39715	.91775	.41310	.91068	36
25	.34884	.93718	.36515	.93095	.38134	.92444	.39741	.91764	.41337	.91056	35
26	.34912	.93708	.36542	.93084	.38161	.92432	.39768	.91752	.41363	.91044	34
27	.34939	.93698	.36569	.93074	.38188	.92421	.39795	.91741	.41390	.91032	33
28	.34966	.93688	.36596	.93063	.38215	.92410	.39822	.91729	.41416	.91020	32
29	.34993	.93677	.36623	.93052	.38241	.92399	.39848	.91718	.41443	.91008	31
30	.35021	.93667	.36650	.93042	.38268	.92388	.39875	.91706	.41469	.90996	30
31	.35048	.93657	.36677	.93031	.38295	.92377	.39902	.91694	.41496	.90984	29
32	.35075	.93647	.36704	.93020	.38322	.92366	.39928	.91683	.41522	.90972	28
33	.35102	.93637	.36731	.93010	.38349	.92355	.39955	.91671	.41549	.90960	27
34	.35130	.93626	.36758	.92999	.38376	.92343	.39982	.91660	.41575	.90948	26
35	.35157	.93616	.36785	.92988	.38403	.92332	.40008	.91648	.41602	.90936	25
36	.35184	.93606	.36812	.92978	.38430	.92321	.40035	.91636	.41628	.90924	24
37	.35211	.93596	.36839	.92967	.38456	.92310	.40062	.91625	.41655	.90911	23
38	.35239	.93585	.36867	.92956	.38483	.92299	.40088	.91613	.41681	.90899	22
39	.35266	.93575	.36894	.92945	.38510	.92287	.40115	.91601	.41707	.90887	21
40	.35293	.93565	.36921	.92935	.38537	.92276	.40141	.91590	.41734	.90875	20
41	.35320	.93555	.36948	.92924	.38564	.92265	.40168	.91578	.41760	.90863	19
42	.35347	.93544	.36975	.92913	.38591	.92254	.40195	.91566	.41787	.90851	18
43	.35375	.93534	.37002	.92902	.38617	.92243	.40221	.91555	.41813	.90839	17
44	.35402	.93524	.37029	.92892	.38644	.92231	.40248	.91543	.41840	.90826	16
45	.35429	.93514	.37056	.92881	.38671	.92220	.40275	.91531	.41866	.90814	15
46	.35456	.93503	.37083	.92870	.38698	.92209	.40301	.91519	.41892	.90802	14
47	.35484	.93493	.37110	.92859	.38725	.92198	.40328	.91508	.41919	.90790	13
48	.35511	.93483	.37137	.92849	.38752	.92186	.40355	.91496	.41945	.90778	12
49	.35538	.93472	.37164	.92838	.38778	.92175	.40381	.91484	.41972	.90766	11
50	.35565	.93462	.37191	.92827	.38805	.92164	.40408	.91472	.41998	.90753	10
51	.35592	.93452	.37218	.92816	.38832	.92152	.40434	.91461	.42024	.90741	9
52	.35619	.93441	.37245	.92805	.38859	.92141	.40461	.91449	.42051	.90729	8
53	.35647	.93431	.37272	.92794	.38886	.92130	.40488	.91437	.42077	.90717	7
54	.35674	.93420	.37299	.92784	.38912	.92119	.40514	.91425	.42104	.90704	6
55	.35701	.93410	.37326	.92773	.38939	.92107	.40541	.91414	.42130	.90692	5
56	.35728	.93400	.37353	.92762	.38966	.92096	.40567	.91402	.42156	.90680	4
57	.35755	.93389	.37380	.92751	.38993	.92085	.40594	.91390	.42183	.90668	3
58	.35782	.93379	.37407	.92740	.39020	.92073	.40621	.91378	.42209	.90655	2
59	.35810	.93368	.37434	.92729	.39046	.92062	.40647	.91366	.42235	.90643	1
60	.35837	.93358	.37461	.92718	.39073	.92050	.40674	.91355	.42262	.90631	0
′	Cosin	Sine	Cosin	Sine	Cosin	Sine	Cosin	Sine	Cosin	Sine	′
	69°		68°		67°		66°		65°		

Appendix 7

′	25° Sine	25° Cosin	26° Sine	26° Cosin	27° Sine	27° Cosin	28° Sine	28° Cosin	29° Sine	29° Cosin	′
0	.42262	.90631	.43837	.89879	.45399	.89101	.46947	.88295	.48481	.87462	60
1	.42288	.90618	.43863	.89867	.45425	.89087	.46973	.88281	.48506	.87448	59
2	.42315	.90606	.43889	.89854	.45451	.89074	.46999	.88267	.48532	.87434	58
3	.42341	.90594	.43916	.89841	.45477	.89061	.47024	.88254	.48557	.87420	57
4	.42367	.90582	.43942	.89828	.45503	.89048	.47050	.88240	.48583	.87406	56
5	.42394	.90569	.43968	.89816	.45529	.89035	.47076	.88226	.48608	.87391	55
6	.42420	.90557	.43994	.89803	.45554	.89021	.47101	.88213	.48634	.87377	54
7	.42446	.90545	.44020	.89790	.45580	.89008	.47127	.88199	.48659	.87363	53
8	.42473	.90532	.44046	.89777	.45606	.88995	.47153	.88185	.48684	.87349	52
9	.42499	.90520	.44072	.89764	.45632	.88981	.47178	.88172	.48710	.87335	51
10	.42525	.90507	.44098	.89752	.45658	.88968	.47204	.88158	.48735	.87321	50
11	.42552	.90495	.44124	.89739	.45684	.88955	.47229	.88144	.48761	.87306	49
12	.42578	.90483	.44151	.89726	.45710	.88942	.47255	.88130	.48786	.87292	48
13	.42604	.90470	.44177	.89713	.45736	.88928	.47281	.88117	.48811	.87278	47
14	.42631	.90458	.44203	.89700	.45762	.88915	.47306	.88103	.48837	.87264	46
15	.42657	.90446	.44229	.89687	.45787	.88902	.47332	.88089	.48862	.87250	45
16	.42683	.90433	.44255	.89674	.45813	.88888	.47358	.88075	.48888	.87235	44
17	.42709	.90421	.44281	.89662	.45839	.88875	.47383	.88062	.48913	.87221	43
18	.42736	.90408	.44307	.89649	.45865	.88862	.47409	.88048	.48938	.87207	42
19	.42762	.90396	.44333	.89636	.45891	.88848	.47434	.88034	.48964	.87193	41
20	.42788	.90383	.44359	.89623	.45917	.88835	.47460	.88020	.48989	.87178	40
21	.42815	.90371	.44385	.89610	.45942	.88822	.47486	.88006	.49014	.87164	39
22	.42841	.90358	.44411	.89597	.45968	.88808	.47511	.87993	.49040	.87150	38
23	.42867	.90346	.44437	.89584	.45994	.88795	.47537	.87979	.49065	.87136	37
24	.42894	.90334	.44464	.89571	.46020	.88782	.47562	.87965	.49090	.87121	36
25	.42920	.90321	.44490	.89558	.46046	.88768	.47588	.87951	.49116	.87107	35
26	.42946	.90309	.44516	.89545	.46072	.88755	.47614	.87937	.49141	.87093	34
27	.42972	.90296	.44542	.89532	.46097	.88741	.47639	.87923	.49166	.87079	33
28	.42999	.90284	.44568	.89519	.46123	.88728	.47665	.87909	.49192	.87064	32
29	.43025	.90271	.44594	.89506	.46149	.88715	.47690	.87896	.49217	.87050	31
30	.43051	.90259	.44620	.89493	.46175	.88701	.47716	.87882	.49242	.87036	30
31	.43077	.90246	.44646	.89480	.46201	.88688	.47741	.87868	.49268	.87021	29
32	.43104	.90233	.44672	.89467	.46226	.88674	.47767	.87854	.49293	.87007	28
33	.43130	.90221	.44698	.89454	.46252	.88661	.47793	.87840	.49318	.86993	27
34	.43156	.90208	.44724	.89441	.46278	.88647	.47818	.87826	.49344	.86978	26
35	.43182	.90196	.44750	.89428	.46304	.88634	.47844	.87812	.49369	.86964	25
36	.43209	.90183	.44776	.89415	.46330	.88620	.47869	.87798	.49394	.86949	24
37	.43235	.90171	.44802	.89402	.46355	.88607	.47895	.87784	.49419	.86935	23
38	.43261	.90158	.44828	.89389	.46381	.88593	.47920	.87770	.49445	.86921	22
39	.43287	.90146	.44854	.89376	.46407	.88580	.47946	.87756	.49470	.86906	21
40	.43313	.90133	.44880	.89363	.46433	.88566	.47971	.87743	.49495	.86892	20
41	.43340	.90120	.44906	.89350	.46458	.88553	.47997	.87729	.49521	.86878	19
42	.43366	.90108	.44932	.89337	.46484	.88539	.48022	.87715	.49546	.86863	18
43	.43392	.90095	.44958	.89324	.46510	.88526	.48048	.87701	.49571	.86849	17
44	.43418	.90082	.44984	.89311	.46536	.88512	.48073	.87687	.49596	.86834	16
45	.43445	.90070	.45010	.89298	.46561	.88499	.48099	.87673	.49622	.86820	15
46	.43471	.90057	.45036	.89285	.46587	.88485	.48124	.87659	.49647	.86805	14
47	.43497	.90045	.45062	.89272	.46613	.88472	.48150	.87645	.49672	.86791	13
48	.43523	.90032	.45088	.89259	.46639	.88458	.48175	.87631	.49697	.86777	12
49	.43549	.90019	.45114	.89245	.46664	.88445	.48201	.87617	.49723	.86762	11
50	.43575	.90007	.45140	.89232	.46690	.88431	.48226	.87603	.49748	.86748	10
51	.43602	.89994	.45166	.89219	.46716	.88417	.48252	.87589	.49773	.86733	9
52	.43628	.89981	.45192	.89206	.46742	.88404	.48277	.87575	.49798	.86719	8
53	.43654	.89968	.45218	.89193	.46767	.88390	.48303	.87561	.49824	.86704	7
54	.43680	.89956	.45243	.89180	.46793	.88377	.48328	.87546	.49849	.86690	6
55	.43706	.89943	.45269	.89167	.46819	.88363	.48354	.87532	.49874	.86675	5
56	.43733	.89930	.45295	.89153	.46844	.88349	.48379	.87518	.49899	.86661	4
57	.43759	.89918	.45321	.89140	.46870	.88336	.48405	.87504	.49924	.86646	3
58	.43785	.89905	.45347	.89127	.46896	.88322	.48430	.87490	.49950	.86632	2
59	.43811	.89892	.45373	.89114	.46921	.88308	.48456	.87476	.49975	.86617	1
60	.43837	.89879	.45399	.89101	.46947	.88295	.48481	.87462	.50000	.86603	0
′	Cosin 64°	Sine 64°	Cosin 63°	Sine 63°	Cosin 62°	Sine 62°	Cosin 61°	Sine 61°	Cosin 60°	Sine 60°	′

Manual of Field Geology

′	30° Sine	30° Cosin	31° Sine	31° Cosin	32° Sine	32° Cosin	33° Sine	33° Cosin	34° Sine	34° Cosin	′
0	.50000	.86603	.51504	.85717	.52992	.84805	.54464	.83867	.55919	.82904	60
1	.50025	.86588	.51529	.85702	.53017	.84789	.54488	.83851	.55943	.82887	59
2	.50050	.86573	.51554	.85687	.53041	.84774	.54513	.83835	.55968	.82871	58
3	.50076	.86559	.51579	.85672	.53066	.84759	.54537	.83819	.55992	.82855	57
4	.50101	.86544	.51604	.85657	.53091	.84743	.54561	.83804	.56016	.82839	56
5	.50126	.86530	.51628	.85642	.53115	.84728	.54586	.83788	.56040	.82822	55
6	.50151	.86515	.51653	.85627	.53140	.84712	.54610	.83772	.56064	.82806	54
7	.50176	.86501	.51678	.85612	.53164	.84697	.54635	.83756	.56088	.82790	53
8	.50201	.86486	.51703	.85597	.53189	.84681	.54659	.83740	.56112	.82773	52
9	.50227	.86471	.51728	.85582	.53214	.84666	.54683	.83724	.56136	.82757	51
10	.50252	.86457	.51753	.85567	.53238	.84650	.54708	.83708	.56160	.82741	50
11	.50277	.86442	.51778	.85551	.53263	.84635	.54732	.83692	.56184	.82724	49
12	.50302	.86427	.51803	.85536	.53288	.84619	.54756	.83676	.56208	.82708	48
13	.50327	.86413	.51828	.85521	.53312	.84604	.54781	.83660	.56232	.82692	47
14	.50352	.86398	.51852	.85506	.53337	.84588	.54805	.83645	.56256	.82675	46
15	.50377	.86384	.51877	.85491	.53361	.84573	.54829	.83629	.56280	.82659	45
16	.50403	.86369	.51902	.85476	.53386	.84557	.54854	.83613	.56305	.82643	44
17	.50428	.86354	.51927	.85461	.53411	.84542	.54878	.83597	.56329	.82626	43
18	.50453	.86340	.51952	.85446	.53435	.84526	.54902	.83581	.56353	.82610	42
19	.50478	.86325	.51977	.85431	.53460	.84511	.54927	.83565	.56377	.82593	41
20	.50503	.86310	.52002	.85416	.53484	.84495	.54951	.83549	.56401	.82577	40
21	.50528	.86295	.52026	.85401	.53509	.84480	.54975	.83533	.56425	.82561	39
22	.50553	.86281	.52051	.85385	.53534	.84464	.54999	.83517	.56449	.82544	38
23	.50578	.86266	.52076	.85370	.53558	.84448	.55024	.83501	.56473	.82528	37
24	.50603	.86251	.52101	.85355	.53583	.84433	.55048	.83485	.56497	.82511	36
25	.50628	.86237	.52126	.85340	.53607	.84417	.55072	.83469	.56521	.82495	35
26	.50654	.86222	.52151	.85325	.53632	.84402	.55097	.83453	.56545	.82478	34
27	.50679	.86207	.52175	.85310	.53656	.84386	.55121	.83437	.56569	.82462	33
28	.50704	.86192	.52200	.85294	.53681	.84370	.55145	.83421	.56593	.82446	32
29	.50729	.86178	.52225	.85279	.53705	.84355	.55169	.83405	.56617	.82429	31
30	.50754	.86163	.52250	.85264	.53730	.84339	.55194	.83389	.56641	.82413	30
31	.50779	.86148	.52275	.85249	.53754	.84324	.55218	.83373	.56665	.82396	29
32	.50804	.86133	.52299	.85234	.53779	.84308	.55242	.83356	.56689	.82380	28
33	.50829	.86119	.52324	.85218	.53804	.84292	.55266	.83340	.56713	.82363	27
34	.50854	.86104	.52349	.85203	.53828	.84277	.55291	.83324	.56736	.82347	26
35	.50879	.86089	.52374	.85188	.53853	.84261	.55315	.83308	.56760	.82330	25
36	.50904	.86074	.52399	.85173	.53877	.84245	.55339	.83292	.56784	.82314	24
37	.50929	.86059	.52423	.85157	.53902	.84230	.55363	.83276	.56808	.82297	23
38	.50954	.86045	.52448	.85142	.53926	.84214	.55388	.83260	.56832	.82281	22
39	.50979	.86030	.52473	.85127	.53951	.84198	.55412	.83244	.56856	.82264	21
40	.51004	.86015	.52498	.85112	.53975	.84182	.55436	.83228	.56880	.82248	20
41	.51029	.86000	.52522	.85096	.54000	.84167	.55460	.83212	.56904	.82231	19
42	.51054	.85985	.52547	.85081	.54024	.84151	.55484	.83195	.56928	.82214	18
43	.51079	.85970	.52572	.85066	.54049	.84135	.55509	.83179	.56952	.82198	17
44	.51104	.85956	.52597	.85051	.54073	.84120	.55533	.83163	.56976	.82181	16
45	.51129	.85941	.52621	.85035	.54097	.84104	.55557	.83147	.57000	.82165	15
46	.51154	.85926	.52646	.85020	.54122	.84088	.55581	.83131	.57024	.82148	14
47	.51179	.85911	.52671	.85005	.54146	.84072	.55605	.83115	.57047	.82132	13
48	.51204	.85896	.52696	.84989	.54171	.84057	.55630	.83098	.57071	.82115	12
49	.51229	.85881	.52720	.84974	.54195	.84041	.55654	.83082	.57095	.82098	11
50	.51254	.85866	.52745	.84959	.54220	.84025	.55678	.83066	.57119	.82082	10
51	.51279	.85851	.52770	.84943	.54244	.84009	.55702	.83050	.57143	.82065	9
52	.51304	.85836	.52794	.84928	.54269	.83994	.55726	.83034	.57167	.82048	8
53	.51329	.85821	.52819	.84913	.54293	.83978	.55750	.83017	.57191	.82032	7
54	.51354	.85806	.52844	.84897	.54317	.83962	.55775	.83001	.57215	.82015	6
55	.51379	.85792	.52869	.84882	.54342	.83946	.55799	.82985	.57238	.81999	5
56	.51404	.85777	.52893	.84866	.54366	.83930	.55823	.82969	.57262	.81982	4
57	.51429	.85762	.52918	.84851	.54391	.83915	.55847	.82953	.57286	.81965	3
58	.51454	.85747	.52943	.84836	.54415	.83899	.55871	.82936	.57310	.81949	2
59	.51479	.85732	.52967	.84820	.54440	.83883	.55895	.82920	.57334	.81932	1
60	.51504	.85717	.52992	.84805	.54464	.83867	.55919	.82904	.57358	.81915	0
′	Cosin	Sine	Cosin	Sine	Cosin	Sine	Cosin	Sine	Cosin	Sine	′
	59°		58°		57°		56°		55°		

Appendix 7

′	35° Sine	35° Cosin	36° Sine	36° Cosin	37° Sine	37° Cosin	38° Sine	38° Cosin	39° Sine	39° Cosin	′
0	.57358	.81915	.58779	.80902	.60182	.79864	.61566	.78801	.62932	.77715	60
1	.57381	.81899	.58802	.80885	.60205	.79846	.61589	.78783	.62955	.77696	59
2	.57405	.81882	.58826	.80867	.60228	.79829	.61612	.78765	.62977	.77678	58
3	.57429	.81865	.58849	.80850	.60251	.79811	.61635	.78747	.63000	.77660	57
4	.57453	.81848	.58873	.80833	.60274	.79793	.61658	.78729	.63022	.77641	56
5	.57477	.81832	.58896	.80816	.60298	.79776	.61681	.78711	.63045	.77623	55
6	.57501	.81815	.58920	.80799	.60321	.79758	.61704	.78694	.63068	.77605	54
7	.57524	.81798	.58943	.80782	.60344	.79741	.61726	.78676	.63090	.77586	53
8	.57548	.81782	.58967	.80765	.60367	.79723	.61749	.78658	.63113	.77568	52
9	.57572	.81765	.58990	.80748	.60390	.79706	.61772	.78640	.63135	.77550	51
10	.57596	.81748	.59014	.80730	.60414	.79688	.61795	.78622	.63158	.77531	50
11	.57619	.81731	.59037	.80713	.60437	.79671	.61818	.78604	.63180	.77513	49
12	.57643	.81714	.59061	.80696	.60460	.79653	.61841	.78586	.63203	.77494	48
13	.57667	.81698	.59084	.80679	.60483	.79635	.61864	.78568	.63225	.77476	47
14	.57691	.81681	.59108	.80662	.60506	.79618	.61887	.78550	.63248	.77458	46
15	.57715	.81664	.59131	.80644	.60529	.79600	.61909	.78532	.63271	.77439	45
16	.57738	.81647	.59154	.80627	.60553	.79583	.61932	.78514	.63293	.77421	44
17	.57762	.81631	.59178	.80610	.60576	.79565	.61955	.78496	.63316	.77402	43
18	.57786	.81614	.59201	.80593	.60599	.79547	.61978	.78478	.63338	.77384	42
19	.57810	.81597	.59225	.80576	.60622	.79530	.62001	.78460	.63361	.77366	41
20	.57833	.81580	.59248	.80558	.60645	.79512	.62024	.78442	.63383	.77347	40
21	.57857	.81563	.59272	.80541	.60668	.79494	.62046	.78424	.63406	.77329	39
22	.57881	.81546	.59295	.80524	.60691	.79477	.62069	.78405	.63428	.77310	38
23	.57904	.81530	.59318	.80507	.60714	.79459	.62092	.78387	.63451	.77292	37
24	.57928	.81513	.59342	.80489	.60738	.79441	.62115	.78369	.63473	.77273	36
25	.57952	.81496	.59365	.80472	.60761	.79424	.62138	.78351	.63496	.77255	35
26	.57976	.81479	.59389	.80455	.60784	.79406	.62160	.78333	.63518	.77236	34
27	.57999	.81462	.59412	.80438	.60807	.79388	.62183	.78315	.63540	.77218	33
28	.58023	.81445	.59436	.80420	.60830	.79371	.62206	.78297	.63563	.77199	32
29	.58047	.81428	.59459	.80403	.60853	.79353	.62229	.78279	.63585	.77181	31
30	.58070	.81412	.59482	.80386	.60876	.79335	.62251	.78261	.63608	.77162	30
31	.58094	.81395	.59506	.80368	.60899	.79318	.62274	.78243	.63630	.77144	29
32	.58118	.81378	.59529	.80351	.60922	.79300	.62297	.78225	.63653	.77125	28
33	.58141	.81361	.59552	.80334	.60945	.79282	.62320	.78206	.63675	.77107	27
34	.58165	.81344	.59576	.80316	.60968	.79264	.62342	.78188	.63698	.77088	26
35	.58189	.81327	.59599	.80299	.60991	.79247	.62365	.78170	.63720	.77070	25
36	.58212	.81310	.59622	.80282	.61015	.79229	.62388	.78152	.63742	.77051	24
37	.58236	.81293	.59646	.80264	.61038	.79211	.62411	.78134	.63765	.77033	23
38	.58260	.81276	.59669	.80247	.61061	.79193	.62433	.78116	.63787	.77014	22
39	.58283	.81259	.59693	.80230	.61084	.79176	.62456	.78098	.63810	.76996	21
40	.58307	.81242	.59716	.80212	.61107	.79158	.62479	.78079	.63832	.76977	20
41	.58330	.81225	.59739	.80195	.61130	.79140	.62502	.78061	.63854	.76959	19
42	.58554	.81208	.59763	.80178	.61153	.79122	.62524	.78043	.63877	.76940	18
43	.58378	.81191	.59786	.80160	.61176	.79105	.62547	.78025	.63899	.76921	17
44	.58401	.81174	.59809	.80143	.61199	.79087	.62570	.78007	.63922	.76903	16
45	.58425	.81157	.59832	.80125	.61222	.79069	.62592	.77988	.63944	.76884	15
46	.58449	.81140	.59856	.80108	.61245	.79051	.62615	.77970	.63966	.76866	14
47	.58472	.81123	.59879	.80091	.61268	.79033	.62638	.77952	.63989	.76847	13
48	.58496	.81106	.59902	.80073	.61291	.79016	.62660	.77934	.64011	.76828	12
49	.58519	.81089	.59926	.80056	.61314	.78998	.62683	.77916	.64033	.76810	11
50	.58543	.81072	.59949	.80038	.61337	.78980	.62706	.77897	.64056	.76791	10
51	.58567	.81055	.59972	.80021	.61360	.78962	.62728	.77879	.64078	.76772	9
52	.58590	.81038	.59995	.80003	.61383	.78944	.62751	.77861	.64100	.76754	8
53	.58614	.81021	.60019	.79986	.61406	.78926	.62774	.77843	.64123	.76735	7
54	.58637	.81004	.60042	.79968	.61429	.78908	.62796	.77824	.64145	.76717	6
55	.58661	.80987	.60065	.79951	.61451	.78891	.62819	.77806	.64167	.76698	5
56	.58684	.80970	.60089	.79934	.61474	.78873	.62842	.77788	.64190	.76679	4
57	.58708	.80953	.60112	.79916	.61497	.78855	.62864	.77769	.64212	.76661	3
58	.58731	.80936	.60135	.79899	.61520	.78837	.62887	.77751	.64234	.76642	2
59	.58755	.80919	.60158	.79881	.61543	.78819	.62909	.77733	.64256	.76623	1
60	.58779	.80902	.60182	.79864	.61566	.78801	.62932	.77715	.64279	.76604	0
′	Cosin	Sine	Cosin	Sine	Cosin	Sine	Cosin	Sine	Cosin	Sine	′
	54°		53°		52°		51°		50°		

′	40° Sine	40° Cosin	41° Sine	41° Cosin	42° Sine	42° Cosin	43° Sine	43° Cosin	44° Sine	44° Cosin	′
0	.64279	.76604	.65606	.75471	.66913	.74314	.68200	.73135	.69466	.71934	60
1	.64301	.76586	.65628	.75452	.66935	.74295	.68221	.73116	.69487	.71914	59
2	.64323	.76567	.65650	.75433	.66956	.74276	.68242	.73096	.69508	.71894	58
3	.64346	.76548	.65672	.75414	.66978	.74256	.68264	.73076	.69529	.71873	57
4	.64368	.76530	.65694	.75395	.66999	.74237	.68285	.73056	.69549	.71853	56
5	.64390	.76511	.65716	.75375	.67021	.74217	.68306	.73036	.69570	.71833	55
6	.64412	.76492	.65738	.75356	.67043	.74198	.68327	.73016	.69591	.71813	54
7	.64435	.76473	.65759	.75337	.67064	.74178	.68349	.72996	.69612	.71792	53
8	.64457	.76455	.65781	.75318	.67086	.74159	.68370	.72976	.69633	.71772	52
9	.64479	.76436	.65803	.75299	.67107	.74139	.68391	.72957	.69654	.71752	51
10	.64501	.76417	.65825	.75280	.67129	.74120	.68412	.72937	.69675	.71732	50
11	.64524	.76398	.65847	.75261	.67151	.74100	.68434	.72917	.69696	.71711	49
12	.64546	.76380	.65869	.75241	.67172	.74080	.68455	.72897	.69717	.71691	48
13	.64568	.76361	.65891	.75222	.67194	.74061	.68476	.72877	.69737	.71671	47
14	.64590	.76342	.65913	.75203	.67215	.74041	.68497	.72857	.69758	.71650	46
15	.64612	.76323	.65935	.75184	.67237	.74022	.68518	.72837	.69779	.71630	45
16	.64635	.76304	.65956	.75165	.67258	.74002	.68539	.72817	.69800	.71610	44
17	.64657	.76286	.65978	.75146	.67280	.73983	.68561	.72797	.69821	.71590	43
18	.64679	.76267	.66000	.75126	.67301	.73963	.68582	.72777	.69842	.71569	42
19	.64701	.76248	.66022	.75107	.67323	.73944	.68603	.72757	.69862	.71549	41
20	.64723	.76229	.66044	.75088	.67344	.73924	.68624	.72737	.69883	.71529	40
21	.64746	.76210	.66066	.75069	.67366	.73904	.68645	.72717	.69904	.71508	39
22	.64768	.76192	.66088	.75050	.67387	.73885	.68666	.72697	.69925	.71488	38
23	.64790	.76173	.66109	.75030	.67409	.73865	.68688	.72677	.69946	.71468	37
24	.64812	.76154	.66131	.75011	.67430	.73846	.68709	.72657	.69966	.71447	36
25	.64834	.76135	.66153	.74992	.67452	.73826	.68730	.72637	.69987	.71427	35
26	.64856	.76116	.66175	.74973	.67473	.73806	.68751	.72617	.70008	.71407	34
27	.64878	.76097	.66197	.74953	.67495	.73787	.68772	.72597	.70029	.71386	33
28	.64901	.76078	.66218	.74934	.67516	.73767	.68793	.72577	.70049	.71366	32
29	.64923	.76059	.66240	.74915	.67538	.73747	.68814	.72557	.70070	.71345	31
30	.64945	.76041	.66262	.74896	.67559	.73728	.68835	.72537	.70091	.71325	30
31	.64967	.76022	.66284	.74876	.67580	.73708	.68857	.72517	.70112	.71305	29
32	.64989	.76003	.66306	.74857	.67602	.73688	.68878	.72497	.70132	.71284	28
33	.65011	.75984	.66327	.74838	.67623	.73669	.68899	.72477	.70153	.71264	27
34	.65033	.75965	.66349	.74818	.67645	.73649	.68920	.72457	.70174	.71243	26
35	.65055	.75946	.66371	.74799	.67666	.73629	.68941	.72437	.70195	.71223	25
36	.65077	.75927	.66393	.74780	.67688	.73610	.68962	.72417	.70215	.71203	24
37	.65100	.75908	.66414	.74760	.67709	.73590	.68983	.72397	.70236	.71182	23
38	.65122	.75889	.66436	.74741	.67730	.73570	.69004	.72377	.70257	.71162	22
39	.65144	.75870	.66458	.74722	.67752	.73551	.69025	.72357	.70277	.71141	21
40	.65166	.75851	.66480	.74703	.67773	.73531	.69046	.72337	.70298	.71121	20
41	.65188	.75832	.66501	.74683	.67795	.73511	.69067	.72317	.70319	.71100	19
42	.65210	.75813	.66523	.74664	.67816	.73491	.69088	.72297	.70339	.71080	18
43	.65232	.75794	.66545	.74644	.67837	.73472	.69109	.72277	.70360	.71059	17
44	.65254	.75775	.66566	.74625	.67859	.73452	.69130	.72257	.70381	.71039	16
45	.65276	.75756	.66588	.74606	.67880	.73432	.69151	.72236	.70401	.71019	15
46	.65298	.75738	.66610	.74586	.67901	.73413	.69172	.72216	.70422	.70998	14
47	.65320	.75719	.66632	.74567	.67923	.73393	.69193	.72196	.70443	.70978	13
48	.65342	.75700	.66653	.74548	.67944	.73373	.69214	.72176	.70463	.70957	12
49	.65364	.75680	.66675	.74528	.67965	.73353	.69235	.72156	.70484	.70937	11
50	.65386	.75661	.66697	.74509	.67987	.73333	.69256	.72136	.70505	.70916	10
51	.65408	.75642	.66718	.74489	.68008	.73314	.69277	.72116	.70525	.70896	9
52	.65430	.75623	.66740	.74470	.68029	.73294	.69298	.72095	.70546	.70875	8
53	.65452	.75604	.66762	.74451	.68051	.73274	.69319	.72075	.70567	.70855	7
54	.65474	.75585	.66783	.74431	.68072	.73254	.69340	.72055	.70587	.70834	6
55	.65496	.75566	.66805	.74412	.68093	.73234	.69361	.72035	.70608	.70813	5
56	.65518	.75547	.66827	.74392	.68115	.73215	.69382	.72015	.70628	.70793	4
57	.65540	.75528	.66848	.74373	.68136	.73195	.69403	.71995	.70649	.70772	3
58	.65562	.75509	.66870	.74353	.68157	.73175	.69424	.71974	.70670	.70752	2
59	.65584	.75490	.66891	.74334	.68179	.73155	.69445	.71954	.70690	.70731	1
60	.65606	.75471	.66913	.74314	.68200	.73135	.69466	.71934	.70711	.70711	0
′	Cosin	Sine	Cosin	Sine	Cosin	Sine	Cosin	Sine	Cosin	Sine	′
	49°		48°		47°		46°		45°		

Appendix 7

′	0° Tang	0° Cotang	1° Tang	1° Cotang	2° Tang	2° Cotang	3° Tang	3° Cotang	′
0	.00000	Infinite	.01746	57.2900	.03492	28.6363	.05241	19.0811	60
1	.00029	3437.75	.01775	56.3506	.03521	28.3994	.05270	18.9755	59
2	.00058	1718.87	.01804	55.4415	.03550	28.1664	.05299	18.8711	58
3	.00087	1145.92	.01833	54.5613	.03579	27.9372	.05328	18.7678	57
4	.00116	859.436	.01862	53.7086	.03609	27.7117	.05357	18.6656	56
5	.00145	687.549	.01891	52.8821	.03638	27.4899	.05387	18.5645	55
6	.00175	572.957	.01920	52.0807	.03667	27.2715	.05416	18.4645	54
7	.00204	491.106	.01949	51.3032	.03696	27.0566	.05445	18.3655	53
8	.00233	429.718	.01978	50.5485	.03725	26.8450	.05474	18.2677	52
9	.00262	381.971	.02007	49.8157	.03754	26.6367	.05503	18.1708	51
10	.00291	343.774	.02036	49.1039	.03783	26.4316	.05533	18.0750	50
11	.00320	312.521	.02066	48.4121	.03812	26.2296	.05562	17.9802	49
12	.00349	286.478	.02095	47.7395	.03842	26.0307	.05591	17.8863	48
13	.00378	264.441	.02124	47.0853	.03871	25.8348	.05620	17.7934	47
14	.00407	245.552	.02153	46.4489	.03900	25.6418	.05649	17.7015	46
15	.00436	229.182	.02182	45.8294	.03929	25.4517	.05678	17.6106	45
16	.00465	214.858	.02211	45.2261	.03958	25.2644	.05708	17.5205	44
17	.00495	202.219	.02240	44.6386	.03987	25.0798	.05737	17.4314	43
18	.00524	190.984	.02269	44.0661	.04016	24.8978	.05766	17.3432	42
19	.00553	180.932	.02298	43.5081	.04046	24.7185	.05795	17.2558	41
20	.00582	171.885	.02328	42.9641	.04075	24.5418	.05824	17.1693	40
21	.00611	163.700	.02357	42.4335	.04104	24.3675	.05854	17.0837	39
22	.00640	156.259	.02386	41.9158	.04133	24.1957	.05883	16.9990	38
23	.00669	149.465	.02415	41.4106	.04162	24.0263	.05912	16.9150	37
24	.00698	143.237	.02444	40.9174	.04191	23.8593	.05941	16.8319	36
25	.00727	137.507	.02473	40.4358	.04220	23.6945	.05970	16.7496	35
26	.00756	132.219	.02502	39.9655	.04250	23.5321	.05999	16.6681	34
27	.00785	127.321	.02531	39.5059	.04279	23.3718	.06029	16.5874	33
28	.00815	122.774	.02560	39.0568	.04308	23.2137	.06058	16.5075	32
29	.00844	118.540	.02589	38.6177	.04337	23.0577	.06087	16.4283	31
30	.00873	114.589	.02619	38.1885	.04366	22.9038	.06116	16.3499	30
31	.00902	110.892	.02648	37.7686	.04395	22.7519	.06145	16.2722	29
32	.00931	107.426	.02677	37.3579	.04424	22.6020	.06175	16.1952	28
33	.00960	104.171	.02706	36.9560	.04454	22.4541	.06204	16.1190	27
34	.00989	101.107	.02735	36.5627	.04483	22.3081	.06233	16.0435	26
35	.01018	98.2179	.02764	36.1776	.04512	22.1640	.06262	15.9687	25
36	.01047	95.4895	.02793	35.8006	.04541	22.0217	.06291	15.8945	24
37	.01076	92.9085	.02822	35.4313	.04570	21.8813	.06321	15.8211	23
38	.01105	90.4633	.02851	35.0695	.04599	21.7426	.06350	15.7483	22
39	.01135	88.1436	.02881	34.7151	.04628	21.6056	.06379	15.6762	21
40	.01164	85.9398	.02910	34.3678	.04658	21.4704	.06408	15.6048	20
41	.01193	83.8435	.02939	34.0273	.04687	21.3369	.06437	15.5340	19
42	.01222	81.8470	.02968	33.6935	.04716	21.2049	.06467	15.4638	18
43	.01251	79.9434	.02997	33.3662	.04745	21.0747	.06496	15.3943	17
44	.01280	78.1263	.03026	33.0452	.04774	20.9460	.06525	15.3254	16
45	.01309	76.3900	.03055	32.7303	.04803	20.8188	.06554	15.2571	15
46	.01338	74.7292	.03084	32.4213	.04833	20.6932	.06584	15.1893	14
47	.01367	73.1390	.03114	32.1181	.04862	20.5691	.06613	15.1222	13
48	.01396	71.6151	.03143	31.8205	.04891	20.4465	.06642	15.0557	12
49	.01425	70.1533	.03172	31.5284	.04920	20.3253	.06671	14.9898	11
50	.01455	68.7501	.03201	31.2416	.04949	20.2056	.06700	14.9244	10
51	.01484	67.4019	.03230	30.9599	.04978	20.0872	.06730	14.8596	9
52	.01513	66.1055	.03259	30.6833	.05007	19.9702	.06759	14.7954	8
53	.01542	64.8580	.03288	30.4116	.05037	19.8546	.06788	14.7317	7
54	.01571	63.6567	.03317	30.1446	.05066	19.7403	.06817	14.6685	6
55	.01600	62.4992	.03346	29.8823	.05095	19.6273	.06847	14.6059	5
56	.01629	61.3829	.03376	29.6245	.05124	19.5156	.06876	14.5438	4
57	.01658	60.3058	.03405	29.3711	.05153	19.4051	.06905	14.4823	3
58	.01687	59.2659	.03434	29.1220	.05182	19.2959	.06934	14.4212	2
59	.01716	58.2612	.03463	28.8771	.05212	19.1879	.06963	14.3607	1
60	.01746	57.2900	.03492	28.6363	.05241	19.0811	.06993	14.3007	0
′	Cotang	Tang	Cotang	Tang	Cotang	Tang	Cotang	Tang	′
	89°		88°		87°		86°		

′	4° Tang	4° Cotang	5° Tang	5° Cotang	6° Tang	6° Cotang	7° Tang	7° Cotang	′
0	.06993	14.3007	.08749	11.4301	.10510	9.51436	.12278	8.14435	60
1	.07022	14.2411	.08778	11.3919	.10540	9.48781	.12308	8.12481	59
2	.07051	14.1821	.08807	11.3540	.10569	9.46141	.12338	8.10536	58
3	.07080	14.1235	.08837	11.3163	.10599	9.43515	.12367	8.08600	57
4	.07110	14.0655	.08866	11.2789	.10628	9.40904	.12397	8.06674	56
5	.07139	14.0079	.08895	11.2417	.10657	9.38307	.12426	8.04756	55
6	.07168	13.9507	.08925	11.2048	.10687	9.35724	.12456	8.02848	54
7	.07197	13.8940	.08954	11.1681	.10716	9.33155	.12485	8.00948	53
8	.07227	13.8378	.08983	11.1316	.10746	9.30599	.12515	7.99058	52
9	.07256	13.7821	.09013	11.0954	.10775	9.28058	.12544	7.97176	51
10	.07285	13.7267	.09042	11.0594	.10805	9.25530	.12574	7.95302	50
11	.07314	13.6719	.09071	11.0237	.10834	9.23016	.12603	7.93438	49
12	.07344	13.6174	.09101	10.9882	.10863	9.20516	.12633	7.91582	48
13	.07373	13.5634	.09130	10.9529	.10893	9.18028	.12662	7.89734	47
14	.07402	13.5098	.09159	10.9178	.10922	9.15554	.12692	7.87895	46
15	.07431	13.4566	.09189	10.8829	.10952	9.13093	.12722	7.86064	45
16	.07461	13.4039	.09218	10.8483	.10981	9.10646	.12751	7.84242	44
17	.07490	13.3515	.09247	10.8139	.11011	9.08211	.12781	7.82428	43
18	.07519	13.2996	.09277	10.7797	.11040	9.05789	.12810	7.80622	42
19	.07548	13.2480	.09306	10.7457	.11070	9.03379	.12840	7.78825	41
20	.07578	13.1969	.09335	10.7119	.11099	9.00983	.12869	7.77035	40
21	.07607	13.1461	.09365	10.6783	.11128	8.98598	.12899	7.75254	39
22	.07636	13.0958	.09394	10.6450	.11158	8.96227	.12929	7.73480	38
23	.07665	13.0458	.09423	10.6118	.11187	8.93867	.12958	7.71715	37
24	.07695	12.9962	.09453	10.5789	.11217	8.91520	.12988	7.69957	36
25	.07724	12.9469	.09482	10.5462	.11246	8.89185	.13017	7.68208	35
26	.07753	12.8981	.09511	10.5136	.11276	8.86862	.13047	7.66466	34
27	.07782	12.8496	.09541	10.4813	.11305	8.84551	.13076	7.64732	33
28	.07812	12.8014	.09570	10.4491	.11335	8.82252	.13106	7.63005	32
29	.07841	12.7536	.09600	10.4172	.11364	8.79964	.13136	7.61287	31
30	.07870	12.7062	.09629	10.3854	.11394	8.77689	.13165	7.59575	30
31	.07899	12.6591	.09658	10.3538	.11423	8.75425	.13195	7.57872	29
32	.07929	12.6124	.09688	10.3224	.11452	8.73172	.13224	7.56176	28
33	.07958	12.5660	.09717	10.2913	.11482	8.70931	.13254	7.54487	27
34	.07987	12.5199	.09746	10.2602	.11511	8.68701	.13284	7.52806	26
35	.08017	12.4742	.09776	10.2294	.11541	8.66482	.13313	7.51132	25
36	.08046	12.4288	.09805	10.1988	.11570	8.64275	.13343	7.49465	24
37	.08075	12.3838	.09834	10.1683	.11600	8.62078	.13372	7.47806	23
38	.08104	12.3390	.09864	10.1381	.11629	8.59893	.13402	7.46154	22
39	.08134	12.2946	.09893	10.1080	.11659	8.57718	.13432	7.44509	21
40	.08163	12.2505	.09923	10.0780	.11688	8.55555	.13461	7.42871	20
41	.08192	12.2067	.09952	10.0483	.11718	8.53402	.13491	7.41240	19
42	.08221	12.1632	.09981	10.0187	.11747	8.51259	.13521	7.39616	18
43	.08251	12.1201	.10011	9.98931	.11777	8.49128	.13550	7.37999	17
44	.08280	12.0772	.10040	9.96007	.11806	8.47007	.13580	7.36389	16
45	.08309	12.0346	.10069	9.93101	.11836	8.44896	.13609	7.34786	15
46	.08339	11.9923	.10099	9.90211	.11865	8.42795	.13639	7.33190	14
47	.08368	11.9504	.10128	9.87338	.11895	8.40705	.13669	7.31600	13
48	.08397	11.9087	.10158	9.84482	.11924	8.38625	.13698	7.30018	12
49	.08427	11.8673	.10187	9.81641	.11954	8.36555	.13728	7.28442	11
50	.08456	11.8262	.10216	9.78817	.11983	8.34496	.13758	7.26873	10
51	.08485	11.7853	.10246	9.76009	.12013	8.32446	.13787	7.25310	9
52	.08514	11.7448	.10275	9.73217	.12042	8.30406	.13817	7.23754	8
53	.08544	11.7045	.10305	9.70441	.12072	8.28376	.13846	7.22204	7
54	.08573	11.6645	.10334	9.67680	.12101	8.26355	.13876	7.20661	6
55	.08602	11.6248	.10363	9.64935	.12131	8.24345	.13906	7.19125	5
56	.08632	11.5853	.10393	9.62205	.12160	8.22344	.13935	7.17594	4
57	.08661	11.5461	.10422	9.59490	.12190	8.20352	.13965	7.16071	3
58	.08690	11.5072	.10452	9.56791	.12219	8.18370	.13995	7.14553	2
59	.08720	11.4685	.10481	9.54106	.12249	8.16398	.14024	7.13042	1
60	.08749	11.4301	.10510	9.51436	.12278	8.14435	.14054	7.11537	0
′	Cotang	Tang	Cotang	Tang	Cotang	Tang	Cotang	Tang	′
	85°		84°		83°		82°		

Appendix 7

′	8° Tang	8° Cotang	9° Tang	9° Cotang	10° Tang	10° Cotang	11° Tang	11° Cotang	′
0	.14054	7.11537	.15838	6.31375	.17633	5.67128	.19438	5.14455	60
1	.14084	7.10038	.15868	6.30189	.17663	5.66165	.19468	5.13658	59
2	.14113	7.08546	.15898	6.29007	.17693	5.65205	.19498	5.12862	58
3	.14143	7.07059	.15928	6.27829	.17723	5.64248	.19529	5.12069	57
4	.14173	7.05579	.15958	6.26655	.17753	5.63295	.19559	5.11279	56
5	.14202	7.04105	.15988	6.25486	.17783	5.62344	.19589	5.10490	55
6	.14232	7.02637	.16017	6.24321	.17813	5.61397	.19619	5.09704	54
7	.14262	7.01174	.16047	6.23160	.17843	5.60452	.19649	5.08921	53
8	.14291	6.99718	.16077	6.22003	.17873	5.59511	.19680	5.08139	52
9	.14321	6.98268	.16107	6.20851	.17903	5.58573	.19710	5.07360	51
10	.14351	6.96823	.16137	6.19703	.17933	5.57638	.19740	5.06584	50
11	.14381	6.95385	.16167	6.18559	.17963	5.56706	.19770	5.05809	49
12	.14410	6.93952	.16196	6.17419	.17993	5.55777	.19801	5.05037	48
13	.14440	6.92525	.16226	6.16283	.18023	5.54851	.19831	5.04267	47
14	.14470	6.91104	.16256	6.15151	.18053	5.53927	.19861	5.03499	46
15	.14499	6.89688	.16286	6.14023	.18083	5.53007	.19891	5.02734	45
16	.14529	6.88278	.16316	6.12899	.18113	5.52090	.19921	5.01971	44
17	.14559	6.86874	.16346	6.11779	.18143	5.51176	.19952	5.01210	43
18	.14588	6.85475	.16376	6.10664	.18173	5.50264	.19982	5.00451	42
19	.14618	6.84082	.16405	6.09552	.18203	5.49356	.20012	4.99695	41
20	.14648	6.82694	.16435	6.08444	.18233	5.48451	.20042	4.98940	40
21	.14678	6.81312	.16465	6.07340	.18263	5.47548	.20073	4.98188	39
22	.14707	6.79936	.16495	6.06240	.18293	5.46648	.20103	4.97438	38
23	.14737	6.78564	.16525	6.05143	.18323	5.45751	.20133	4.96690	37
24	.14767	6.77199	.16555	6.04051	.18353	5.44857	.20164	4.95945	36
25	.14796	6.75838	.16585	6.02962	.18384	5.43966	.20194	4.95201	35
26	.14826	6.74483	.16615	6.01878	.18414	5.43077	.20224	4.94460	34
27	.14856	6.73133	.16645	6.00797	.18444	5.42192	.20254	4.93721	33
28	.14886	6.71789	.16674	5.99720	.18474	5.41309	.20285	4.92984	32
29	.14915	6.70450	.16704	5.98646	.18504	5.40429	.20315	4.92249	31
30	.14945	6.69116	.16734	5.97576	.18534	5.39552	.20345	4.91516	30
31	.14975	6.67787	.16764	5.96510	.18564	5.38677	.20376	4.90785	29
32	.15005	6.66463	.16794	5.95448	.18594	5.37805	.20406	4.90056	28
33	.15034	6.65144	.16824	5.94390	.18624	5.36936	.20436	4.89330	27
34	.15064	6.63831	.16854	5.93335	.18654	5.36070	.20466	4.88605	26
35	.15094	6.62523	.16884	5.92283	.18684	5.35206	.20497	4.87882	25
36	.15124	6.61219	.16914	5.91236	.18714	5.34345	.20527	4.87162	24
37	.15153	6.59921	.16944	5.90191	.18745	5.33487	.20557	4.86444	23
38	.15183	6.58627	.16974	5.89151	.18775	5.32631	.20588	4.85727	22
39	.15213	6.57339	.17004	5.88114	.18805	5.31778	.20618	4.85013	21
40	.15243	6.56055	.17033	5.87080	.18835	5.30928	.20648	4.84300	20
41	.15272	6.54777	.17063	5.86051	.18865	5.30080	.20679	4.83590	19
42	.15302	6.53503	.17093	5.85024	.18895	5.29235	.20709	4.82882	18
43	.15332	6.52234	.17123	5.84001	.18925	5.28393	.20739	4.82175	17
44	.15362	6.50970	.17153	5.82982	.18955	5.27553	.20770	4.81471	16
45	.15391	6.49710	.17183	5.81966	.18986	5.26715	.20800	4.80769	15
46	.15421	6.48456	.17213	5.80953	.19016	5.25880	.20830	4.80068	14
47	.15451	6.47206	.17243	5.79944	.19046	5.25048	.20861	4.79370	13
48	.15481	6.45961	.17273	5.78938	.19076	5.24218	.20891	4.78673	12
49	.15511	6.44720	.17303	5.77936	.19106	5.23391	.20921	4.77978	11
50	.15540	6.43484	.17333	5.76937	.19136	5.22566	.20952	4.77286	10
51	.15570	6.42253	.17363	5.75941	.19166	5.21744	.20982	4.76595	9
52	.15600	6.41026	.17393	5.74949	.19197	5.20925	.21013	4.75906	8
53	.15630	6.39804	.17423	5.73960	.19227	5.20107	.21043	4.75219	7
54	.15660	6.38587	.17453	5.72974	.19257	5.19293	.21073	4.74534	6
55	.15689	6.37374	.17483	5.71992	.19287	5.18480	.21104	4.73851	5
56	.15719	6.36165	.17513	5.71013	.19317	5.17671	.21134	4.73170	4
57	.15749	6.34961	.17543	5.70037	.19347	5.16863	.21164	4.72490	3
58	.15779	6.33761	.17573	5.69064	.19378	5.16058	.21195	4.71813	2
59	.15809	6.32566	.17603	5.68094	.19408	5.15256	.21225	4.71137	1
60	.15838	6.31375	.17633	5.67128	.19438	5.14455	.21256	4.70463	0
′	Cotang	Tang	Cotang	Tang	Cotang	Tang	Cotang	Tang	′
	81°		80°		79°		78°		

′	12° Tang	12° Cotang	13° Tang	13° Cotang	14° Tang	14° Cotang	15° Tang	15° Cotang	′
0	.21256	4.70463	.23087	4.33148	.24933	4.01078	.26795	3.73205	60
1	.21286	4.69791	.23117	4.32573	.24964	4.00582	.26826	3.72771	59
2	.21316	4.69121	.23148	4.32001	.24995	4.00086	.26857	3.72338	58
3	.21347	4.68452	.23179	4.31430	.25026	3.99592	.26888	3.71907	57
4	.21377	4.67786	.23209	4.30860	.25056	3.99099	.26920	3.71476	56
5	.21408	4.67121	.23240	4.30291	.25087	3.98607	.26951	3.71046	55
6	.21438	4.66458	.23271	4.29724	.25118	3.98117	.26982	3.70616	54
7	.21469	4.65797	.23301	4.29159	.25149	3.97627	.27013	3.70188	53
8	.21499	4.65138	.23332	4.28595	.25180	3.97139	.27044	3.69761	52
9	.21529	4.64480	.23363	4.28032	.25211	3.96651	.27076	3.69335	51
10	.21560	4.63825	.23393	4.27471	.25242	3.96165	.27107	3.68909	50
11	.21590	4.63171	.23424	4.26911	.25273	3.95680	.27138	3.68485	49
12	.21621	4.62518	.23455	4.26352	.25304	3.95196	.27169	3.68061	48
13	.21651	4.61868	.23485	4.25795	.25335	3.94713	.27201	3.67638	47
14	.21682	4.61219	.23516	4.25239	.25366	3.94232	.27232	3.67217	46
15	.21712	4.60572	.23547	4.24685	.25397	3.93751	.27263	3.66796	45
16	.21743	4.59927	.23578	4.24132	.25428	3.93271	.27294	3.66376	44
17	.21773	4.59283	.23608	4.23580	.25459	3.92793	.27326	3.65957	43
18	.21804	4.58641	.23639	4.23030	.25490	3.92316	.27357	3.65538	42
19	.21834	4.58001	.23670	4.22481	.25521	3.91839	.27388	3.65121	41
20	.21864	4.57363	.23700	4.21933	.25552	3.91364	.27419	3.64705	40
21	.21895	4.56726	.23731	4.21387	.25583	3.90890	.27451	3.64289	39
22	.21925	4.56091	.23762	4.20842	.25614	3.90417	.27482	3.63874	38
23	.21956	4.55458	.23793	4.20298	.25645	3.89945	.27513	3.63461	37
24	.21986	4.54826	.23823	4.19756	.25676	3.89474	.27545	3.63048	36
25	.22017	4.54196	.23854	4.19215	.25707	3.89004	.27576	3.62636	35
26	.22047	4.53568	.23885	4.18675	.25738	3.88536	.27607	3.62224	34
27	.22078	4.52941	.23916	4.18137	.25769	3.88068	.27638	3.61814	33
28	.22108	4.52316	.23946	4.17600	.25800	3.87601	.27670	3.61405	32
29	.22139	4.51693	.23977	4.17064	.25831	3.87136	.27701	3.60996	31
30	.22169	4.51071	.24008	4.16530	.25862	3.86671	.27732	3.60588	30
31	.22200	4.50451	.24039	4.15997	.25893	3.86208	.27764	3.60181	29
32	.22231	4.49832	.24069	4.15465	.25924	3.85745	.27795	3.59775	28
33	.22261	4.49215	.24100	4.14934	.25955	3.85284	.27826	3.59370	27
34	.22292	4.48600	.24131	4.14405	.25986	3.84824	.27858	3.58966	26
35	.22322	4.47986	.24162	4.13877	.26017	3.84364	.27889	3.58562	25
36	.22353	4.47374	.24193	4.13350	.26048	3.83906	.27921	3.58160	24
37	.22383	4.46764	.24223	4.12825	.26079	3.83449	.27952	3.57758	23
38	.22414	4.46155	.24254	4.12301	.26110	3.82992	.27983	3.57357	22
39	.22444	4.45548	.24285	4.11778	.26141	3.82537	.28015	3.56957	21
40	.22475	4.44942	.24316	4.11256	.26172	3.82083	.28046	3.56557	20
41	.22505	4.44338	.24347	4.10736	.26203	3.81630	.28077	3.56159	19
42	.22536	4.43735	.24377	4.10216	.26235	3.81177	.28109	3.55761	18
43	.22567	4.43134	.24408	4.09699	.26266	3.80726	.28140	3.55364	17
44	.22597	4.42534	.24439	4.09182	.26297	3.80276	.28172	3.54968	16
45	.22628	4.41936	.24470	4.08666	.26328	3.79827	.28203	3.54573	15
46	.22658	4.41340	.24501	4.08152	.26359	3.79378	.28234	3.54179	14
47	.22689	4.40745	.24532	4.07639	.26390	3.78931	.28266	3.53785	13
48	.22719	4.40152	.24562	4.07127	.26421	3.78485	.28297	3.53393	12
49	.22750	4.39560	.24593	4.06616	.26452	3.78040	.28329	3.53001	11
50	.22781	4.38969	.24624	4.06107	.26483	3.77595	.28360	3.52609	10
51	.22811	4.38381	.24655	4.05599	.26515	3.77152	.28391	3.52219	9
52	.22842	4.37793	.24686	4.05092	.26546	3.76709	.28423	3.51829	8
53	.22872	4.37207	.24717	4.04586	.26577	3.76268	.28454	3.51441	7
54	.22903	4.36623	.24747	4.04081	26608	3.75828	.28486	3.51053	6
55	.22934	4.36040	.24778	4.03578	.26639	3.75388	.28517	3.50666	5
56	.22964	4.35459	.24809	4.03076	.26670	3.74950	.28549	3.50279	4
57	.22995	4.34879	.24840	4.02574	.26701	3.74512	.28580	3.49894	3
58	.23026	4.34300	.24871	4.02074	.26733	3.74075	.28612	3.49509	2
59	.23056	4.33723	.24902	4.01576	.26764	3.73640	.28643	3.49125	1
60	.23087	4.33148	.24933	4.01078	.26795	3.73205	.28675	3.48741	0
′	Cotang	Tang	Cotang	Tang	Cotang	Tang	Cotang	Tang	′
	77°		76°		75°		74°		

Appendix 7

′	16° Tang	16° Cotang	17° Tang	17° Cotang	18° Tang	18° Cotang	19° Tang	19° Cotang	′
0	.28675	3.48741	.30573	3.27085	.32492	3.07768	.34433	2.90421	60
1	.28706	3.48359	.30605	3.26745	.32524	3.07464	.34465	2.90147	59
2	.28738	3.47977	.30637	3.26406	.32556	3.07160	.34498	2.89873	58
3	.28769	3.47596	.30669	3.26067	.32588	3.06857	.34530	2.89600	57
4	.28800	3.47216	.30700	3.25729	.32621	3.06554	.34563	2.89327	56
5	.28832	3.46837	.30732	3.25392	.32653	3.06252	.34596	2.89055	55
6	.28864	3.46458	.30764	3.25055	.32685	3.05950	.34628	2.88783	54
7	.28895	3.46080	.30796	3.24719	.32717	3.05649	.34661	2.88511	53
8	.28927	3.45703	.30828	3.24383	.32749	3.05349	.34693	2.88240	52
9	.28958	3.45327	.30860	3.24049	.32782	3.05049	.34726	2.87970	51
10	.28990	3.44951	.30891	3.23714	.32814	3.04749	.34758	2.87700	50
11	.29021	3.44576	.30923	3.23381	.32846	3.04450	.34791	2.87430	49
12	.29053	3.44202	.30955	3.23048	.32878	3.04152	.34824	2.87161	48
13	.29084	3.43829	.30987	3.22715	.32911	3.03854	.34856	2.86892	47
14	.29116	3.43456	.31019	3.22384	.32943	3.03556	.34889	2.86624	46
15	.29147	3.43084	.31051	3.22053	.32975	3.03260	.34922	2.86356	45
16	.29179	3.42713	.31083	3.21722	.33007	3.02963	.34954	2.86089	44
17	.29210	3.42343	.31115	3.21392	.33040	3.02667	.34987	2.85822	43
18	.29242	3.41973	.31147	3.21063	.33072	3.02372	.35020	2.85555	42
19	.29274	3.41604	.31178	3.20734	.33104	3.02077	.35052	2.85289	41
20	.29305	3.41236	.31210	3.20406	.33136	3.01783	.35085	2.85023	40
21	.29337	3.40869	.31242	3.20079	.33169	3.01489	.35118	2.84758	39
22	.29368	3.40502	.31274	3.19752	.33201	3.01196	.35150	2.84494	38
23	.29400	3.40136	.31306	3.19426	.33233	3.00903	.35183	2.84229	37
24	.29432	3.39771	.31338	3.19100	.33266	3.00611	.35216	2.83965	36
25	.29463	3.39406	.31370	3.18775	.33298	3.00319	.35248	2.83702	35
26	.29495	3.39042	.31402	3.18451	.33330	3.00028	.35281	2.83439	34
27	.29526	3.38679	.31434	3.18127	.33363	2.99738	.35314	2.83176	33
28	.29558	3.38317	.31466	3.17804	.33395	2.99447	.35346	2.82914	32
29	.29590	3.37955	.31498	3.17481	.33427	2.99158	.35379	2.82653	31
30	.29621	3.37594	.31530	3.17159	.33460	2.98868	.35412	2.82391	30
31	.29653	3.37234	.31562	3.16838	.33492	2.98580	.35445	2.82130	29
32	.29685	3.36875	.31594	3.16517	.33524	2.98292	.35477	2.81870	28
33	.29716	3.36516	.31626	3.16197	.33557	2.98004	.35510	2.81610	27
34	.29748	3.36158	.31658	3.15877	.33589	2.97717	.35543	2.81350	26
35	.29780	3.35800	.31690	3.15558	.33621	2.97430	.35576	2.81091	25
36	.29811	3.35443	.31722	3.15240	.33654	2.97144	.35608	2.80833	24
37	.29843	3.35087	.31754	3.14922	.33686	2.96858	.35641	2.80574	23
38	.29875	3.34732	.31786	3.14605	.33718	2.96573	.35674	2.80316	22
39	.29906	3.34377	.31818	3.14288	.33751	2.96288	.35707	2.80059	21
40	.29938	3.34023	.31850	3.13972	.33783	2.96004	.35740	2.79802	20
41	.29970	3.33670	.31882	3.13656	.33816	2.95721	.35772	2.79545	19
42	.30001	3.33317	.31914	3.13341	.33848	2.95437	.35805	2.79289	18
43	.30033	3.32965	.31946	3.13027	.33881	2.95155	.35838	2.79033	17
44	.30065	3.32614	.31978	3.12713	.33913	2.94872	.35871	2.78778	16
45	.30097	3.32264	.32010	3.12400	.33945	2.94591	.35904	2.78523	15
46	.30128	3.31914	.32042	3.12087	.33978	2.94309	.35937	2.78269	14
47	.30160	3.31565	.32074	3.11775	.34010	2.94028	.35969	2.78014	13
48	.30192	3.31216	.32106	3.11464	.34043	2.93748	.36002	2.77761	12
49	.30224	3.30868	.32139	3.11153	.34075	2.93468	.36035	2.77507	11
50	.30255	3.30521	.32171	3.10842	.34108	2.93189	.36068	2.77254	10
51	.30287	3.30174	.32203	3.10532	.34140	2.92910	.36101	2.77002	9
52	.30319	3.29829	.32235	3.10223	.34173	2.92632	.36134	2.76750	8
53	.30351	3.29483	.32267	3.09914	.34205	2.92354	.36167	2.76498	7
54	.30382	3.29139	.32299	3.09606	.34238	2.92076	.36199	2.76247	6
55	.30414	3.28795	.32331	3.09298	.34270	2.91799	.36232	2.75996	5
56	.30446	3.28452	.32363	3.08991	.34303	2.91523	.36265	2.75746	4
57	.30478	3.28109	.32396	3.08685	.34335	2.91246	.36298	2.75496	3
58	.30509	3.27767	.32428	3.08379	.34368	2.90971	.36331	2.75246	2
59	.30541	3.27426	.32460	3.08073	.34400	2.90696	.36364	2.74997	1
60	.30573	3.27085	.32492	3.07768	.34433	2.90421	.36397	2.74748	0
′	Cotang	Tang	Cotang	Tang	Cotang	Tang	Cotang	Tang	′
	73°		72°		71°		70°		

′	20° Tang	20° Cotang	21° Tang	21° Cotang	22° Tang	22° Cotang	23° Tang	23° Cotang	′
0	.36397	2.74748	.38386	2.60509	.40403	2.47509	.42447	2.35585	60
1	.36430	2.74499	.38420	2.60283	.40436	2.47302	.42482	2.35395	59
2	.36463	2.74251	.38453	2.60057	.40470	2.47095	.42516	2.35205	58
3	.36496	2.74004	.38487	2.59831	.40504	2.46888	.42551	2.35015	57
4	.36529	2.73756	.38520	2.59606	.40538	2.46682	.42585	2.34825	56
5	.36562	2.73509	.38553	2.59381	.40572	2.46476	.42619	2.34636	55
6	.36595	2.73263	.38587	2.59156	.40606	2.46270	.42654	2.34447	54
7	.36628	2.73017	.38620	2.58932	.40640	2.46065	.42688	2.34258	53
8	.36661	2.72771	.38654	2.58708	.40674	2.45860	.42722	2.34069	52
9	.36694	2.72526	.38687	2.58484	.40707	2.45655	.42757	2.33881	51
10	.36727	2.72281	.38721	2.58261	.40741	2.45451	.42791	2.33693	50
11	.36760	2.72036	.38754	2.58038	.40775	2.45246	.42826	2.33505	49
12	.36793	2.71792	.38787	2.57815	.40809	2.45043	.42860	2.33317	48
13	.36826	2.71548	.38821	2.57593	.40843	2.44839	.42894	2.33130	47
14	.36859	2.71305	.38854	2.57371	.40877	2.44636	.42929	2.32943	46
15	.36892	2.71062	.38888	2.57150	.40911	2.44433	.42963	2.32756	45
16	.36925	2.70819	.38921	2.56928	.40945	2.44230	.42998	2.32570	44
17	.36958	2.70577	.38955	2.56707	.40979	2.44027	.43032	2.32383	43
18	.36991	2.70335	.38988	2.56487	.41013	2.43825	.43067	2.32197	42
19	.37024	2.70094	.39022	2.56266	.41047	2.43623	.43101	2.32012	41
20	.37057	2.69853	.39055	2.56046	.41081	2.43422	.43136	2.31826	40
21	.37090	2.69612	.39089	2.55827	.41115	2.43220	.43170	2.31641	39
22	.37123	2.69371	.39122	2.55608	.41149	2.43019	.43205	2.31456	38
23	.37157	2.69131	.39156	2.55389	.41183	2.42819	.43239	2.31271	37
24	.37190	2.68892	.39190	2.55170	.41217	2.42618	.43274	2.31086	36
25	.37223	2.68653	.39223	2.54952	.41251	2.42418	.43308	2.30902	35
26	.37256	2.68414	.39257	2.54734	.41285	2.42218	.43343	2.30718	34
27	.37289	2.68175	.39290	2.54516	.41319	2.42019	.43378	2.30534	33
28	.37322	2.67937	.39324	2.54299	.41353	2.41819	.43412	2.30351	32
29	.37355	2.67700	.39357	2.54082	.41387	2.41620	.43447	2.30167	31
30	.37388	2.67462	.39391	2.53865	.41421	2.41421	.43481	2.29984	30
31	.37422	2.67225	.39425	2.53648	.41455	2.41223	.43516	2.29801	29
32	.37455	2.66989	.39458	2.53432	.41490	2.41025	.43550	2.29619	28
33	.37488	2.66752	.39492	2.53217	.41524	2.40827	.43585	2.29437	27
34	.37521	2.66516	.39526	2.53001	.41558	2.40629	.43620	2.29254	26
35	.37554	2.66281	.39559	2.52786	.41592	2.40432	.43654	2.29073	25
36	.37588	2.66046	.39593	2.52571	.41626	2.40235	.43689	2.28891	24
37	.37621	2.65811	.39626	2.52357	.41660	2.40038	.43724	2.28710	23
38	.37654	2.65576	.39660	2.52142	.41694	2.39841	.43758	2.28528	22
39	.37687	2.65342	.39694	2.51929	.41728	2.39645	.43793	2.28348	21
40	.37720	2.65109	.39727	2.51715	.41763	2.39449	.43828	2.28167	20
41	.37754	2.64875	.39761	2.51502	.41797	2.39253	.43862	2.27987	19
42	.37787	2.64642	.39795	2.51289	.41831	2.39058	.43897	2.27806	18
43	.37820	2.64410	.39829	2.51076	.41865	2.38863	.43932	2.27626	17
44	.37853	2.64177	.39862	2.50864	.41899	2.38668	.43966	2.27447	16
45	.37887	2.63945	.39896	2.50652	.41933	2.38473	.44001	2.27267	15
46	.37920	2.63714	.39930	2.50440	.41968	2.38279	.44036	2.27088	14
47	.37953	2.63483	.39963	2.50229	.42002	2.38084	.44071	2.26909	13
48	.37986	2.63252	.39997	2.50018	.42036	2.37891	.44105	2.26730	12
49	.38020	2.63021	.40031	2.49807	.42070	2.37697	.44140	2.26552	11
50	.38053	2.62791	.40065	2.49597	.42105	2.37504	.44175	2.26374	10
51	.38086	2.62561	.40098	2.49386	.42139	2.37311	.44210	2.26196	9
52	.38120	2.62332	.40132	2.49177	.42173	2.37118	.44244	2.26018	8
53	.38153	2.62103	.40166	2.48967	.42207	2.36925	.44279	2.25840	7
54	.38186	2.61874	.40200	2.48758	.42242	2.36733	.44314	2.25663	6
55	.38220	2.61646	.40234	2.48549	.42276	2.36541	.44349	2.25486	5
56	.38253	2.61418	.40267	2.48340	.42310	2.36349	.44384	2.25309	4
57	.38286	2.61190	.40301	2.48132	.42345	2.36158	.44418	2.25132	3
58	.38320	2.60963	.40335	2.47924	.42379	2.35967	.44453	2.24956	2
59	.38353	2.60736	.40369	2.47716	.42413	2.35776	.44488	2.24780	1
60	.38386	2.60509	.40403	2.47509	.42447	2.35585	.44523	2.24604	0
′	Cotang	Tang	Cotang	Tang	Cotang	Tang	Cotang	Tang	′
	69°		68°		67°		66°		

Appendix 7

′	24° Tang	24° Cotang	25° Tang	25° Cotang	26° Tang	26° Cotang	27° Tang	27° Cotang	′
0	.44523	2.24604	.46631	2.14451	.48773	2.05030	.50953	1.96261	60
1	.44558	2.24428	.46666	2.14288	.48809	2.04879	.50989	1.96120	59
2	.44593	2.24252	.46702	2.14125	.48845	2.04728	.51026	1.95979	58
3	.44627	2.24077	.46737	2.13963	.48881	2.04577	.51063	1.95838	57
4	.44662	2.23902	.46772	2.13801	.48917	2.04426	.51099	1.95698	56
5	.44697	2.23727	.46808	2.13639	.48953	2.04276	.51136	1.95557	55
6	.44732	2.23553	.46843	2.13477	.48989	2.04125	.51173	1.95417	54
7	.44767	2.23378	.46879	2.13316	.49026	2.03975	.51209	1.95277	53
8	.44802	2.23204	.46914	2.13154	.49062	2.03825	.51246	1.95137	52
9	.44837	2.23030	.46950	2.12993	.49098	2.03675	.51283	1.94997	51
10	.44872	2.22857	.46985	2.12832	.49134	2.03526	.51319	1.94858	50
11	.44907	2.22683	.47021	2.12671	.49170	2.03376	.51356	1.94718	49
12	.44942	2.22510	.47056	2.12511	.49206	2.03227	.51393	1.94579	48
13	.44977	2.22337	.47092	2.12350	.49242	2.03078	.51430	1.94440	47
14	.45012	2.22164	.47128	2.12190	.49278	2.02929	.51467	1.94301	46
15	.45047	2.21992	.47163	2.12030	.49315	2.02780	.51503	1.94162	45
16	.45082	2.21819	.47199	2.11871	.49351	2.02631	.51540	1.94023	44
17	.45117	2.21647	.47234	2.11711	.49387	2.02483	.51577	1.93885	43
18	.45152	2.21475	.47270	2.11552	.49423	2.02335	.51614	1.93746	42
19	.45187	2.21304	.47305	2.11392	.49459	2.02187	.51651	1.93608	41
20	.45222	2.21132	.47341	2.11233	.49495	2.02039	.51688	1.93470	40
21	.45257	2.20961	.47377	2.11075	.49532	2.01891	.51724	1.93332	39
22	.45292	2.20790	.47412	2.10916	.49568	2.01743	.51761	1.93195	38
23	.45327	2.20619	.47448	2.10758	.49604	2.01596	.51798	1.93057	37
24	.45362	2.20449	.47483	2.10600	.49640	2.01449	.51835	1.92920	36
25	.45397	2.20278	.47519	2.10442	.49677	2.01302	.51872	1.92782	35
26	.45432	2.20108	.47555	2.10284	.49713	2.01155	.51909	1.92645	34
27	.45467	2.19938	.47590	2.10126	.49749	2.01008	.51946	1.92508	33
28	.45502	2.19769	.47626	2.09969	.49786	2.00862	.51983	1.92371	32
29	.45538	2.19599	.47662	2.09811	.49822	2.00715	.52020	1.92235	31
30	.45573	2.19430	.47698	2.09654	.49858	2.00569	.52057	1.92098	30
31	.45608	2.19261	.47733	2.09498	.49894	2.00423	.52094	1.91962	29
32	.45643	2.19092	.47769	2.09341	.49931	2.00277	.52131	1.91826	28
33	.45678	2.18923	.47805	2.09184	.49967	2.00131	.52168	1.91690	27
34	.45713	2.18755	.47840	2.09028	.50004	1.99986	.52205	1.91554	26
35	.45748	2.18587	.47876	2.08872	.50040	1.99841	.52242	1.91418	25
36	.45784	2.18419	.47912	2.08716	.50076	1.99695	.52279	1.91282	24
37	.45819	2.18251	.47948	2.08560	.50113	1.99550	.52316	1.91147	23
38	.45854	2.18084	.47984	2.08405	.50149	1.99406	.52353	1.91012	22
39	.45889	2.17916	.48019	2.08250	.50185	1.99261	.52390	1.90876	21
40	.45924	2.17749	.48055	2.08094	.50222	1.99116	.52427	1.90741	20
41	.45960	2.17582	.48091	2.07939	.50258	1.98972	.52464	1.90607	19
42	.45995	2.17416	.48127	2.07785	.50295	1.98828	.52501	1.90472	18
43	.46030	2.17249	.48163	2.07630	.50331	1.98684	.52538	1.90337	17
44	.46065	2.17083	.48198	2.07476	.50368	1.98540	.52575	1.90203	16
45	.46101	2.16917	.48234	2.07321	.50404	1.98396	.52613	1.90069	15
46	.46136	2.16751	.48270	2.07167	.50441	1.98253	.52650	1.89935	14
47	.46171	2.16585	.48306	2.07014	.50477	1.98110	.52687	1.89801	13
48	.46206	2.16420	.48342	2.06860	.50514	1.97966	.52724	1.89667	12
49	.46242	2.16255	.48378	2.06706	.50550	1.97823	.52761	1.89533	11
50	.46277	2.16090	.48414	2.06553	.50587	1.97681	.52798	1.89400	10
51	.46312	2.15925	.48450	2.06400	.50623	1.97538	.52836	1.89266	9
52	.46348	2.15760	.48486	2.06247	.50660	1.97395	.52873	1.89133	8
53	.46383	2.15596	.48521	2.06094	.50696	1.97253	.52910	1.89000	7
54	.46418	2.15432	.48557	2.05942	.50733	1.97111	.52947	1.88867	6
55	.46454	2.15268	.48593	2.05790	.50769	1.96969	.52985	1.88734	5
56	.46489	2.15104	.48629	2.05637	.50806	1.96827	.53022	1.88602	4
57	.46525	2.14940	.48665	2.05485	.50843	1.96685	.53059	1.88469	3
58	.46560	2.14777	.48701	2.05333	.50879	1.96544	.53096	1.88337	2
59	.46595	2.14614	.48737	2.05182	.50916	1.96402	.53134	1.88205	1
60	.46631	2.14451	.48773	2.05030	.50953	1.96261	.53171	1.88073	0
′	Cotang	Tang	Cotang	Tang	Cotang	Tang	Cotang	Tang	′
	65°		64°		63°		62°		

′	28° Tang	28° Cotang	29° Tang	29° Cotang	30° Tang	30° Cotang	31° Tang	31° Cotang	′
0	.53171	1.88073	.55431	1.80405	.57735	1.73205	.60086	1.66428	60
1	.53208	1.87941	.55469	1.80281	.57774	1.73089	.60126	1.66318	59
2	.53246	1.87809	.55507	1.80158	.57813	1.72973	.60165	1.66209	58
3	.53283	1.87677	.55545	1.80034	.57851	1.72857	.60205	1.66099	57
4	.53320	1.87546	.55583	1.79911	.57890	1.72741	.60245	1.65990	56
5	.53358	1.87415	.55621	1.79788	.57929	1.72625	.60284	1.65881	55
6	.53395	1.87283	.55659	1.79665	.57968	1.72509	.60324	1.65772	54
7	.53432	1.87152	.55697	1.79542	.58007	1.72393	.60364	1.65663	53
8	.53470	1.87021	.55736	1.79419	.58046	1.72278	.60403	1.65554	52
9	.53507	1.86891	.55774	1.79296	.58085	1.72163	.60443	1.65445	51
10	.53545	1.86760	.55812	1.79174	.58124	1.72047	.60483	1.65337	50
11	.53582	1.86630	.55850	1.79051	.58162	1.71932	.60522	1.65228	49
12	.53620	1.86499	.55888	1.78929	.58201	1.71817	.60562	1.65120	48
13	.53657	1.86369	.55926	1.78807	.58240	1.71702	.60602	1.65011	47
14	.53694	1.86239	.55964	1.78685	.58279	1.71588	.60642	1.64903	46
15	.53732	1.86109	.56003	1.78563	.58318	1.71473	.60681	1.64795	45
16	.53769	1.85979	.56041	1.78441	.58357	1.71358	.60721	1.64687	44
17	.53807	1.85850	.56079	1.78319	.58396	1.71244	.60761	1.64579	43
18	.53844	1.85720	.56117	1.78198	.58435	1.71129	.60801	1.64471	42
19	.53882	1.85591	.56156	1.78077	.58474	1.71015	.60841	1.64363	41
20	.53920	1.85462	.56194	1.77955	.58513	1.70901	.60883	1.64256	40
21	.53957	1.85333	.56232	1.77834	.58552	1.70787	.60921	1.64148	39
22	.53995	1.85204	.56270	1.77713	.58591	1.70673	.60960	1.64041	38
23	.54032	1.85075	.56309	1.77592	.58631	1.70560	.61000	1.63934	37
24	.54070	1.84946	.56347	1.77471	.58670	1.70446	.61040	1.63826	36
25	.54107	1.84818	.56385	1.77351	.58709	1.70332	.61080	1.63719	35
26	.54145	1.84689	.56424	1.77230	.58748	1.70219	.61120	1.63612	34
27	.54183	1.84561	.56462	1.77110	.58787	1.70106	.61160	1.63505	33
28	.54220	1.84433	.56501	1.76990	.58826	1.69992	.61200	1.63398	32
29	.54258	1.84305	.56539	1.76869	.58865	1.69879	.61240	1.63292	31
30	.54296	1.84177	.56577	1.76749	.58905	1.69766	.61280	1.63185	30
31	.54333	1.84049	.56616	1.76629	.58944	1.69653	.61320	1.63079	29
32	.54371	1.83922	.56654	1.76510	.58983	1.69541	.61360	1.62972	28
33	.54409	1.83794	.56693	1.76390	.59022	1.69428	.61400	1.62866	27
34	.54446	1.83667	.56731	1.76271	.59061	1.69316	.61440	1.62760	26
35	.54484	1.83540	.56769	1.76151	.59101	1.69203	.61480	1.62654	25
36	.54522	1.83413	.56808	1.76032	.59140	1.69091	.61520	1.62548	24
37	.54560	1.83286	.56846	1.75913	.59179	1.68979	.61561	1.62442	23
38	.54597	1.83159	.56885	1.75794	.59218	1.68866	.61601	1.62336	22
39	.54635	1.83033	.56923	1.75675	.59258	1.68754	.61641	1.62230	21
40	.54673	1.82906	.56962	1.75556	.59297	1.68643	.61681	1.62125	20
41	.54711	1.82780	.57000	1.75437	.59336	1.68531	.61721	1.62019	19
42	.54748	1.82654	.57039	1.75319	.59376	1.68419	.61761	1.61914	18
43	.54786	1.82528	.57078	1.75200	.59415	1.68308	.61801	1.61808	17
44	.54824	1.82402	.57116	1.75082	.59454	1.68196	.61842	1.61703	16
45	.54862	1.82276	.57155	1.74964	.59494	1.68085	.61882	1.61598	15
46	.54900	1.82150	.57193	1.74846	.59533	1.67974	.61922	1.61493	14
47	.54938	1.82025	.57232	1.74728	.59573	1.67863	.61962	1.61388	13
48	.54975	1.81899	.57271	1.74610	.59612	1.67752	.62003	1.61283	12
49	.55013	1.81774	.57309	1.74492	.59651	1.67641	.62043	1.61179	11
50	.55051	1.81649	.57348	1.74375	.59691	1.67530	.62083	1.61074	10
51	.55089	1.81524	.57386	1.74257	.59730	1.67419	.62124	1.60970	9
52	.55127	1.81399	.57425	1.74140	.59770	1.67309	.62164	1.60865	8
53	.55165	1.81274	.57464	1.74022	.59809	1.67198	.62204	1.60761	7
54	.55203	1.81150	.57503	1.73905	.59849	1.67088	.62245	1.60657	6
55	.55241	1.81025	.57541	1.73788	.59888	1.66978	.62285	1.60553	5
56	.55279	1.80901	.57580	1.73671	.59928	1.66867	.62325	1.60449	4
57	.55317	1.80777	.57619	1.73555	.59967	1.66757	.62366	1.60345	3
58	.55355	1.80653	.57657	1.73438	.60007	1.66647	.62406	1.60241	2
59	.55393	1.80529	.57696	1.73321	.60046	1.66538	.62446	1.60137	1
60	.55431	1.80405	.57735	1.73205	.60086	1.66428	.62487	1.60033	0
′	Cotang	Tang	Cotang	Tang	Cotang	Tang	Cotang	Tang	′
	61°		60°		59°		58°		

Appendix 7

′	32° Tang	32° Cotang	33° Tang	33° Cotang	34° Tang	34° Cotang	35° Tang	35° Cotang	′
0	.62487	1.60033	.64941	1.53986	.67451	1.48256	.70021	1.42815	60
1	.62527	1.59930	.64982	1.53888	.67493	1.48163	.70064	1.42726	59
2	.62568	1.59826	.65024	1.53791	.67536	1.48070	.70107	1.42638	58
3	.62608	1.59723	.65065	1.53693	.67578	1.47977	.70151	1.42550	57
4	.62649	1.59620	.65106	1.53595	.67620	1.47885	.70194	1.42462	56
5	.62689	1.59517	.65148	1.53497	.67663	1.47792	.70238	1.42374	55
6	.62730	1.59414	.65189	1.53400	.67705	1.47699	.70281	1.42286	54
7	.62770	1.59311	.65231	1.53302	.67748	1.47607	.70325	1.42198	53
8	.62811	1.59208	.65272	1.53205	.67790	1.47514	.70368	1.42110	52
9	.62852	1.59105	.65314	1.53107	.67832	1.47422	.70412	1.42022	51
10	.62892	1.59002	.65355	1.53010	.67875	1.47330	.70455	1.41934	50
11	.62933	1.58900	.65397	1.52913	.67917	1.47238	.70499	1.41847	49
12	.62973	1.58797	.65438	1.52816	.67960	1.47146	.70542	1.41759	48
13	.63014	1.58695	.65480	1.52719	.68002	1.47053	.70586	1.41672	47
14	.63055	1.58593	.65521	1.52622	.68045	1.46962	.70629	1.41584	46
15	.63095	1.58490	.65563	1.52525	.68088	1.46870	.70673	1.41497	45
16	.63136	1.58388	.65604	1.52429	.68130	1.46778	.70717	1.41409	44
17	.63177	1.58286	.65646	1.52332	.68173	1.46686	.70760	1.41322	43
18	.63217	1.58184	.65688	1.52235	.68215	1.46595	.70804	1.41235	42
19	.63258	1.58083	.65729	1.52139	.68258	1.46503	.70848	1.41148	41
20	.63299	1.57981	.65771	1.52043	.68301	1.46411	.70891	1.41061	40
21	.63340	1.57879	.65813	1.51946	.68343	1.46320	.70935	1.40974	39
22	.63380	1.57778	.65854	1.51850	.68386	1.46229	.70979	1.40887	38
23	.63421	1.57676	.65896	1.51754	.68429	1.46137	.71023	1.40800	37
24	.63462	1.57575	.65938	1.51658	.68471	1.46046	.71066	1.40714	36
25	.63503	1.57474	.65980	1.51562	.68514	1.45955	.71110	1.40627	35
26	.63544	1.57372	.66021	1.51466	.68557	1.45864	.71154	1.40540	34
27	.63584	1.57271	.66063	1.51370	.68600	1.45773	.71198	1.40454	33
28	.63625	1.57170	.66105	1.51275	.68642	1.45682	.71242	1.40367	32
29	.63666	1.57069	.66147	1.51179	.68685	1.45592	.71285	1.40281	31
30	.63707	1.56969	.66189	1.51084	.68728	1.45501	.71329	1.40195	30
31	.63748	1.56868	.66230	1.50988	.68771	1.45410	.71373	1.40109	29
32	.63789	1.56767	.66272	1.50893	.68814	1.45320	.71417	1.40022	28
33	.63830	1.56667	.66314	1.50797	.68857	1.45229	.71461	1.39936	27
34	.63871	1.56566	.66356	1.50702	.68900	1.45139	.71505	1.39850	26
35	.63912	1.56466	.66398	1.50607	.68942	1.45049	.71549	1.39764	25
36	.63953	1.56366	.66440	1.50512	.68985	1.44958	.71593	1.39679	24
37	.63994	1.56265	.66482	1.50417	.69028	1.44868	.71637	1.39593	23
38	.64035	1.56165	.66524	1.50322	.69071	1.44778	.71681	1.39507	22
39	.64076	1.56065	.66566	1.50228	.69114	1.44688	.71725	1.39421	21
40	.64117	1.55966	.66608	1.50133	.69157	1.44598	.71769	1.39336	20
41	.64158	1.55866	.66650	1.50038	.69200	1.44508	.71813	1.39250	19
42	.64199	1.55766	.66692	1.49944	.69243	1.44418	.71857	1.39165	18
43	.64240	1.55666	.66734	1.49849	.69286	1.44329	.71901	1.39079	17
44	.64281	1.55567	.66776	1.49755	.69329	1.44239	.71946	1.38994	16
45	.64322	1.55467	.66818	1.49661	.69372	1.44149	.71990	1.38909	15
46	.64363	1.55368	.66860	1.49566	.69416	1.44060	.72034	1.38824	14
47	.64404	1.55269	.66902	1.49472	.69459	1.43970	.72078	1.38738	13
48	.64446	1.55170	.66944	1.49378	.69502	1.43881	.72122	1.38653	12
49	.64487	1.55071	.66986	1.49284	.69545	1.43792	.72167	1.38568	11
50	.64528	1.54972	.67028	1.49190	.69588	1.43703	.72211	1.38484	10
51	.64569	1.54873	.67071	1.49097	.69631	1.43614	.72255	1.38399	9
52	.64610	1.54774	.67113	1.49003	.69675	1.43525	.72299	1.38314	8
53	.64652	1.54675	.67155	1.48909	.69718	1.43436	.72344	1.38229	7
54	.64693	1.54576	.67197	1.48816	.69761	1.43347	.72388	1.38145	6
55	.64734	1.54478	.67239	1.48722	.69804	1.43258	.72432	1.38060	5
56	.64775	1.54379	.67282	1.48629	.69847	1.43169	.72477	1.37976	4
57	.64817	1.54281	.67324	1.48536	.69891	1.43080	.72521	1.37891	3
58	.64858	1.54183	.67366	1.48442	.69934	1.42992	.72565	1.37807	2
59	.64899	1.54085	.67409	1.48349	.69977	1.42903	.72610	1.37722	1
60	.64941	1.53986	.67451	1.48256	.70021	1.42815	.72654	1.37638	0
′	Cotang	Tang	Cotang	Tang	Cotang	Tang	Cotang	Tang	′
	57°		56°		55°		54°		

′	36° Tang	36° Cotang	37° Tang	37° Cotang	38° Tang	38° Cotang	39° Tang	39° Cotang	′
0	.72654	1.37638	.75355	1.32704	.78129	1.27994	.80978	1.23490	60
1	.72699	1.37554	.75401	1.32624	.78175	1.27917	.81027	1.23416	59
2	.72743	1.37470	.75447	1.32544	.78222	1.27841	.81075	1.23343	58
3	.72788	1.37386	.75492	1.32464	.78269	1.27764	.81123	1.23270	57
4	.72832	1.37302	.75538	1.32384	.78316	1.27688	.81171	1.23196	56
5	.72877	1.37218	.75584	1.32304	.78363	1.27611	.81220	1.23123	55
6	.72921	1.37134	.75629	1.32224	.78410	1.27535	.81268	1.23050	54
7	.72966	1.37050	.75675	1.32144	.78457	1.27458	.81316	1.22977	53
8	.73010	1.36967	.75721	1.32064	.78504	1.27382	.81364	1.22904	52
9	.73055	1.36883	.75767	1.31984	.78551	1.27306	.81413	1.22831	51
10	.73100	1.36800	.75812	1.31904	.78598	1.27230	.81461	1.22758	50
11	.73144	1.36716	.75858	1.31825	.78645	1.27153	.81510	1.22685	49
12	.73189	1.36633	.75904	1.31745	.78692	1.27077	.81558	1.22612	48
13	.73234	1.36549	.75950	1.31666	.78739	1.27001	.81606	1.22539	47
14	.73278	1.36466	.75996	1.31586	.78786	1.26925	.81655	1.22467	46
15	.73323	1.36383	.76042	1.31507	.78834	1.26849	.81703	1.22394	45
16	.73368	1.36300	.76088	1.31427	.78881	1.26774	.81752	1.22321	44
17	.73413	1.36217	.76134	1.31348	.78928	1.26698	.81800	1.22249	43
18	.73457	1.36134	.76180	1.31269	.78975	1.26622	.81849	1.22176	42
19	.73502	1.36051	.76226	1.31190	.79022	1.26546	.81898	1.22104	41
20	.73547	1.35968	.76272	1.31110	.79070	1.26471	.81946	1.22031	40
21	.73592	1.35885	.76318	1.31031	.79117	1.26395	.81995	1.21959	39
22	.73637	1.35802	.76364	1.30952	.79164	1.26319	.82044	1.21886	38
23	.73681	1.35719	.76410	1.30873	.79212	1.26244	.82092	1.21814	37
24	.73726	1.35637	.76456	1.30795	.79259	1.26169	.82141	1.21742	36
25	.73771	1.35554	.76502	1.30716	.79306	1.26093	.82190	1.21670	35
26	.73816	1.35472	.76548	1.30637	.79354	1.26018	.82238	1.21598	34
27	.73861	1.35389	.76594	1.30558	.79401	1.25943	.82287	1.21526	33
28	.73906	1.35307	.76640	1.30480	.79449	1.25867	.82336	1.21454	32
29	.73951	1.35224	.76686	1.30401	.79496	1.25792	.82385	1.21382	31
30	.73996	1.35142	.76733	1.30323	.79544	1.25717	.82434	1.21310	30
31	.74041	1.35060	.76779	1.30244	.79591	1.25642	.82483	1.21238	29
32	.74086	1.34978	.76825	1.30166	.79639	1.25567	.82531	1.21166	28
33	.74131	1.34896	.76871	1.30087	.79686	1.25492	.82580	1.21094	27
34	.74176	1.34814	.76918	1.30009	.79734	1.25417	.82629	1.21023	26
35	.74221	1.34732	.76964	1.29931	.79781	1.25343	.82678	1.20951	25
36	.74267	1.34650	.77010	1.29853	.79829	1.25268	.82727	1.20879	24
37	.74312	1.34568	.77057	1.29775	.79877	1.25193	.82776	1.20808	23
38	.74357	1.34487	.77103	1.29696	.79924	1.25118	.82825	1.20736	22
39	.74402	1.34405	.77149	1.29618	.79972	1.25044	.82874	1.20665	21
40	.74447	1.34323	.77196	1.29541	.80020	1.24969	.82923	1.20593	20
41	.74492	1.34242	.77242	1.29463	.80067	1.24895	.82972	1.20522	19
42	.74538	1.34160	.77289	1.29385	.80115	1.24820	.83022	1.20451	18
43	.74583	1.34079	.77335	1.29307	.80163	1.24746	.83071	1.20379	17
44	.74628	1.33998	.77382	1.29229	.80211	1.24672	.83120	1.20308	16
45	.74674	1.33916	.77428	1.29152	.80258	1.24597	.83169	1.20237	15
46	.74719	1.33835	.77475	1.29074	.80306	1.24523	.83218	1.20166	14
47	.74764	1.33754	.77521	1.28997	.80354	1.24449	.83268	1.20095	13
48	.74810	1.33673	.77568	1.28919	.80402	1.24375	.83317	1.20024	12
49	.74855	1.33592	.77615	1.28842	.80450	1.24301	.83366	1.19953	11
50	.74900	1.33511	.77661	1.28764	.80498	1.24227	.83415	1.19882	10
51	.74946	1.33430	.77708	1.28687	.80546	1.24153	.83465	1.19811	9
52	.74991	1.33349	.77754	1.28610	.80594	1.24079	.83514	1.19740	8
53	.75037	1.33268	.77801	1.28533	.80642	1.24005	.83564	1.19669	7
54	.75082	1.33187	.77848	1.28456	.80690	1.23931	.83613	1.19599	6
55	.75128	1.33107	.77895	1.28379	.80738	1.23858	.83662	1.19528	5
56	.75173	1.33026	.77941	1.28302	.80786	1.23784	.83712	1.19457	4
57	.75219	1.32946	.77988	1.28225	.80834	1.23710	.83761	1.19387	3
58	.75264	1.32865	.78035	1.28148	.80882	1.23637	.83811	1.19316	2
59	.75310	1.32785	.78082	1.28071	.80930	1.23563	.83860	1.19246	1
60	.75355	1.32704	.78129	1.27994	.80978	1.23490	.83910	1.19175	0
′	Cotang	Tang	Cotang	Tang	Cotang	Tang	Cotang	Tang	′
	53°	53°	52°	52°	51°	51°	50°	50°	

Appendix 7

′	40° Tang	40° Cotang	41° Tang	41° Cotang	42° Tang	42° Cotang	43° Tang	43° Cotang	′
0	.83910	1.19175	.86929	1.15037	.90040	1.11061	.93252	1.07237	60
1	.83960	1.19105	.86980	1.14969	.90093	1.10996	.93306	1.07174	59
2	.84009	1.19035	.87031	1.14902	.90146	1.10931	.93360	1.07112	58
3	.84059	1.18964	.87082	1.14834	.90199	1.10867	.93415	1.07049	57
4	.84108	1.18894	.87133	1.14767	.90251	1.10802	.93469	1.06987	56
5	.84158	1.18824	.87184	1.14699	.90304	1.10737	.93524	1.06925	55
6	.84208	1.18754	.87236	1.14632	.90357	1.10672	.93578	1.06862	54
7	.84258	1.18684	.87287	1.14565	.90410	1.10607	.93633	1.06800	53
8	.84307	1.18614	.87338	1.14498	.90463	1.10543	.93688	1.06738	52
9	.84357	1.18544	.87389	1.14430	.90516	1.10478	.93742	1.06676	51
10	.84407	1.18474	.87441	1.14363	.90569	1.10414	.93797	1.06613	50
11	.84457	1.18404	.87492	1.14296	.90621	1.10349	.93852	1.06551	49
12	.84507	1.18334	.87543	1.14229	.90674	1.10285	.93906	1.06489	48
13	.84556	1.18264	.87595	1.14162	.90727	1.10220	.93961	1.06427	47
14	.84606	1.18194	.87646	1.14095	.90781	1.10156	.94016	1.06365	46
15	.84656	1.18125	.87698	1.14028	.90834	1.10091	.94071	1.06303	45
16	.84706	1.18055	.87749	1.13961	.90887	1.10027	.94125	1.06241	44
17	.84756	1.17986	.87801	1.13894	.90940	1.09963	.94180	1.06179	43
18	.84806	1.17916	.87852	1.13828	.90993	1.09899	.94235	1.06117	42
19	.84856	1.17846	.87904	1.13761	.91046	1.09834	.94290	1.06056	41
20	.84906	1.17777	.87955	1.13694	.91099	1.09770	.94345	1.05994	40
21	.84956	1.17708	.88007	1.13627	.91153	1.09706	.94400	1.05932	39
22	.85006	1.17638	.88059	1.13561	.91206	1.09642	.94455	1.05870	38
23	.85057	1.17569	.88110	1.13494	.91259	1.09578	.94510	1.05809	37
24	.85107	1.17500	.88162	1.13428	.91313	1.09514	.94565	1.05747	36
25	.85157	1.17430	.88214	1.13361	.91366	1.09450	.94620	1.05685	35
26	.85207	1.17361	.88265	1.13295	.91419	1.09386	.94676	1.05624	34
27	.85257	1.17292	.88317	1.13228	.91473	1.09322	.94731	1.05562	33
28	.85308	1.17223	.88369	1.13162	.91526	1.09258	.94786	1.05501	32
29	.85358	1.17154	.88421	1.13096	.91580	1.09195	.94841	1.05439	31
30	.85408	1.17085	.88473	1.13029	.91633	1.09131	.94896	1.05378	30
31	.85458	1.17016	.88524	1.12963	.91687	1.09067	.94952	1.05317	29
32	.85509	1.16947	.88576	1.12897	.91740	1.09003	.95007	1.05255	28
33	.85559	1.16878	.88628	1.12831	.91794	1.08940	.95062	1.05194	27
34	.85609	1.16809	.88680	1.12765	.91847	1.08876	.95118	1.05133	26
35	.85660	1.16741	.88732	1.12699	.91901	1.08813	.95173	1.05072	25
36	.85710	1.16672	.88784	1.12633	.91955	1.08749	.95229	1.05010	24
37	.85761	1.16603	.88836	1.12567	.92008	1.08686	.95284	1.04949	23
38	.85811	1.16535	.88888	1.12501	.92062	1.08622	.95340	1.04888	22
39	.85862	1.16466	.88940	1.12435	.92116	1.08559	.95395	1.04827	21
40	.85912	1.16398	.88992	1.12369	.92170	1.08496	.95451	1.04766	20
41	.85963	1.16329	.89045	1.12303	.92224	1.08432	.95506	1.04705	19
42	.86014	1.16261	.89097	1.12238	.92277	1.08369	.95562	1.04644	18
43	.86064	1.16192	.89149	1.12172	.92331	1.08306	.95618	1.04583	17
44	.86115	1.16124	.89201	1.12106	.92385	1.08243	.95673	1.04522	16
45	.86166	1.16056	.89253	1.12041	.92439	1.08179	.95729	1.04461	15
46	.86216	1.15987	.89306	1.11975	.92493	1.08116	.95785	1.04401	14
47	.86267	1.15919	.89358	1.11909	.92547	1.08053	.95841	1.04340	13
48	.86318	1.15851	.89410	1.11844	.92601	1.07990	.95897	1.04279	12
49	.86368	1.15783	.89463	1.11778	.92655	1.07927	.95952	1.04218	11
50	.86419	1.15715	.89515	1.11713	.92709	1.07864	.96008	1.04158	10
51	.86470	1.15647	.89567	1.11648	.92763	1.07801	.96064	1.04097	9
52	.86521	1.15579	.89620	1.11582	.92817	1.07738	.96120	1.04036	8
53	.86572	1.15511	.89672	1.11517	.92872	1.07676	.96176	1.03976	7
54	.86623	1.15443	.89725	1.11452	.92926	1.07613	.96232	1.03915	6
55	.86674	1.15375	.89777	1.11387	.92980	1.07550	.96288	1.03855	5
56	.86725	1.15308	.89830	1.11321	.93034	1.07487	.96344	1.03794	4
57	.86776	1.15240	.89883	1.11256	.93088	1.07425	.96400	1.03734	3
58	.86827	1.15172	.89935	1.11191	.93143	1.07362	.96457	1.03674	2
59	.86878	1.15104	.89988	1.11126	.93197	1.07299	.96513	1.03613	1
60	.86929	1.15037	.90040	1.11061	.93252	1.07237	.96569	1.03553	0
′	Cotang	Tang	Cotang	Tang	Cotang	Tang	Cotang	Tang	′
	49°		48°		47°		46°		

′	44° Tang	44° Cotang	′	′	44° Tang	44° Cotang	′	′	44° Tang	44° Cotang	′
0	.96569	1.03553	60	20	.97700	1.02355	40	40	.98843	1.01170	20
1	.96625	1.03493	59	21	.97756	1.02295	39	41	.98901	1.01112	19
2	.96681	1.03433	58	22	.97813	1.02236	38	42	.98958	1.01053	18
3	.96738	1.03372	57	23	.97870	1.02176	37	43	.99016	1.00994	17
4	.96794	1.03312	56	24	.97927	1.02117	36	44	.99073	1.00935	16
5	.96850	1.03252	55	25	.97984	1.02057	35	45	.99131	1.00876	15
6	.96907	1.03192	54	26	.98041	1.01998	34	46	.99189	1.00818	14
7	.96963	1.03132	53	27	.98098	1.01939	33	47	.99247	1.00759	13
8	.97020	1.03072	52	28	.98155	1.01879	32	48	.99304	1.00701	12
9	.97076	1.03012	51	29	.98213	1.01820	31	49	.99362	1.00642	11
10	.97133	1.02952	50	30	.98270	1.01761	30	50	.99420	1.00583	10
11	.97189	1.02892	49	31	.98327	1.01702	29	51	.99478	1.00525	9
12	.97246	1.02832	48	32	.98384	1.01642	28	52	.99536	1.00467	8
13	.97302	1.02772	47	33	.98441	1.01583	27	53	.99594	1.00408	7
14	.97359	1.02713	46	34	.98499	1.01524	26	54	.99652	1.00350	6
15	.97416	1.02653	45	35	.98556	1.01465	25	55	.99710	1.00291	5
16	.97472	1.02593	44	36	.98613	1.01406	24	56	.99768	1.00233	4
17	.97529	1.02533	43	37	.98671	1.01347	23	57	.99826	1.00175	3
18	.97586	1.02474	42	38	.98728	1.01288	22	58	.99884	1.00116	2
19	.97643	1.02414	41	39	.98786	1.01229	21	59	.99942	1.00058	1
20	.97700	1.02355	40	40	.98843	1.01170	20	60	1.00000	1.00000	0
′	Cotang	Tang 45°	′	′	Cotang	Tang 45°	′	′	Cotang	Tang 45°	′

APPENDIX 8. EQUIVALENCE OF SOME UNITS OF WEIGHT AND MEASURE

Underlined figures are exact; others are rounded off. Condensed from Letter Circular 1035 (Jan., 1960) of the U.S. Department of Commerce, National Bureau of Standards, Washington 25, D.C.

1 in. = 0.08333 ft; 0.02778 yd; 2.54 cm.
1 ft = 12 in.; 0.6061 rods; 0.3048 m; 0.0001894 mi
1 yd = 3 ft; 0.9144 m; 0.1818 rods; 0.0005682 mi
1 m = 1000 mm; 100 cm; 10 decimeters 0.1 dekameters; 0.01 hectometers; 0.001 km
1 m = 39.37 in.; 3.2808 ft; 1.0936 yd; 0.0006214 mi
1 fathom = 6 ft; 1.8288 m
1 rod = 198 in.; 16.5 ft; 5.5 yd
1 chain = 100 links; 66 ft; 0.0125 mi; 20.117 m;
1 mi = 5280 ft; 1760 yd; 320 rods; 1609.344 m;
1 nautical mi = 6076.1 ft; 1852 m
1 sq in. = 6.4516 sq cm; 0.00694 sq ft
1 sq ft = 144 sq in.; 0.1111 sq yd; 0.0929 sq m
1 sq yd = 1296 sq in.; 9 sq ft; 0.8361 sq m
1 sq m = 1551 sq in.; 10.76 sq ft; 1.196 sq yd
1 acre = 43560 sq ft; 4840 sq yd; 0.405 hectares; 0.00156 sq mi
1 sq mi = 640 acres; 259 hectares
1 cu cm = 0.0610 cu in.; 0.000001 cu m
1 cu in. = 0.0005787 cu ft; 16.387 cu cm
1 cu ft = 1728 cu in.; 0.03704 cu yd; 0.0283 cu m; 7.480 gal (U.S.)
1 cu yd = 46656 cu in.; 27 cu ft; 0.7645 cu m
1 cu m = 35.315 cu ft; 1.3079 cu yd
1 gal (U.S.) = 231 cu in; 128 fl oz; 0.1337 cu ft; 3.785 liters
1 liter = 61.025 cu in.; 0.2642 gal (U.S.); 0.0353 cu ft
1 acre ft = 43560 cu ft; 325851 gal (U.S.); 1233.5 cu m
1 oz (avoir.) = 437.5 grains; 28.350 grams; 0.0625 lbs (avoir.)
1 gram = 15.432 grains; 0.03527 oz (avoir.); 0.002205 lbs (avoir.)
1 short (net) ton = 2000 lbs; 0.9072 metric ton; 0.8929 long (gross) ton

APPENDIX 9. PROTRACTOR FOR INTERCONVERSION OF TRUE DIP AND APPARENT DIP

DIRECTIONS

True Dip = dip measured in the field.

Direction or Divergence Angle = angle between strike and direction of cross section.

Apparent Dip = inclination in the plane of cross section.

To determine the apparent dip, set a straightedge (or fine line drawn on a strip of celluloid) to pass through the lower right corner of the grid (point 0) and the point of intersection of the lines representing the true dip and the direction angle. Read apparent dip on the graduated arc.

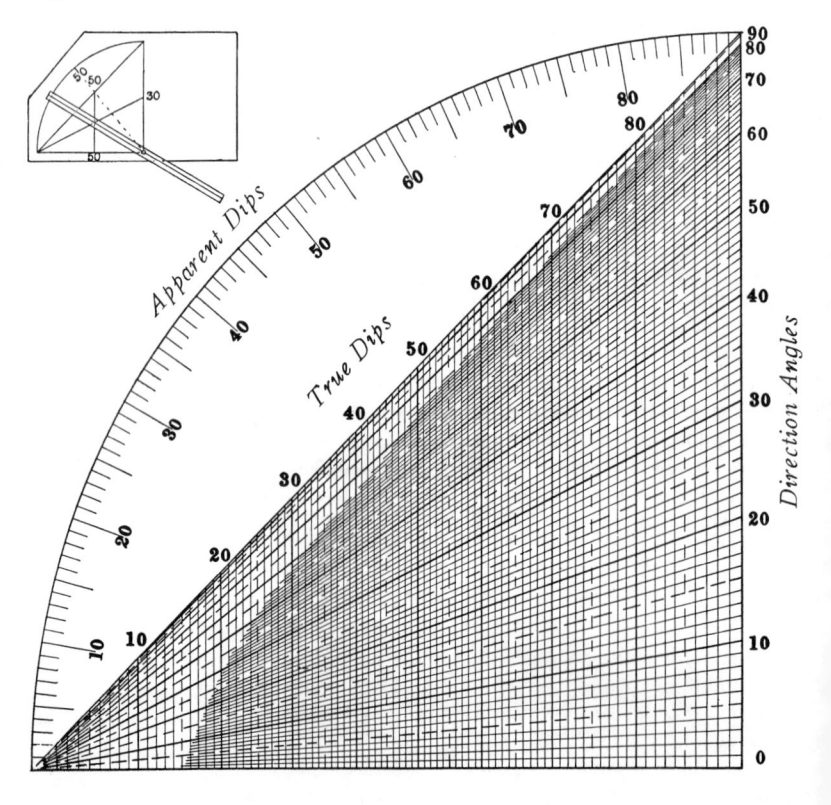

Copyright, 1925, by W. S. Tangier Smith; copyright renewed, 1953; reprinted by permission.

Index

Page numbers in **boldface type** refer to figures.

Aa lava, 259
Abbreviations for field notes, 8, 329
Abstracts of reports, 192
Adamellite, **276**, 277
Adit, of mine, 175
Adjustments, of alidade, 93, 94
   of barometer (altimeter) readings, 56
   of Brunton compass, 34
   of control networks, 125
   of radial line plots, 166
   of traverses, 54
Aerial photograph compilations, 154–169
   from controlled photographs, 163
   cross sections from, 157, 169
   marking and transferring points for, 161
   materials for, 160
   plotting control for, 162
   radial line method for, 159
   transferring details for, 167–169
   from uncontrolled photographs, 165
Aerial photograph projects, preparations for, 154
Aerial photographs, compilations from, 154–169
   computing scale at any elevation, 83–84
   cross sections from, 157, 169
   determining average scale of, 81
   determining exact scales of, 81
   distortion of, by relief, 76
   distortion of, by tilt, 75, 165, 166
   geologic mapping on, 84, 155
   inking, 85, 167
   kinds, 73
   locating points on, by bearings and pacing, 83

Aerial photographs, locating points on, by inspection, 82
   marking points and lines on, 84
   overlaps of, 74, 155
   plane table surveys on, 150
   plotting north arrows on, 82
   rectified to correct tilt, 150
   in selecting control stations, 116
   stereoscopic viewing of, 77
   in transferring features to maps, 85, 150
Age (of geologic time), 211
Age relations, 3, *see also* Intrusive sequence, Stratigraphic sequence
Agglomerate, 255
Agglutinate, 255
Ahrens, L. H., 181
Alidade, *see* Peep-sight alidade, Telescopic alidade
Alluvial deposits, mapping, 245
Alterations of plutonic igneous rocks, 291, 292, 293
   of volcanic rocks, 257, 261, 265, 269
Alteration zones as map units, 274, 293
Altimeter, *see* Barometer
American Commission on Stratigraphic Nomenclature, 210
American Geological Institute, 188, 332
American Society of Photogrammetry, 74, 154
Amphibolite, 300, 302
Amygdales, 260
Anderson, C. A., 266
Andesite, 254
Anorthosite, **276**
Aplite, 275
Apparent dip, used in sections, 46, 362
Areal geologic mapping, *see* Mapping

363

Index

Arenite, 216, **217**
Argillite, 219
Ash, volcanic, 256
Ash-flow deposits, 266
Autoliths, 285
Axial plane cleavage, 311

Back, in mine workings, 175
Backsight orientation, 100, 146
Badgley, P. C., 45, 227, 317, 318
Bailey, E. B., and W. J. McCallien, 269
Balk, Robert, 289, 290
Banded gneiss, 300
Banding, igneous flow, 263, **279**, 287–289
Barometer (altimeter)
  control surveys with, 123
  correcting readings from, 56
  locating points on base map with, 55
  testing, 55
Basalt, 254, 257–262, 268
Basanite, 255
Base lines, 90, 114, 116
  measuring, 119
Base maps, 49
  enlarging, 49
  features plotted on, 57
  locating points on, 51–57, 88–90
  planimetric, 49, 50
  preparing for field, 50
  scales of, 49
  scribing techniques on, 50
  sources of, 50
  topographic, 49
Beaman arc, 93, 104
  procedure with, 105
Beaman's rules for three-point locations, 145
Bearing lines, 22
  on aerial photographs, 83
  for locating points on maps, 51–53
Bedding, from bedding fabric, 226
  identifying in outcrop, 33
  measuring attitude of, 28
  in metamorphic rocks, 302
  symbols for, 235, 335
Bedding cleavage, compared to beds, 222

Bedding cleavage, in shale, 219
Beds, 222
  animal markings on, 231
  current directions in, 225, 226, 227, 228–230
  cut-and-fill structures in, 228, 233
  cyclic and rhythmic repetitions of, 223
  desiccation marks in, 231
  fabric of, 226
  internal features of, 225
  lava flows, 251–253, 257–262
  in metamorphic rocks, 302
  of pyroclastic rocks, 265–267
  sets of, 223
  shapes of, 223
  stratigraphic tops of, see Stratigraphic sequence, determining
  surfaces between, 228
Bentonite beds, 266
Bibliography in reports, 193
Billings, M. P., 62
Bioherm, 15
Biostratigraphic units, 211
Biostrome, 59
Birdseye, C. H., 132, **144**, 145
Bishop, M. S., 207
Blocks, volcanic, 255
Bombs, volcanic, 255, **256**
Bouchard, Harry, and F. H. Moffitt, 109, 123, 125, 134
Boudins, 313, **324**
Bradley, W. C., 246
Breccia, sedimentary, 215, 219
  volcanic, 255
Breccia pipes and necks, 269
Brunton Pocket Transit, 21, see also Compass
Bureau of Land Management, lines on aerial photographs, 81, 82
  township system, **19**
Burnham, C. W., 322
Busk, H. G., 45

Calcarenite, 220, 221
Carbonate rocks, 220, 300
Cartographic units, 58, 209
  of metamorphic rocks, 297, 319, 322

Cartographic units, of plutonic igneous
    rocks, 273
  of surficial deposits, 245
  of volcanic rocks, 252
Cataclastic rocks, 299, 301
Cements in sedimentary rocks, 213, 216,
    220–221
Chamberlin, T. C., 2
Chart, isogonic, 339
Charts for estimating compositions of
    rocks, 332
Chert, **217**
Claystone, 218
Cleavages in metamorphic rocks, 306,
    311–313
  in sedimentary rocks, 219, 222
Clinometer, measuring dip with, 28–31
  measuring vertical angles with, 25
Cloos, Ernst, 307
Cloos, Hans, 290
Coast and Geodetic Survey, U.S., 115,
    131, 132
Cognate ejecta, plutonic, 257
Cognate xenoliths, 285
Coleman, R. G., 277
Collimation marks, 161
Collins, J. J., 173
Colluvium, 245
Color charts, 216, 247
Colored inks for aerial photographs, 160
Colored pencils, on aerial photographs,
    80
  uses of, on base maps, 66
Color index, 277
  charts for estimating, 332
Coloring illustrations, 200
Colors of sedimentary rocks, 216
  in logging drill holes, 244
Colors of soils, 247
Columnar joints, in basic lavas, 260, **261**
  in ignimbrites, 266
  in intrusions, 268, 269
Columnar sections
  compiling, 48
  in reports, 205
  thicknesses for, 47, 240
Compass bearings, 22–24

Compass (especially Brunton compass),
    21
  care and adjustments of, 34
  as hand level, 27
  locating points on aerial photographs
    with, 83
  locating points on base maps with,
    51–57
  measuring bearings with, 23
  measuring stratigraphic sections with,
    236, 239
  measuring strike and dip with, 28–32
  measuring trend and plunge of linear
    features with, 33
  measuring vertical angles with, 25
  repairing, 34
  sketch map of control network with,
    118
  traversing with, 36–44, 54, 236
  used on traverse board (as peep-
    sight alidade), 88–90
Compass traverse, 36–44, 54, 236
Compilations from aerial photographs,
    see Aerial photograph compilations
Complex geologic relations, interpreting,
    2
Computations, of aerial photograph
    scales, 81, 83–84
  of bearings, 128
  of coordinates, 129
  of differences in elevations, 26, 27, **90**,
    127
  of gradienter readings, 110
  of stadia measurements, 103, 108
  of stratigraphic thicknesses, 240, **241**
  of transit traverse data, 134
  of triangles, 125
Conglomerates, 215, 219
Contact metamorphic zones, 321
Contacts, 58
  described in notes, 12
  flagging for plane table mapping, 141
  gradational, 64
  of hybrid magmas, 280
  mapping on base maps, 60–70, 90
  of plutons, 277, 280, 281
  within plutons, 279, **284**, **288**
  poorly exposed, 61

Index

Contacts, projection of, 62
Contour interval, choosing, 141
Contours, accuracy of on quadrangle maps, 53
  drawing, 149
  on aerial photographs, 151
  in detailed surveys, 171
Control networks, 113, 116
Control points, plotting on plane table sheets, 138
Control signals, for locations with a compass, 53
  for precise surveys, 118
Control stations, in detailed surveys, 171
  in mines, 175, 177
  selection of, 116
Control survey data, sources of, 115
Control surveys, 113
  for aerial photograph projects, 155
  by alidade and plane table, 132
  for detailed studies, 171
  for stratigraphic measurements, 238
  by transit-tape traverses, 133
  by triangulation, 114
Conversion of units of weight and measure, 361
Coordinate systems, 129–132
Coquinite, 221
Cores, drill-hole, 173, 243
Corrections, of barometric readings, 56
  for curvature and refraction, 127
  for magnetic declination, 22, 339
  of taped distances, 120–122
Correlation, lithic (of rock units), 59
  using fossils, 59, 211–212
Craftint, 200, 204
Creep of soil and mantle,
  effect on attitude of beds, 33
  in mapping concealed contacts, 61
Cross-bedding, **222**, 223, 225, 235
  in metamorphic rocks, 304
Crosscut, in mine workings, 175
Cross sections, geologic
  from aerial photographs, 157, 169
  composing on plate with map, 202
  drawing final copy of, 44–47, 204
  for field summaries, 71
  from plane table maps, 152

Cross sections, at quarry faces, 172
  rules for orientation of, 45
  selection of, 45
  from traverse maps, 44–47
Crowder, D. F., 324
Crowell, J. C., 229, 230, **231**
Current directions from sedimentary features, 225, 226, 227, 228, 229, 230
Current ripples, 228
Curvature and refraction, corrections for, 127
Cyclic bedding, 223

Dacite, 254
Declination, magnetic, 22, 339
Delineating features on aerial photographs, 167
Desiccation features in sediments, 231
Diagrams, fence, 205
  isometric (axonometric), 197
  panel, 205
  perspective, 198
  stereographic, 199, 317, **318**
Diamond drilling, 173, 184
Difference in elevation,
  with barometer (altimeter), 55, 123
  by leveling with hand level, 27
  by leveling with transit, 123
  by stadia, 103, 105
  by stepping method, 110
  from vertical angles, 26, **90**, 127
Differential leveling, 123
Dikes, plutonic, 278, 279, **281**, **284**, **288**, 291, 292, **293**, 323–324
  volcanic, 268
Diorite, **276**
Dip, initial, 222, 250
Dip of beds, 28–33
  plotting, 42
Dip protractor, 362
Dixon, W. J., and F. J. Massey, 178
Dolomitic rocks, classifying, 221
Domes, volcanic, 263
Drafting illustrations, 195, 199, 201, 205
Drag on faults, **68**
Drag folds, 314

Index

Drawing detailed maps and sections, 201
Drawing methods, 199
Drawings, duplication of, 196
  in field notes, 10
  in reports, 197, 198
Drift, in mine workings, 175
Drilling, in detailed studies, 172
  diamond, 173
Dunbar, C. O., and John Rodgers, 209, 220
Dunite, **276**
Duplication processes for illustrations, 196

Eardley, A. J., 74
Elevation, difference in, *see* Difference in elevation
Elevation of instrument, 106, 108
Entrenching tool, folding, 173
Epoch (of geologic time), 211
Equipment, field
  for aerial photograph compilations, 160
  for compass traverse, 38
  for control surveys with transit, 328
  for general geologic work, 5, 327
  for mapping on photographs, 80, 327
  for mine mapping, 174
  for plane table mapping, 328
  for taping base line, 119, 121
  for trimming fossils, 16
Error of closure, 54
Errors permissible in plane table mapping, 137
Errors of plotting on plane table, 133
Estimating positions on a map, 53, 145
Excavating and cleaning outcrops, 172
Extrusive rocks, *see* Volcanic rocks

Fabric, of metamorphic rocks, 283, 298, 305–307, 309–313
  of plutonic igneous rocks, 282–285, **286**, 287–289
  of sedimentary rocks, 215, 226
  of volcanic rocks, 257–260, 262–265, 266
Face (breast) in mine workings, 175

Facies, in detecting unconformities, 233
  in establishing stratigraphic units, 212
  in plutons, 274
Fairbairn, H. W., et al., 181
Faults, classification of, by attitude, 292
  details in outcrops, 67
  distinguishing from unconformities, 232
  evidence for, 67
  examining in mines, 177
  mapping, 67
  in plutons, 279, 290–292
  symbols for, 68, **293**, 334
Faunizones, 211
Fence diagrams, 207
Field notes, 7
  on backs of photographs, 85
  descriptions in, 9
  organizing for report writing, 186
  in stadia mapping, 148
  summaries of, 71
Field work, scheduling of, 4, 70, 137
Fine-grained detrital rocks, naming, 218
Fisher, R. V., 266
Flaser structure, 299
Flinn, Derek, 179
Flow, solid (metamorphic), 282, 283, 308, 316
Flow casts, *see* Flute casts, etc.
Flow layers, igneous, 286–289
Flow structures, of basic lavas, 257–259
  of plutonic igneous rocks, 282, 285, 286–290
  of silicic lavas, 262
  in volcanic domes, 264–265
  in volcanic intrusions, 268
Flute casts, 229
Flysch figures, *see* Load casts, etc.
Folds, drag, 314
  foliations and lineations in, 309–315
  geometric styles of, 307
  kinds of, 307, **311**, 324
  primary, in sediments, 225, 230
  profiles of, 57
  refolded, 308, 313
  slip (shear), 311, **315**, **324**
  symbols for, 57, 334, 335

Index

Foliations, metamorphic, 305–306, 309–313
Foraminifera, 17, 242
Formal stratigraphic units, 209
Formations, 209, *see also* Rock units
Fossils, correlating with, 59
   in establishing stratigraphic units, 211
   finding, 15
   as indicators of tops of beds, **234**, 235
   as indicators of unconformities, 233
   labeling, 17
   in metamorphic rocks, 303
   names for, 190
   new species of, 191
Fracture cleavage, 306
Fractures, complementary (conjugate), 316
   in metamorphic rocks, 316
   in plutons
      classifying by age, 291
      classifying by origin, 292
      Cloos' system of naming, 290
Frye, J. C., and A. R. Leonard, 246
Fuller, R. E., 262

Gabbro, **276**
Gair, J. E., **314**
Gannett, S. S., 127
Geodes, stratified, 260
Geographic names, in reports, 189
   for rock units, 190, 210
Geographic positions, 115
Geological Survey, U.S.,
   control data, 115
   Geologic Names Committee, 190
   Map Information Office, 50
   names on quadrangle maps, 189
   Suggestions to authors–, 188
   symbols used on topographic maps, 337 (in *note*)
   topographic (quadrangle) maps, 50
Geologic Names Committee, 190
Gilbert, C. M., 216, **217**, **218**, 266
Gilbert, G. K., 2
Gilluly, James, 294
Glacial drift, 245
Glauconite as indicator of unconformity, 233

Gneiss, 300
Gneissose structure, **282**, 306
Goldsmith, Richard, 300
Gradational contacts, 64
   mapping, 64, 66, 280
   new symbol for, 64, 334
   in plutons, 278, 280, **284**
Graded beds, **222**, 225, **227**, 230, **231**, 234
   in metamorphic rocks, 304
Gradienter screw, procedure with, 109
Grain sizes, cards for determining, 213
   of metamorphic rocks, 299
   of plutonic igneous rocks, 275
   of sedimentary rocks, 213
Granite (potassic granite), **276**
Granitic rocks, 275, **276**, **279**, 281, **282**, **284**, **288**, 294
Granodiorite, **276**, 277
Granofels, 300
Granulite, 300
Graphic symbols for rocks, 338
Graphs and curves, 198
Gravel, classification of, 213
Graywacke, 218
Great circle, in three-point locations, 144
Greenly, Edward, and Howel Williams, 63
Grids, plotting on illustrations, 203
   polyconic, 132
   rectangular, constructing, 131
   in transferring map points, 138
Groove casts, 229
Groups, 209
   of plutonic igneous rocks, 274
Grout, F. F., 178

Hammer, geologist's, 7
Hand lens, 7
Hand level, difference in elevation with 27
Hewett, D. F., 240
H. I., in measuring difference in elevation, 101, **103**, 104
Hodgson, R. A., 56, 123
Hoel, P. G., 178

Index

Hoelscher, R. P., and C. H. Springer, 197, 198, 199
Hoover, T. J., 184
Horizons, stratigraphic (time), 59
Hornfels, 300, 321
Howard, A. D., 246
Hutchinson, R. M., 290
Hutton, C. O., and F. J. Turner, 321
Hydrographic symbols for maps, 337 (in *note*)
Hydrothermal alteration, 291, 292, 294, 301
Hypotheses, in field studies, 2
multiple working, 2

Igneous rocks, *see* Volcanic rocks, Plutonic igneous rocks
Igneous rock series, plutonic, 274
volcanic, 253
Ignimbrite, 257, 266
Ijolite, **276**
Illustrations, columnar sections, 47–48, 205
duplication of, 196
enlarging and reducing, 201
geologic maps, detailed, 201
geologic sections, 47, 201
kinds of, 196, 201, 205
planning for a report, 195
preparation of, 47–48, 195–207
stratigraphic, 205
Inclusions in plutonic rocks, 279, 280, **281**, 285
Index maps, 187, 197
Index minerals, metamorphic, 320
Ingram, R. L., 224
Initial dip, of sedimentary beds, 222
of volcanic rocks, 251
Inking, field sheets, 71
final maps, 204
plane table sheets, 151, 153
Instrumentman, duties at plane table, 147, 148
Interpolation of contours, 149
Interpretations at outcrops, 2–4
Intersection, location by, with compass and base map, 52
with peep-sight alidade, 89

Intersection, location by, of plane table stations, 132, 139, 142
precision of, with alidade, 133
by radial lines on photographs, 159
Intrusive bodies, metamorphosed, 303
Intrusive rocks, *see* Volcanic rocks, Plutonic igneous rocks
Intrusive sequence, in plutonic rocks, 272, 279, 280–281, **288**, 289–290, 291
in volcanic rocks, 269
Inverse position computations, 115
Isogonic chart, 339
Isometric (axonometric) drawings, 197

Jackson, C. F., and J. B. Knaebel, 174, 178
Jacob's staff, 236
Johannsen, Albert, 275
Johnson tripod head, 96
Joints, columnar, 260, 266, 268, 269
on detailed maps, 171
by erosional unloading, 291–292
in metamorphic rocks, 316
in plutonic igneous rocks, 290–292

Keratophyre, 254
Key beds, as rock units, 59, 209
of volcanic rocks, 252–253, 265
Kleinpell, R. M., 211
Krauskopf, K. B., **260**, 270
Krumbein, W. C., and J. W. Tukey, 180
Krumbein, W. C., and L. L. Sloss, 209
Kuenen, P. H., 229

Labeling, drill cores, 243
hand specimens, 17–18
oriented specimens, 318, 319
well cuttings, 244
Lahars, 266
Laminations, **222**, 224
Lamprophyre, 275
Landforms, in reports, 187
in studying surficial deposits, 245
Landslides, effect on beds, 33
Lapilli, 256
Lapilli tuff, 256
Latite, 254

Latitudes and departures, computation of, 130
Latitude and longitude, 113, 115
Lava flows, subaqueous, 261
Lava movement, determining direction of, 258, 259, 261, 262
Lavas, structures of, 257–265
Lava tubes, 258
Layer structures in plutonic rocks, 286
Legends (explanations) on maps, 153, 202
Lenses in bedded rocks, 223
Lentils as rock units, 209
Leopold, L. B., and J. P. Miller, 246
LeRoy, L. W., 243
Lettering illustrations, 200
Leucitite, 255
Level, in mine workings, 175
Leveling with transit, 123
Limestones, classifying and describing, 217, 220
Linear features (lineations),
  in lavas, 258, 261, 262
  measuring attitude of, 33
  in metamorphic rocks, 305, 306, 309–315, 317
  plotting on stereographic nets, 317
  in plutonic igneous rocks, 280–285, 287, 289, 290
  reorienting in field, 227
  in sedimentary rocks, 225–230
  trend and plunge of, 33
Lineated schist, 298
List of illustrations in report, 192
Lists, use of in reports, 193
Lithogenetic units, 58
Lithologic symbols, 338
  in columnar sections, 47
  in cross sections, **205**
  printed patterns for, 200
Lithologic units, *see* Rock units
Load casts, 230
Lobeck, A. K., 198
Locality descriptions, 18
Locating points on a base map, 51–57, 89 (*see also* Traversing)
Logging cores, 243
Logging wells and drill holes, 242

Logs from cuttings, 244
Low, J. W., 135
Lowell, J. D., 224, 225
Lueder, D. R., 74

McKee, E. D., and G. W. Weir, 223, 224
McKinstry, H. E., 173
Mackin, J. H., 265, 294
Magnetic declination, 22, 339
Magnetic dip of compass needle, 35, 94
Magnetic disturbances of compass bearings, 24, 178
Magnetic orientation of plane table, 100, 146, 151
Mallory, V. S., 211
Map case, 6
Map grids, constructing, 131, 203
Map Information Office, 50
Mapping, on aerial photographs, 73, 84, 150, 155 [171
  with alidade and plane table, 135,
  alteration zones, 274, 293
  on base maps, 49
  with colored pencils, 66
  by color index, 277
  covered units, 62
  current structures, 225–227, 228–229
  daily routine of, 70
  in detailed studies, 171
  faults, 67
  fractures in plutons, 290–293
  glacial deposits, 245
  gradational contacts, 64
  inclusions, 286
  landforms, 245
  layers and schlieren in plutons, 286
  by matching standardized rock chips, 280
  metamorphic zones, 298, 319
  migmatites, 322
  by outcrop method, 63
  with peep-sight alidade, 88
  plutonic fabrics, 282, 284
  rates of, 62
  by reconnaissance methods, 69
  sharp contacts, 60
  by soils, 61, 246

Index

Mapping, by specific gravities, 277
 surficial deposits, 245
 underground, 174
 volcanic units, 250–253, 267
Maps, base, 49
 drafting, 201
 duplicating, 196
 enlarging, 49, 201
 ground control for, 113
 mine, 174
 planimetric, 49, 50
 outcrop or exposure, 63
 reconnaissance, 69
 for scribing, 50
 sketch, 69, 148, 197
 sources of, 50
 topographic, 49
Map symbols, 57, 61, 64, 68, 293, 334–337
Marble, 300, 302
Martin, N. R., 290
Melteigite, 276
Members, 209, *see also* Rock units
Mertie, J. B., 240
Metalavas, 303, 315
Metamorphic deformation, 301, 305, 307–317, 321, 323
 in massive rocks, 315
 in migmatites, 323
 zones of, 321
Metamorphic foliations and lineations, *see* Metamorphic rocks
Metamorphic rocks, cartographic units of, 297
 cleavages in, 306, 309–313
 deformation of, 301, 305, 307–317, 321, 323
 folds in, 307, 309–314
 foliations in, 305, 306, 309–313
 joints and veins in, 316
 lineations in, 306, 309–313, 316
 naming, 298
 oriented samples of, 318
 premetamorphic lithology of, 302
 refolded, 308, 313
 stratigraphic sequence in, 303
 studies of, 296, 305
 textures of, 283, 298

Metamorphic rocks, zones of, 298, 319
Metasomatic dikes and veins, 278, 284, 286, 322
Metasomatic layering, 288, 289
Metasomatic rocks, 272, 278, 283, 284, 292, 294, 301, 322, 323
Metasomatic zones, 274, 293, 322
Microfossils
 from cores and cuttings, 243
 kinds and occurrences, 16
 sampling for, 240, 242
Migmatites, 278, 322
Mine mapping, 174
 safety rules for, 178
 by traverses, 177
 work sheets for, 174
Mine workings, terminology of, 175
Moffitt, F. H., 74, 150, 154, 172
Monzonite, 276
Mortar structure, 299
Mudstone, 219
Muller, S. W., 211, 234
Munsell Color Company, 216, 247
Mylonite, 299, 301

Naming rocks, metamorphic, 298, 322
 plutonic igneous, 274
 in reports, 191
 sedimentary, 212
 volcanic, 253
Naming rock units, 190, 209–211, 253, 273–274, 297–298
Nepheline syenite, 276
Nephelinite, 255
Nevin, C. M., 62
Nickelsen, R. P., 313
Norite, 276
North, magnetic and true, 22, 339
North arrows on aerial photographs, 82
Notebook, field, 5, 6
 for samples, 183
 for stadia mapping, 106–107, 148
 for vertical angle measurements, 101
Notes, geologic, 7–13, 71
 descriptions in, 9
 for plane table mapping, 148
 summaries of, 71
Note numbers entered on maps, 18, 58

Oblique aerial photographs, 73
Observations at outcrops, 1–3, 7–9, 71, 208, 272, 297
Offset of instrument station, 142
Orbicules, 285
Ore sampling, 179–184
Oriented samples, 14, 285, 318
Orienting plane table, 100
Oscillation ripples, 228
Outcrop maps, 63
Outcrops, descriptions of, 7–9
  drawings of in notes, **10–11**, 197
  finding in covered areas, 62
  observations of, 1–3
  photographs of, **10–11**, 197
Outline for description of rock unit, 9
Outline for geologic report, 186
Overlays for color work on aerial photographs, 80
Overlays for compiling aerial photographs, 162

Pace, determining, 37
Pacing distances, 37
Pahoehoe lava, 258
Palagonite tuff, 261
Panel diagrams, 207
Parallax correction, 93
Parks, R. D., 178
Pass points for radial line compilation, 161
Pebble counts, 220
Pebble gravel, classified, 213
Pedometer, use in pacing, 37
Peep-sight alidade, 88
  mapping with, 89
Pegmatites, 275
Pelletite, 221
Pencils, choice of, 6, 80
Peridotite, **276**
Period (of geologic time), 211
Perrin, P. G., 188
Perspective drawings, 198
Petrofabric studies, 305, 318
Pettijohn, F. J., 209, 220
Phacoidal structure, **299**
Phillips, F. C., 227, 317, 318
Phonolite, 255

Photogeology, 74
Photogrammetry, 74
Photographs, *see also* Aerial photographs
  described in notes, 8, **10**
  as illustrations in reports, 197
  for plotting data at cliffs and cuts, 172
Photo indexes, 75
Phyllite, 300, **312**, **314**
Phyllonite, 300, 301, **315**, 323
Pillow lava, 261, **262**
Pipe amygdales, 258
Pitcher, W. S., 286
Pitcher, W. S., and R. C. Sinha, 178
Planar fabrics, in metamorphic rocks, 282–283, 298, 305, 309–313
  in plutonic igneous rocks, 282–285, **286**, 287–289
  in sedimentary rocks, 215, 226
  in volcanic rocks, 257–260, 262–265, 266
Plane table, 96
  in aerial photograph projects, 156, 159
  detailed mapping with, 171
  geologic mapping with, 135–153
  measuring stratigraphic sections with, 238
  precision of plotting on, 133
  setting up and orienting, 99
  surveying steep cuts with, 172
  traversing with, 146
  tripods for, 96
Plane table maps, inking and completing, 151–152
Plane table projects, 135, 170
  scales for, 135, 137, 154
Plane table sheets, 97–98
  selecting layout of, 137
Plane table traverses, 146
Planning field studies, 4
  detailed examinations, 170
  of metamorphic rocks, 296, 305
  plane table projects, 136
  of plutonic igneous rocks, 272
  sampling, 178–182
  of sedimentary rocks, 208

Index

Planning field studies, triangulation surveys, 114
Plant fossils, 16
Plats, township, 82
Plotting, control networks, 131, 138, 162
 strike and dip, 42
 traverse data, 40, 44
Plunge of linear features, 33
Plutonic igneous rocks,
 age relations in, 272, 273, 279, 280–282, **288**, 289–290, 291, 294
 alterations of, 291, 292, 293
 contacts of, 277
 hybrid zones of, 280
 inclusions in, 285
 layers and schlieren in, 286
 naming, 274
 planar and linear fabrics in, 282
 studies of, 272
 units of, 273
Pocket stereoscope, 77
Polaris at elongation, observing, 123
Polyconic grids, 132
Polymetamorphic rocks, 301
Porcellanite, **217**
Porosity of rocks, 215
Porphyries, 275
Powers, M. C., **215**
Precision of plotting on plane table, 133
Prentice, J. E., **226**
Primary structures in metamorphic rocks, 296, 302–304, 315
Prints from illustrations, 196
Printed patterns, 200
Protoclastic textures, 281, **282**, 283, 291, **293**
Protractor, for interconversion of true and apparent dip, 362
Pseudoscopic image, 80
 in delineating streams, 167
Pseudotachylite, 301
Pyroclastic cones, 265
Pyroclastic rocks, classification of, 255
 deposits of, 264
 grains in, 255, 256–257
Pyroxenite, **276**

Quadrilaterals in triangulation, 117, 122, 125
Quartz diorite, **276**
Quartzite, 300, 302, **312**, **313**
Quartz keratophyre, 254
Quartz monzonite, **276**

Radial line method, 159
Rain, effects on aerial photographs, 85
 preparations for, 7
 protecting plane table sheets from, 98
Raise, in mine workings, 175, **177**
Ramsay, J. G., 318
Ray, R. G., 74
Read, H. H., 323
Reconnaissance mapping, 69
Reconnoitering, control systems, 116, 118, 137
 for detailed studies, 170
 for plane table projects, 137
 to select rock units in plutons, 273
 traverse courses, 38, 54
Recorder, in survey parties, 136, 147
References in reports, system for making, 194
Refolding, 308, 313, 318
Refraction in measuring vertical angles, 123, 127
Relict structures in metamorphic rocks, 302–305
Reports, geologic, 185
 abstracts of, 192
 bibliography in, 193
 clarity of, 188
 form of, 193
 front matter for, 191
 illustrations for, 195–207, *see also* Illustrations
 lists and tables in, 193
 organizing and starting, 185
 outline for, 186
 purposes of, 185
 special terms in, 189
 use of references in, 193–195
Resection method, 142
Retrograde metamorphism, 301, 320
Reynolds, W. F., 125
Rhyolite, 254, **263**

Rhythmic bedding, 223
Ripple marks, 228
Ridgway, J. L., 197
Rittmann, A., **264**
Rock chips, used in mapping, 280
Rock names, *see* Naming rocks
Rock units, compared to time-stratigraphic units, 212
  correlating, 59
  descriptions of in notes, 9
  descriptions of in reports, 186, 187, 190
  general nature of, 58
  informal names for, 210
  key beds, 59, 209
  of metamorphic rocks, 297, 319–323
  naming, 190, 209, 210
  of plutonic igneous rocks, 273
  of sedimentary rocks, 209
  selecting new units, 60
  surficial deposits, 62, 63, 245
  of volcanic rocks, 252
Rod, stadia, 112
Rods (rodding structure), 313, **324**
Rodman, duties, 112, 147–149
Rounding of sedimentary grains, 215
Rubble, sedimentary, 215
  volcanic, 255

Safety in mines, 178
Safety in using hammer, 14
Samples, of altered plutonic rocks, 293
  bias of, 183
  for bulk composition, 180
  channel, 182, 183
  of cores, 243
  of cuttings, 244
  for distribution of compositions, 181
  fossil, 14, 240
  marking and numbering, 17, 243, 244
  microfossil, 16, 240
  packing, 18
  for petrofabric studies, 318–319
  oriented, 14, 318
  rock, 13
  serial, 179
  specifications for, 13, 179, 183
  spot, 179

Samples, stratified, 182
Sampling methods, 13, 183
Sampling plans, 178, 179, 180, 181
Sand-size cards, 213
Sand sizes, classification of, 213
Sandstones, classifying and describing 216
Schenck, H. G., and S. W. Muller, 211
Schenk, E. T., and J. H. McMasters, 191
Schist, 300, 302, **310, 311**
  knobby, 301, **309**
Schistosity (schistose texture), 298, 30
Schlieren in plutonic rocks, 286
Schmidt projection, 317
Scientific method, 2
Scoria bomb, **256**
Scoria flows, 259
Scribing materials and methods,
  on base maps, 50
  on plane table sheets, 98
Secondary control points, 140
Sederholm, J. J., 323
Sedimentary rocks, beds and related small-scale structures in, 208, 22; 235
  colors of, 216, 247
  fabric of, 215
  grain sizes of, 213
  logging wells and drill holes in, 24;
  measuring stratigraphic sections of 235
  naming and describing, 212
  porosity and permeability of, 215
  sorting of, 214
  studies of, 208
  surficial deposits of, 245
  unconformities in, 232
Semischist, 300
Serial (columnar) sections, 205
Serial samples, 179
Series, of plutonic igneous rocks, 274
  in stratigraphic sequences, 211
  of volcanic rocks, 253
Sets of beds, **222**, 223
Shaft, in mine workings, 175
Shale, 218, 219

Shaw, D. M., 178
Shear fractures in plutons, 291–292
Shells, fabric of in sediments, 226
Shelton, J. S., **267**
Shrock, R. R., 235
Sills, 268
Siltstone, 218
Skarn, 300
Sketch maps, of control networks, 118
   for plane table mapping, 137, 141, **148**
   by reconnaissance, 69
Sketches in notes, **10–11**, 309, 318
Skialiths, 285
Slate, 300, **312**
Slaty cleavage, 306, 311
Slip (shear) cleavage, 306
Slip (shear) folds, 311, **315**, 324
Slope direction in sediments, determining, 225, 230
Slump structures in sediments, 225, 230
Smith, H. T. U., 74
Smith, R. L., 266
Soils, color charts for, 247
   in detecting unconformities, 233, 252
   in geologic mapping, 61, 246
   horizons of, 246
   profiles of, 246
Soil Survey Staff (U.S. Department of Agriculture), 246
Solar ephemeris, 124
Sorting in sedimentary rocks, 214, 219
Spads, use in mine mapping, 177
Specific gravity, used in mapping rocks, 277
Specimens, *see* Samples
Spilite, 255
Spiracles in lavas, 261
Spot samples, 179
Stadia constant, 109
Stadia intercept, 103
Stadia interval factor, 109
Stadia mapping, 147, 151, 171
   instrument stations for, 139
   by moving instrument around rod, 151
   by traversing, 146

Stadia measurements,
   computations for, 104, 108
   with gradienter screw, 109
   by half intercept, 107
   by level sight, 107–108
   notebook for, 105, **106–107**
   principles of, 102
   steps in, 106
Stadia rod, uses of, 112, 151, 238
Stages, 211
Stanley, G. M., 246
State plane coordinate systems, 130
Statistical methods in sampling, 178, 179, 180, 181, 182
Stearns, H. T., **251**, **259**
Stearns, H. T., and G. A. Macdonald, 253, 259
Stebinger screw, 109
Stepping method with alidade, 110
Stereographic projection, uses of, 227, 285, 293, 317
Stereo pairs, 77
Stereoscope, mirror, 78
   pocket, 77
Stereoscopic image, 73, 77
   obtaining with stereoscope, 77
   obtaining without stereoscope, 79
Stokes, W. L., **229**
Stopes, 175, 177
Strata, *see* Beds
Stratigraphic horizons on key beds, 59, 265
Stratigraphic sections, measuring, 235
   by Brunton-tape method, 239
   by compass-pace surveys, 36, 236
   corrections for slope and oblique measures, 240, **241**
   by Jacob's staff, 236
   from maps or photographs, 236
   by plane table methods, 238
   requirements for measurements, 235
   by transit-tape traverses, 238
Stratigraphic sequence, determining,
   in metamorphic rocks, 303, 312–313, 314
   in sedimentary rocks, 225, 226, 228, 229, 230, 231, 232, 233

Stratigraphic sequence, determining, in volcanic rocks, 258, 259, 260, 261, 262, 266
Stratigraphic tops of beds, see Stratigraphic sequence, determining
Stratigraphic units, 58, 209–212
  measuring, 235
  in metamorphic terrains, 297, 303
  naming, 190, 209–211
  of pyroclastic rocks, 265–267
  in volcanic terrains, 250–253
Striding level, care and adjustment of, 92, 94, 95
Strike and dip, 28
  measuring with compass, 28–32
  plotting and checking, 42
  by three-point method, 31
  where to take, 32
Structural zones in metamorphic rocks, 321
Structure maps, 188, 197, 202
Structures, see Beds, Folds, Faults, Metamorphic rocks, etc.
Structures, described in reports, 187, 188
Sun compass, 25
Surficial deposits
  in general mapping, 62
  kinds of, 245
  relations to landforms, 245
Survey data, sources of, 115
Sutton, John, and Janet Watson, 321
Swanson, L. W., 74
Syenite, **276**
Syenodiorite, **276**
System (as a stratigraphic term), 211

Table of contents in reports, 192
Tables, use of in reports, 193
Tactite, 300
Tape, standardization of, 121
Taped distance, corrections of, 120, 122
Taping, on slope, 121
  stratigraphic sections, 238, 239
  with tape held level, 119
Telescopic alidade, 91
  adjustments of, 93, 94
  care of in field, 95

Telescopic alidade, geologic mapping with, 135
  intersection with, 142
  kinds of, 93
  line of collimation of, 95
  measuring vertical angles with, 101
  resection with, 142
  stadia mapping with, 147, 151
  stadia measurements with, 102–110
  three-point locations with, 143
  triangulation with, 132
Temperature correction, for barometric readings, 56
  for taped distances, 120
Tension fractures in plutons, **278**, **279**, 291, 292
Tephrite, 255
Terraces, tracing, 245–246
Textures, metamorphic, 282, 283, 298
  plutonic igneous, 275, 282
  relict, in metamorphic rocks, 302–303
Thickness of beds, 224
Thickness of units, 235–240
Three-point locations with plane table, 143
Three-point method for strike and dip, 31
Tilt distortion of aerial photographs, 75
Time-stratigraphic units, 210–212
Time units (intervals), 211
Tonalite, **276**
Tongues, as rock units, 209
Topographic mapping, see Plane table
Topographic profiles
  from aerial photographs, 158
  for cross sections, 43
Topographic symbols for maps, 337 (in note)
Tops and bottoms of beds, see Stratigraphic sequence, determining
Township-range cadastral system, **19**
Trachyte, 254
Transit, measuring vertical angles with, 122
  for taping on slope, 121
  in traversing, 133, 238
  in triangulation surveys, 113, 122
Traverse board and tripod, 88

Traverse, error of closure of, 54
  general scheme of, 36
  planning, 38, 54
Traversing, with alidade and plane table, 146, 238
  with barometer (altimeter), 56
  on a base map, 54
  with compass, 36, 54, 236
  in mines, 177, 178
  with transit and tape, 133, 238
  by turning angles, 41
Trend and plunge of linear features, 33
Triangle of error, in intersecting, 52
  in radial line plots, 165, 166–167
  in three-point (plane table) locations, 144–146
Triangles, computations from, 125
Triangulation networks, 116
Triangulation surveys, for aerial photograph projects, 156
  with alidade and plane table, 132
  computations for, 125
  with peep-sight alidade, 90
  plans for, 114
  for stratigraphic measurements, 238
  with transit, 122
Tripods, plane table, 96
True north by observation on Polaris, 123
Tuff-breccia, 255
Tuff, classification of, 256, 257
Turner, F. J., 321
Type locality (area), for rock unit, 210
Type section of rock unit, 210

Ultramylonite, 301
Unconformities, in columnar sections, 47
  evidence for, 232, 252
  stratigraphic significance of, 232
  in surficial deposits, 245
  value as contacts, 59, 70
  in volcanic rocks, 252
Underground mapping, see Mine mapping
Units of weight and measure, 361
Urtite, **276**

Varves, 234
Vegetation, used in mapping, 62, 84
Veins, in metamorphic rocks, **314, 315,** 316, 322, **324**
  in plutonic igneous rocks, 290–293, 294
  sampling, 181–183
Vertebrate fossils, 15
Vertical aerial photographs, 73
Vertical angles, see Difference in elevation
Vertical control for surveys, 113, 123
Volcanic complexes, 253
Volcanic intrusions
  identifying, 267–269
  mapping, 253, 267
  naming, 253
  superficial, 270
Volcanic mudflow deposits, 266
Volcanic necks, 253, 269
Volcanic rocks, alterations of, 252, 257, 261, 265
  cartographic units of, 252
  flows of, 257–265
  initial dip of, 250
  intrusive bodies of, 267
  pyroclastic deposits of, 265
  stratigraphic relations of, 250–252
  unconformities in, 252
Volcanic vents, indications of, 265, 267, 269

Wacke, 216, **217**
Walking contacts, 60
Wallis, W. A., and H. V. Roberts, 178
Waters, A. C., 253, 261, 262
Waters, A. C., and Konrad Krauskopf, 274
Weiss, L. E., and D. B. McIntyre, 318
Welded tuff, 257, 266
Well logs, 242
Wentworth, C. K., and G. A. Macdonald, **259**
Wentworth, C. K., and Howel Williams, 255
Wentworth scale of grain sizes, 213
Wet weather, preparations for, 7
Williams, Howel, **264,** 266, 269

Williams, Howel, and Helmut Meyer-Abich, 270
Williams, Howel, F. J. Turner, and C. M. Gilbert, 217
Wilmarth, M. G., 190
Wilson, Druid, G. C. Keroher, and B. E. Hansen, 190
Wilson, Druid, W. J. Sando, and R. W. Kopf, 190
Wilson, Gilbert, 313

Winze, in mine workings, 175
Wood, Alan, and A. J. Smith, 229

Xenocrysts, 285
Xenoliths, **279, 281**, 285

Zip-a-tone, 200, 201, 204
Zoned inclusions, 286
Zones, alteration, 274, 293
   of hybrid magmas, 280
   metamorphic, 298, 319